Archaeoastronomy: Current Research

Archaeoastronomy: Current Research

Edited by
Hudson Pearce

STATES
ACADEMIC PRESS
www.statesacademicpress.com

Published by States Academic Press,
109 South 5th Street,
Brooklyn, NY 11249, USA

ISBN: 978-1-63989-057-6

Cataloging-in-Publication Data

Archaeoastronomy : current research / edited by Hudson Pearce.
p. cm.
Includes bibliographical references and index.
ISBN 978-1-63989-057-6
1. Archaeoastronomy. 2. Astronomy. 3. Astronomy--History. I. Pearce, Hudson.
GN799.A8 H36 2022
520.93--dc23

For information on all States Academic Press publications
visit our website at www.statesacademicpress.com

STATES
ACADEMIC PRESS

Contents

Preface

The study of the phenomena of the sky and its use in culture by the people of the past is known as archaeoastronomy. The complementary niches in cognitive archaeology and landscape archaeology are fulfilled by archaeoastronomy. It uses methods from diverse fields such as anthropology, archaeology, astronomy, history, etc. This field studies material evidence to explore its relation to the sky. Such studies help in exploring how a wider landscape is integrated into the beliefs regarding the cycles of nature. Archaeoastronomy can be divided into green archaeoastronomy and brown archaeoastronomy. The basis of green archaeoastronomy is statistics. It is used for evaluation of prehistoric sites with extremely less social evidence. Brown archaeoastronomy utilizes ethnographic and historical records. This helps in establishing a link between early astronomies, calendars and rituals. This book covers in detail some existent theories and innovative concepts revolving around this discipline. Different approaches, evaluations, methodologies and advanced studies on archaeoastronomy have been included herein. This book is a complete source of knowledge on the present status of this important field.

The information shared in this book is based on empirical researches made by veterans in this field of study. The elaborative information provided in this book will help the readers further their scope of knowledge leading to advancements in this field.

Finally, I would like to thank my fellow researchers who gave constructive feedback and my family members who supported me at every step of my research.

Editor

1

ASTRONOMY, ETHICS AND THE PLANETS

Dr Nicholas Campion

Associate Professor in Cosmology and Culture, University of Wales Trinity St. David

Corresponding author: Dr Nicholas Campion (n.campion@uwtsd.ac.uk)

ABSTRACT

This paper looks at the astronomical context for morality and ethics. It explores the ways in which astronomy has inspired theories of good and evil. It will argue that astronomy often has a moral component and that the stars and planets have often been seen as inclining humanity to either good or evil acts, or can be used in order to infer standards of human behaviour. It will also be argued that by influencing culture, astronomy requires or encourages action. The codification of morality in relation to astronomy can be traced to Pharonic Egypt, was developed in the Hellenistic world and reached a full form in the first-century CE in works such as Claudius Ptolemy's *Tetrabiblos*, in which the planets were ascribed personalities. These planetary personalities were embodied in every individual character and could incline people to good or evil depending on the planets' disposition at any one time. The notion of the seven deadly sins - one for each planet - can be traced to this system. The idea is, then, that morality is a dynamic system, moving in space and time as the planets move. While an individual may be inclined to do evil at a particular moment or location in space and time, they might equally be disposed to do good. The system was challenged from a number of perspectives. Hermetic and Gnostic cosmology proposed that, because the entire cosmos was evil, the only solution was to escape from it entirely and the Catholic theologian Thomas Aquinas critiqued the claim that the planets could directly influence moral choice. The paper will conclude with reference to the impact of modern astronomy on morality: the use of Newtonianism to support theories of natural justice and universal human rights, and the application Einsteinian relativity to the theory of cultural relativism, in which no one culture can claim to be superior to any other.

KEYWORDS: Astronomy, Culture, Morality, Plato, Thomas Aquinas, Isaac Newton, Albert Einstein

---

1. INTRODUCTION

The cultural applications of astronomy concern every aspect of human affairs, including morality. This paper will consider the relationship between morality and astronomy in one particular strand represented in Platonic, Hermetic and Catholic thought, concluding with some final thoughts about the modern era. The major secondary source on the topic regarding the classical and Medieval West in English is Morton Bloomfield's *The Seven Deadly Sins* (1952), an expansion of his original paper published in the *Harvard Theological Review* (1941). Morton (1952: 16) used the terms 'Soul-Drama', from Reitzeinstein and Schaeder (1926) and 'Soul Journey' from Bousset (1901) as a description of the doctrine of the soul's descent and ascent through the planetary spheres. In this scheme, each planet and planetary sphere assumes a moral quality which the soul either acquires (as it descends) or abandons (as it ascends) as it makes its journey through the spheres, As Bloomfield argued, the seven planets then become associated with the seven deadly, or cardinal, sins of Roman Catholicism. As listed in the probably second-century CE 'Testament of Reuben', these were promiscuity, insatiability, strife, flattery and trickery, arrogance, lying and theft (Charlesworth, 1983: 'Testament of Reuben', 3). As it arrived in the Greek world, Bloomfield claimed, the doctrine had possibly originated in the redeemer mystery of Persian Zoroastrianism, but it also included significant Egyptian content. Bearing in mind the likelihood of Indian influences, in view of the proximity and close trading relationships between India and Mesopotamia, and from there to the Mediterranean, it is impossible to trace any single origin.

The Soul-Drama finds its earliest surviving, comprehensive literary treatment in Plato's 'Myth of Er', in which the soul descends to Earth through the planetary spheres (Plato, 1935: X.61b B, 621.B) and reascends after death. Plato did not ascribe any significance to the planets in this process but he did add that when souls are judged the righteous ascend to the sky, the unjust descend to the earth (Plato, 1935: X 614 C-E), anticipating Christian doctrines of a Heaven above and a Hell below. The sky, then, is a moral, or ethical arena. As Mark Williamson stated, discussing modern space ethics, we have to consider 'who is asking [ethical] questions and who is answering them?' (Williamson, 2003): no question is ever asked without a prior context, or a set of assumptions about the way the world works. This paper looks at the astronomical context for morality and ethics.

2. THE CLASSICAL WORLD

In Babylonian teachings the planetary deities can act as judges of human behaviour (Rochberg, 2004), but the planets appear to have been only endowed with inherent moral qualities between the fifth and first centuries BCE. Early indications are present in the Babylonian nativity omens of the late fifth-century which ascribed good or evil fortune to the planets (Sachs, 1962; Rochberg-Halton, 1988):

> the child born under Jupiter will have a 'regular (life)...will become rich (and) will grow old.
> the child born 'when Venus has come forth', will have a life which is 'exceptionally calm (and) favourable' and marked by longevity.
> the child born with Mars may have a hot temper.
> the child born when Saturn has come forth will live a life which is 'dark, obscure, sick and constrained'.

It follows that, for example, a child born with a hot temper may be inclined to impulsive, violent and therefore potentially immoral acts. These planetary personalities were evident in a more fully developed form in Claudius Ptolemy's second-century CE *Tetrabiblos* in which it was claimed, for example, that Mars may render people kingly, noble and strong when honourably placed but, when difficult, it is 'rash, unruly, indifferent, stubborn, keen, headstrong contemptuous, tyrannical (and) easily angered'(Ptolemy, 1940: III.13). Ptolemy continued, 'If Venus alone takes dominion of the soul in an honourable position, she makes her subjects pleasant, good, eloquent, neat, cheerful, fond of dancing, eager of beauty' and so on (Ptolemy, 1940: III.13). Each planet might have a positive or negative quality at a child's birth, depending on its zodiacal position and angular relationship with the other planets (including the Sun and Moon), but also had an inherent benefic or malefic disposition. As set out by Ptolemy (Ptolemy, 1940: I.5), the Moon, Venus and Jupiter were benefic, Mars and Saturn malefic and the Sun and Mercury were mixed. As the planets move through the zodiac signs and houses, and form aspects to each other, the absolute morality embodied in each planet varies. Morality therefore alters with time, place and motion. Plato had hinted as much in the Republic:

> Not only for plants that grow from the earth but also for animals that live upon it there is a cycle of bearing and barrenness for soul and body as often as the revolutions of their orbs come full circle, in brief courses for the short-lived and oppositely for the opposite; but the laws of pros-

> perous birth of infertility for your race, the men you have bred to be your rulers will not for all their wisdom ascertain by reasoning combined with sensation, but they will escape them, and there will be a time when they will beget children out of season. Now for divine begettings there is a period comprehended by a perfect number (Plato, 1935: VIII, 546 A–B).

Plato's most significant contribution to the question of astronomy and morality was his theory of soul. As set out in *Phaedrus* the human being consisted of four different parts, the body and three-part soul. The highest part, the rational soul, mind or intellect, discerns what is true, judges what is real, makes rational decisions, is related to the movement of the celestial spheres, and can contact the divine:

> Now the divine intelligence, since it is nurtured on mind and pure knowledge, and the intelligence of every soul which is capable of receiving that which befits it, rejoices in seeing reality for a space of time and by gazing upon truth is nourished and made happy until the revolution brings it again to the same place (Plato, 1914: 246 A–D; see also 253 C–D).

Of the other two parts, the 'spirited' soul carries out what reason has decided and the 'appetitive soul' is the seat of emotion or desire. In his well-known analogy Plato describes a chariot pulled by two horses, the charioteer being the rational soul and the two horses the two irrational parts (Plato, 1914: 246 A–D). Plato's scheme of the tripartite soul was codified in the *Corpus Hermeticum*, composed in Hellenistic Egypt from the second-century BCE onwards, in which it was asserted that there are two ways to be freed from planetary fate. That is, astronomical theory requires action. The first way is for the soul to ascend through the planetary spheres, shedding its negative psychological characteristics as it does so (Scott, 1982: Vol. 1, 129, Libellus I.25; see also Bloomfield, 1941). As the soul passes the moon it sheds growth and decay, Mercury trickery, Venus deceit, the sun arrogance, Mars daring and recklessness, Jupiter greed and Saturn falsehood. At the eighth sphere, that of the stars, the soul begins to praise God, encouraged by the sound of voices from the higher levels. Finally the soul ascends to God and, in the logical climax of Platonic theology, it is reunited with God. The second way is to make as direct and spontaneous connection with the light of God, asserting the rational soul's inherent freedom from planetary fate. In this model, as the planets move, they stir up the negative and malefic qualities described by Ptolemy, potentially causing ill-fortune or immoral behaviour, all of which can be neutralised by the spontaneous contact with the divine light:

> the planets replace one another from moment to moment; they do not go on working without change, but succeed one another in rotation. These daemons then make their way through the body, and enter the two irrational parts of the soul; and each daemon perverts the soul in a different way, according to his special mode of action. But the rational part of the soul remains free from the dominion of the daemons, and fit to receive God in itself. If then the rational part of a man's soul is illumined by a ray of light from God, for that [in] man the working of the daemons is brought to nought; for no daemon and no god has power against a single ray of the light of God (Libellus XVI. 15–16 in Scott, 1982).

In one dialogue Poimandres, speaking to Hermes, declares, look 'at what you yourself have in you; for in you too, the word is son, and the mind is father of the word…Now fix your thought upon the Light…and learn to know it' (Libellus I. 6, in Scott, 1982: Vol. 1, 117). Later the statement is made that 'He who has recognised himself [acknowledged the divinity within] enters into the Good' (Libellus I.18, in Scott, 1982: Vol. 1). The need to free one's self from the morally-threatening qualities of the planetary personalities became more pressing with the development of Gnostic theology in the early centuries CE, in which the cosmos was envisaged as a kind of prison. In Gnosticism's cosmophobic (Jonas, 1963) theology, the physical universe inhabited by humanity was created by the evil god Yaldabaoth ('The Apocryphon of John', x–xi, pp. 110–11, Robinson, 1988):

> And he [Yaldabaoth] placed seven kings - each corresponding to the firmaments of heaven - over the seven heavens, and five over the depths of the abyss, that they may reign. And he shared his fire with them, but he did not send forth from the power of the light which he had taken from his mother, for he is ignorant darkness.

Developing the model already set out in the *Corpus Hermeticum* salvation is then possible through Christ:, bi-passing the astronomical universe:

> Live with Christ, and he will save you. For he is the true light and the sun of life. For just as the sun which is visible and makes light for the eyes of the flesh, so Christ illuminates every

> mind and the heart...the light of Christ, which is reason...For everything which is visible is a copy of that which is hidden. For as a fire which burns in a place without being confined to it, so it is with the sun which is in the sky, all of whose rays extend to places on the earth. Similarly, Christ has a single being, and he gives light to every place. This is also the way in which he speaks of our mind, as if it were a lamp which burns up the place. (Being) in a part of the soul, it gives light to all the parts ('The Teachings of Silvanus', 96-98, in Robinson, 1988).

The liberation of the soul from the physical cosmos is facilitated by Christ himself who, at his ascension to Heaven, travelled through the planetary spheres, destroying their malign power:

> It came to pass then, when I saw the mystery of all these words in the vesture which was sent me, that straightway I clothed myself therewith, and I shone most exceedingly and soared into the height. He entereth the firmament. 'I came before the [first] gate of the firmament, shining most exceedingly, and there was no measure for the light which was about me, and the gates of the firmament were shaken one over against another and all opened at once. The powers of the firmament are amazed and fall down and adore him.' And all rulers and all authorities and all angels therein were thrown all together into agitation because of the great light which was on me. And they gazed at the radiant vesture of light with which I was clad, and they saw the mystery which contains their names, and they feared most exceedingly. And all their bonds with which they were bound, were unloosed and every one left his order, and they all fell down before me, adored and said: 'How hath the lord of the universe passed through us without our knowing?' And they all sang praises together to the interiors of the interiors; but me they saw not, but they saw only the light. And they were in great fear and were exceedingly agitated and sang praises to the interiors of the interiors (Mead, 1984: 'Pistis Sophia' I.11; see also Lieven, 2002).

3. THE MEDIEVAL WORLD

The strand of cosmological thought passing from Plato to the *Corpus Hermeticum* and Gnostic Christianity, and taking in Claudius Ptolemy clearly identified a threat to individual fortune, behaviour and morality emerging from the planets' positions at birth as well as the disruptive effects of their daily motions. However, Ptolemaic astrology, Hermetic texts and Gnostic scriptures all but disappeared in Latin Europe following the disintegration of the Western Empire at the end of the fifth-century. The problem of planetary fate then remained dormant for eight hundred years until the translation of a large corpus of astrological material into Latin in the twelfth and thirteenth centuries, largely from Arabic, but including many Greek originals of which the *Tetrabiblos* was the most important. The threat to the individual moral choice required by Christianity was clear from Ptolemy's delineation of the planets' role in the formulation of personality and behaviour. In the context of medieval Christianity it was entirely unacceptable for one's chances of salvation through Christ to be limited by the planets location at one's birth and their subsequent transits. This was offensive both to scripture and to the role of the Church and priesthood: how could a priest offer absolution if the planets had already denied it? The issue was resolved by the Dominican Friar Albertus Magnus (1200–80) who claimed that the planets could influence the body directly, but not the soul (Hendrix 2016). The solution was then set out in much greater detail by Albertus's student, Thomas Aquinas (1225–74), although he was unable to access Plato's *Republic* or *Phaedrus* and the great part of the *Corpus Hermeticum*, and none of the Gnostic scriptures. However, he reached a similar conclusion, namely that the soul (which we may see as equivalent to Plato's rational soul), was free from direct planetary influence (similar to the rational soul in the *Corpus Hemeticum*), by relying on the interpretation of part of the rational soul in Aristotelian thought (Aristotle, 1936, III.V) as the 'Active Intellect'. Thomas considered that the 'Active Intellect' was equivalent to the Christian soul, and if the soul was therefore free from planetary influence, this could only be exerted via the body:

> Of course acts of choice and movements of the will are controlled by God. And human intellectual knowledge is ordered by God through the mediation of the angels. Whereas matters pertinent to bodily things, whether they are internal or external, when they come within the use of man, are governed by God by means of the angels and the celestial bodies (Aquinas, 1975: III.91.2).

Thomas added (1975: III.92.2): 'man is ordered in regard to his body under the celestial bodies, in regard to his intellect under the angels, and in regard to his will under God'. This formula actually provided an explanation for why astrological prediction might work in general circumstances but not specific ones:

> That astrologers not unfrequently forecast the truth by observing the stars may be explained in two ways. First, because a great number of men follow their bodily passions, so that their actions are for the most part disposed in accordance with the inclination of the heavenly bodies: while there are few, namely, the wise alone, who moderate these inclinations by their reason (Aquinas, 1920: 2.2.95.5)

Engagement with the celestial bodies had direct and powerful consequences for moral – or immoral – behaviour. According to Thomas' model the individual's chances of salvation were therefore dependent on resisting the planets' direct influence on the body, promoting baser desires. The Soul Drama was therefore as intense as it was in the Greek world but the Soul Journey was, as in one version described in the *Corpus Hermeticum* and the Gnostic scriptures, directly to God rather than via the planetary spheres.

4. THE MODERN WORLD

If we move forward to the modern world then we can observe how scientific astronomy has abandoned the concept of the Soul Journey, yet still has implications for ethical and moral behaviour. A key example is the influence of Newton's theory of universal law on the argument that, as the universe is governed by one set of laws, so is society, which is contained within the universe (Campion, 2016). Further, every individual is subject to the same law and therefore has equal rights. Newtonianism, the philosophy derived from Newton's work, did not originate human rights theory, but it did contribute the claim that human rights are justified by the laws of physics. Freedom is then the default position of the Newtonian universe and natural rights belong to people through the fact of the existence of the laws of gravity and planetary motion (Paine, 2013). What passes for ethical in one period, though, is not necessarily accepted in another. For example, John Herschel's activities in South Africa from 1833–8 were largely motivated by his belief (along with many other progressive thinkers of the time) that rational progress in science – including astronomy – and imperialism went hand in hand (Musselman, 1998). Setting a wider context, Musselman stated that 'Astronomy has long been recognised as a tool of empire' (Musselman, 1998), referring mainly to those navigational aspects of the discipline which had enabled European explorers to traverse the globe and encounter new territories for conquest and colonisation. More recent influences of astronomy on culture have helped fuel claims that western rationalism may not be superior to the world views of other cultures. Einstein is held particularly responsible for these ideas (Isaacson and Schucking, 2007; Johnson, 1991) as a result of popular equations between scientific relativity on the one hand, and controversial theories of moral and cultural relativism (the idea that no one culture is superior or inferior to another) on the other. The cultural consequences of all three examples, Newton, Herschel and Einstein, are felt in political action: Newtonian natural rights theory encouraged the American revolutionaries of 1776; Herschel was representative of those who wished to civilise colonised peoples by spreading Western values; Einstein has unintentionally encouraged the advocates of cultural relativity and multiculturalism.

5. CONCLUSION

To conclude, while astronomical theory has a constant influence on culture, there appears to be a clear difference between ancient and premodern ideas on the one hand, and modern astronomy on the other. In the former, the entire sky was inherently moral, a notion embodied in theories of the seven deadly sins, the Soul Drama and the Soul Journey. The consequence was that action was necessary in order to navigate the cosmos's moral structure. By contrast modern scientific astronomical theory has no essential moral components, yet influences moral and ethical ideas in wider culture, and consequent action.

REFERENCES

Aquinas, T. (1975) *Summa Contra Gentiles*. 4 Vols., trans. V. J. Bourke. Notre Dame, University of Notre Dame Press.

Aquinas, T. (1920) *Summa Theologica*. 4 Vols., London, Second and Revised Edition, Literally translated by Fathers of the English Dominican Province, London Burns, Oates and Washbourne.

Aristotle (1936) *On the Soul*. trans. W .S. Hett, Cambridge Mass., London, Harvard University Press.

Bloomfield, M. W. (1941) The Origin of the Concept of the Seven Cardinal Sins. *The Harvard Theological Review*, 34.2 (Apr.), 121–128.

Bloomfield, M. W. (1952) *The Seven Deadly Sins; and Introduction to the History of a Religious Concept, with Special Reference to Medieval English Literature*. Ann Arbor, Michigan State College Press.

Reitzeinstein, R., and Schaeder, H. H. (1926) *Studien zun antiken Synkretismus aus Iran und Griechenland.* Studien der Bibliothek Warburg VII, Leipzig and Berlin.
Bousset, W. (1901) Die Himmelreise der Seele. *Archiv für Religionswissenschaft*, IV, 136–169, 229–273.
Campion, N. (2016) Astronomy and Culture in the Eighteenth Century: Isaac Newton's Influence on the Enlightenment and Politics, in J. P. Hernandez, C. G. Garcia, G. Magli, D. Nadali, A. Polcaro and L. Verderame (eds). *Astronomy in Past and Present Cultures, Mediterranean Archaeology & Archaeometry*, 16.4, pp. 497–502.
Charlesworth, J., ed. (1983), The Testament of Reuben, in 'The Testament of the Twelve Patriarchs', in *The Old Testament Pseudepigrapha.* 2 vols., New York, pp. 782–785.
Hendrix, S. (2016) From the Margins to the Image of "The Most Christian Science": Astrology, Theology, and St. Peter's Basilica. *Culture and Cosmos,* 20. 1 and 2, Spring/Summer and Autumn/Winter, pp. 129–46.
Isaacson, W., and Schucking E. L. (2007) *Einstein: His Life and Universe.* New York, Simon and Schuster.
Johnson, P. (1991) *Modern Times: The World from the Twenties to the Nineties.* New York, Harper Collins.
Jonas, H. (1963) *The Gnostic Religion: The Message of the Alien God and the Beginnings of Christianity.* 2nd ed., Boston, Beacon Press.
Lieven, A. von (2002) Gnosis and Astrology: 'Book IV' of the Pistis Sophia, in J. M. Steele and A. Imhausen, *Under One Sky: Astronomy and Mathematics in the Ancient Near East.* Alter Oient und Altes Testament Bsnd 297, Munster: Ugarit Verlag, pp. 223–236.
Mead, G. R. S., trans. (1984) *Pistis Sophia.* Blauvelt, N.Y., Spiritual Science Library [1921].
Musselman, E. G. (1998) Swords into Ploughshares: John Herschel's Progressive View of Astronomical and Imperial Governance. *The British Journal for the History of Science*, 31, 4 (Dec.), pp. 419–435.
Paine, T. (2003) *The Age of Reason.* Mineola, New York, Dover Publications.
Plato (1914) *Phaedrus.* trans H. N. Fowler, Cambridge Mass., London, Harvard University Press.
Plato (1935) *Republic.* 2 Vols., trans. P. Shorey, Cambridge Mass., London, Harvard University Press.
Ptolemy, Claudius (1940) *Tetrabiblos.* trans. F. E. Robbins, Cambridge MA: Harvard University Press.
Reitzeinstein, R., and Schaeder H. H. (1926) *Studien zun antiken Synkretismus aus Iran und Griechenland.* Studien der Bibliothek Warburg VII, Leipzig and Berlin.
Robinson, J. M., ed. (1988) *The Nag Hammadi Library in English.* Leiden, New York: Brill.
Rochberg, F. (2004) *The Heavenly Writing: Divination and Horoscopy, and Astronomy in Mesopotamian Culture,* Cambridge: Cambridge University Press.
Rochberg-Halton, F. (1988) Benefics and Malefics in Babylonian Astrology. in E. Leichty *et al.*, eds., *A Scientific Humanist: Studies in Memory of Abraham Sachs.* Occasional Publications of the Samuel Noah Kramer Fund 9, Philadelphia, Pa., pp. 323–328.
Sachs, A. J. (1952) Babylonian Horoscopes. *Journal of Cuneiform Studies*, Vol. 6, pp. 49–75.
Scott, W. (trans.) (1982) *Hermetica: The Ancient Greek and Latin Writings which contain Religious or Philosophic Teachings Ascribed to Hermes Trismegistus.* 4 vols., Boulder, Co., Shambala.
Williamson, M. (2003) Space ethics and protection of the space environment. *Space Policy* 19, pp. 47–52.

2

PRAILEAITZ I: A MAGDALENIAN LUNAR-SOLAR CAVE AT 15,500 BP IN THE BASQUE COUNTRY

Lionel Sims[1], Xabi Otero[2]

[1]*University of East London*
[2]*Jauzarrea*

Corresponding author: Lionel Sims (lionel.sims@btinternet.com)

ABSTRACT

Hayden et al. (2012) have questioned whether there was a Palaeolithic 'astronomy', yet nevertheless show that ethnographic evidence confirms such knowledge amongst extant 'complex hunter-gatherers'. Included in their critique is a rejection of Marshack's (1972) suggestion that Palaeolithic modern humans made non-decorative lunar notational systems. During fieldwork in the Basque country in March 2015 the authors examined the orientation of the Praileaitz I cave in the valley of the River Deba. The site excavation report (Peñalver, 2014) reveals that the cave was a place of ritual for local hunter-gatherers during the Magdalenian period around 15,500BP. The cave was one of 22 probably inhabited by the same group during the Magdalenian. Since none of the remaining 21 caves displayed any evidence as being places of ritual, then this data-set allows a statistical procedure to identify what particular portfolio of characteristics led to this cave being chosen as a place of ritual. The Praileaitz I cave has a vulva shaped entrance, 29.5 pendants arranged around the cave's chambers, an outward facing vestibular seat, red ochre crayons, a cutting antler baton and abstract 'art'. Many of the pendants displayed marks consistent with Marshack's lunar model. The field archaeoastronomy reveals that the cave entrance had an alignment on summer solstice sunrise. This lunar-solar astronomical finding shows a step-change from tracking discrete months as demonstrated by Marshack to organising 'lunar' rituals abstracted from any particular month displaced onto a solarized time scale. This suggests that the Magdalenian Praileaitz I rituals served as a bridging ritual form between monthly, synodic, rituals in the early Palaeolithic and later monumental alignments on the solarized sidereal moon in the Mesolithic, as for example at Warren Field in Scotland (Gaffney et al., 2013), and in the Neolithic, as for example at Stonehenge (Sims, 2006).

KEYWORDS: Magdalenian, cave, lunar-solar, complexity, vulva, blood, gender, ambiguity.

1. INTRODUCTION

Scholarship remains divided over Marshack's (1971) claims made fifty years ago for Palaeolithic non-decorative lunar notational systems. Hayden and Villeneuve (2012) have called into question not just Marshack's findings but also the suggestion of other archaeoastronomers supportive of Marshack for a Palaeolithic 'astronomy' (d'Errico, 2009; Rappenglück, 2015). Nevertheless from a survey of ethnographic evidence they conclude that 'astronomy' can be found amongst extant 'complex' hunter-gatherers, especially for the solar calendrical timing of rituals, and that this may usefully inform research into prehistoric astronomy.

An element of circularity obtains in their critique, since 'complex counting systems' are one of their criteria for a socially complex culture (Hayden & Villeneuve, 2012, 335). Since the means to establish a sophisticated calendrical system for them defines cultural complexity, then this rules out finding it in a 'simple society'. And as they prefer explanations that link cultural achievements to 'economic, social and political contexts' (Hayden and Villeneuve, 2012, 332), their assumption of a break between egalitarian simplicity to stratified complexity creates a conceptual barrier that obscures the details of this transition.

A deeper engagement with anthropology may suggest a more cautious drawing of the various transformations in early hunting cultures. Wengrow and Graeber (2015) have argued that Palaeolithic cultures may have oscillated between egalitarian and complex modes according to seasonal rhythms. This would suggest that the astronomical knowledge Hayden and Villeneuve associate with complex hunters alone would have been the same people who a season later would be the simple hunters expected to have little or none. And it is a long held understanding in anthropology that while 'economic, social and political contexts' are usually disaggregated from ritual and religion in modern societies they are encapsulated and bound together as a 'total social fact' in a ritual context in early societies (Mauss, 1969). For the present purposes two consequences immediately flow from this understanding. First, all rituals are redundantly multi-dimensional and digital, leading us to expect that symbolic representations will be prescriptively repeated through different performances *and materials* (Rappaport, 1979). Second, that any sophisticated astronomy that could be found in a complex hunting society emerges as a transformation of a prior form linked to a cultural requirement to govern wide areas of foraging life in synchrony with certain heavenly patterns. The BaMbuti (Zeusse, 1979), !Kung (Watts, 2005) and Mescalero Apache (Farrer, 1998), all simple mobile foragers not expected to have any elaborated astronomy according to Hayden and Villeneuve, nevertheless have their own expert knowledge and ritual appropriations of the skyscape. While this knowledge may not be structured as the 'calendars' that Hayden and Villeneuve would recognise, it is these egalitarian understandings that form the infra-structure subsequently transformed into 'astronomical calendars'. If alongside finding differences between egalitarian and stratified hunter gatherers we can also find formal traces of earlier cosmovisions, then once combined this evidence may allow us to locate the reasons for a shift from a cosmology to a religion. If so we would expect that the processes operating on the material plane with substitutions of reversal and estrangement undermining egalitarian relationships would be replicated in the formal properties of the types of astronomy subsequently adopted in complex cultures (Sims, 2006). Following Marshack we might expect the first Palaeolithic astronomy tracking lunar synodic phases, while following Hayden and Villeneuve we might expect the earliest astronomy to emerge from complex cultures expending resources on constructing accurate solarized calendars. Appreciating the thirty thousand years of the European Palaeolithic as a period of rapidly changing seasonal and resource stresses, we might also anticipate a more complex interaction between these two models. Beginning to answer these questions requires enhancing the evidence base and extant theoretical models.

2. LUNAR OR/AND SOLAR

Esteban and Tortosa (2001) have argued that the Parpalló cave in Iberia was chosen by Palaeolithic hunter-gatherers as a place of ritual because of its vulva-shaped entrances and alignment on winter solstice sunrise. The female cave architecture associated with the critical period for survival during winter solstice at the height of the Palaeolithic they suggest attests to an annual sanctuary for conducting a simple naturalistic religion emphasising fertilization of humans, animals and the earth and a possible first step for humans observing the sky. "More evidence of this kind is needed, as only statistical results can give a definitive answer" (Esteban & Tortosa, 2001, 13).

More evidence of another kind is also available from Parpalló cave. Over five thousand inscribed plaquettes dated from the Gravettian to the Magdalenian were found in the cave, with representations of animals and a possible image of a vulva dated to the Gravettian. Not mentioned by the authors is that other Parpalló plaquettes had inscribed serpents and notational systems tracking lunar phases (Villaverde,

1994; Marshack 1971, 102-6, 344-6). This combined evidence suggests that the astronomy of Parpalló cave is therefore not just solar but lunar-solar, and in which the lunar aspect is tracking the synodic phase cycle of the moon. If this interpretation is correct then it suggests that a Palaeolithic lunar-solar 'astronomy' is linked to the social, economic, political *and gender* aspects of some Palaeolithic rituals. But it remains the case that working from single case studies like Parpalló cave requires strengthening by moving towards aggregated data sets.

3. PRAILEAITZ I

In a 5 km stretch of the now lower reaches of the River Deba valley in the Basque country the archaeology of 22 caves reveals occupation remains dating from the Magdalenian, about 15,500 years BP (Peñalver et al., 2006). Some of this archaeology was consistent with stable dwelling places, although they also lived in 'cabins' in the valley. Other caves were complementary and temporary bases for exploiting specific local resources like obtaining raw materials and preparing tools, fishing and gathering molluscs, gathering fruits, tubers and leaves and for specialised hunting of various game animals. During the preceding period of the Solutrean and the Last Glacial Maximum hunting groups from the north retreated south into the Basque country as the northern climate became colder. The archaeology shows a rise in population density during both the Solutrean and Magdalenian in the Basque country.

Figure 1. Praileaitz I entrance

Groups consolidated into single valleys and, with the seashore then at least 10 km further out than today, split into coastal and inland hunter gatherers (Stanford & Bradley, 2012, 194-9). For these reasons the archaeological consensus is that these 22 caves in this small section of the River Deba valley were utilised by the same local group of Magdalenian inland hunter-gatherers. It is also understood that the resource stress consequent from the social circumscription of incomers required continual hunting and foraging with seasonal aggregation once or twice a year for ritual re-bonding of wider coalitions - one of the pre-requisites for 'complexity' (Straus & Clark, 1986).

One of these 22 caves is an exception to this general description. The Praileaitz I cave is located 55 metres above the River Deba on the north facing and very steep valley side. It has no dwelling or resource processing archaeology but does have an elaborate ritual archaeology. Its five metre high entrance is shaped like a vulva (Figure 1). The cave extends into the hillside for about 80 metres and after the vestibule chamber low entrances open serially into two inner chambers each about two metres high and then into a narrowing long low tunnel. Abstract parietal 'art' of red ochre dots are found about halfway along this tunnel. In the vestibule chamber a chocked seat-shaped stone faced outwards to the entrance. Just in front and to the right of this seat were a group of three black stone river-rolled perforated pendants and another single pendant in front and to the left of the seat (Figure 2).

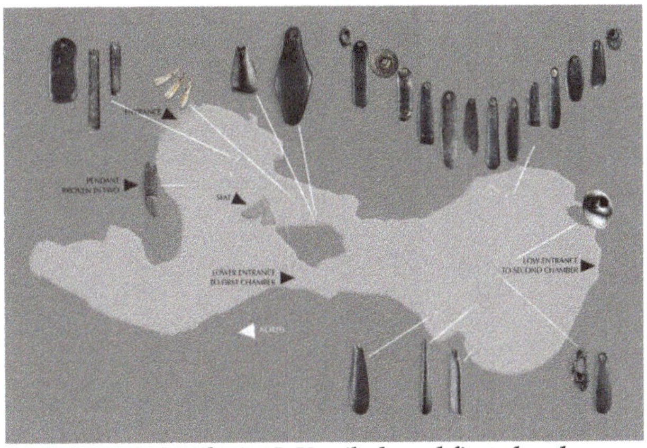

Figure 2. Práileaitz I: Vestibule and first chamber

A further 25 pendants were behind the seat and extending into the first inner chamber. Most of these pendants ranged from 7-12 cms in length. To the right of the seat was a red ochre crayon and a 17 cms long antler sharpened and perforated baton (Figure 3; Peñalver, 2014, 189). Just to the right and behind the seat were three distinctive perforated pendants of *Rupicapra rupicrapra* (wild mountain goat) teeth with the remnants of red ochre upon one. In line with both these two sets of objects and close to the low entrance into the first chamber were two further black stone pendants. One of these, the largest of all the pendants, is by silhouette a lozenge shape. Further, from inspection of the high resolution photographs taken by Otero (Peñalver et al., 2006), 11 of the 26 black stone pendants have inscribed upon them a series of marks that fit a lunar notation. The single stone in front and to the left of the vestibular seat was broken into two pieces, and the broken off part was located 13 metres away inside the second chamber (Figure 2). Since one other stone was broken by roof rock fall but found in articulated arrangement with its parent body, this movement of the part pendant into the inner chamber seems an intentional re-location to register pendants approximating to a total of 29½ . If the cave ritual included alternating the position of this piece with and away from its parent body then this would replicate the 29/30 alternation in the length of the synodic month, indicating a proto-concept of 29½.

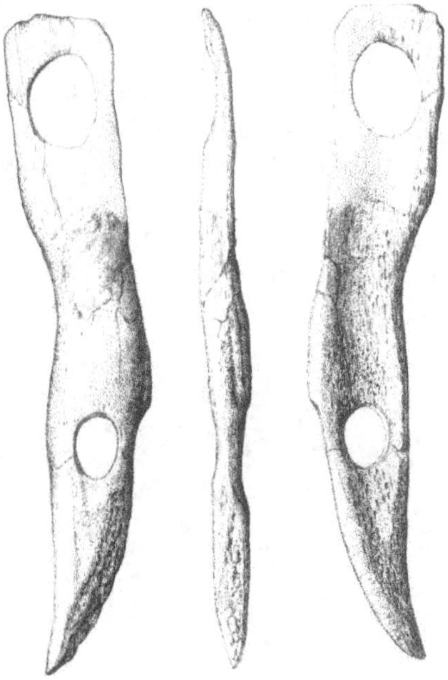

Figure 3 Antler baton from Praileaitiz I

Which, if any, of these fourteen features of Praileaitz I might have been selected by Magdalenian foragers as important for it to be their place of ritual? Using the case study method as a single Palaeolithic cave we might construct a number of narratives consistent with some or many of these features which could well overlap with the commentary provided by Esteban and Tortosa for the Parpalló cave. But since we have here not one but an aggregated data base of 22 caves two questions are made possible. Do the entrance alignments of Praileaitz I and the remaining 21 caves have any astronomical properties similar to Parpalló's, and does the full suite of the cave's properties indicate a specific portfolio distinct from those of the remaining 21 inhabited caves in the lower Deba valley which might add to the understanding provided for Parpalló?

4. RIVER DEBA FIELD WORK AND DATA

During March 2015 five caves, including Praileaitz I, were visited and measurements taken from inside the caves for the azimuths of their entrances and elevations of their outside horizons. Using a Suunto compass and clinometer, both accurate to half a degree, measurements were taken from the three metre inside position of the Praileaitz I seat as a back-sight benchmark. Azimuths were taken of cave entrance edges, the central axial alignment of the cave entrance and of their respective horizon altitudes. Because of very bad weather conditions during the week available for field work, the remaining 17 caves could not be visited, and the azimuth data for them was taken from the Archaeological Resumé (AR) (1995) reports and horizon data from HeyWhatsThat (2015). Corrections were made for the differing magnetic variation between the 1995 AR data and 2015 field data and to the long term variation in the obliquity of the ecliptic since the Magdalenian (Esteban & Tortosa, 2001, 13). Because this long term variation in the obliquity follows a sinusoidal curve at the time of the 15,500BP occupation it measured 23.9° (±0°.1) while today it is 23.4°. This accounts for a difference of just over half a degree, or one solar diameter, from present solar observations.

The azimuths for all 22 caves' entrances, standardised to a zero level horizon, are shown in Figure 4. Five of the caves have a cave entrance edge within ±5° of all four possible solstice alignments. Praileaitz I, the one cave chosen as a place of ritual, is the only one with an entrance edge aligned within 5° of summer solstice sunrise. There is therefore a less than 3% chance that this is a random selection (10/360 = 2.7%). Because all four solstice alignments are possible from the 22 caves in the lower Deba valley, we can test by inter-disciplinary method what is distinctive in the Praileaitz I cave's portfolio of properties.

5. ELABORATING SKYSCAPES

Esteban and Tortosa (2001, 12) showed that the winter sunrise entered the deepest chamber of the Parpalló cave only for the few days around the solstice. Similarly at Praileaitz I the summer solstice sunrise for about seven days enters the vestibule and a ray of sunlight reaches the low narrow opening into the first inner chamber. Into this chamber it will set up a glow of reflected sunlight for a few minutes in the early mornings of the summer solstice week, after which it will return into near total darkness. Considered as a signal, these are digital properties of a long period of 'Off' and a short period of 'On'.

If the Magdalenian foragers had wanted a cave that could have been used as an observing site for calendar purposes, then we would expect them to choose a cave with a wide entrance to track the changing horizon movements of the sun over the course of all or most of the year. Two caves' wide entrances offer such an analogue view of the sun's horizon movements - Arno 1 with a standardised azimuth range of 63°-125° and optimally a wider Santakutz with 47°-317°. Since the standardised solstice azimuths were at 55°, 123°, 237° and 304° for 15,500 years BP, then these two caves were options as solar sites for marking the time of the year. Contrarily Praileaitz I's entrance azimuth amplitude between 16°-56° fails as an observing site to see the sunrise's horizon position for all of the year except at summer solstice. In *not* choosing available caves with entrances entrained on a wide solar horizon we therefore conclude that calendar concerns were not a property sought for this place of ritual but that some digital property of a solstice sun was desired.

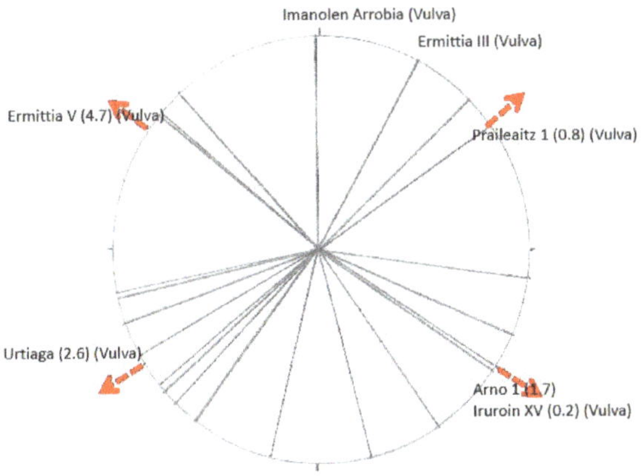

Figure 4 Standardised azimuths of 22 cave entrances in lower River Deba valley (Red dashed arrow indicates standardised solstice azimuth; Name; deviation from solstice azimuth; entrance shape).

If Praileaitz I was chosen for its female architecture to then use as a place for a 'simple fertility ritual', then we need to ask why the six other caves which also had vulva shaped entrances were not chosen. Three, Imanolen Arrobia, Ermittia I and Ermittia III, did not have entrance edges aligned on a solstice azimuth, whereas two others, Ermittia V, and Urtiaga, do have alignments on summer and winter solstice sunsets respectively. Therefore a vulva-shaped entrance alone was not sufficient for a fertility ritual in this cave nor was such an entrance combined with solstice *sunsets*. Iruroin XV has a vulva-shaped entrance and, like Parpalló cave, has an alignment of winter solstice sunrise. Given the choice of two caves with vulva-shaped entrances and summer or winter solstice sunrises, they chose Praileaitz I rather than Iruroin XV. The choice of Praileaitz I cave selected a vulva-shaped entrance with a *summer* solstice *sunrise*. The selection criteria seems to be not the parameter of winter, nor the abstract contemplation of the sky and any of the four of the horizon positions of the solstice sun, nor for a collective sanctuary during the middle of the Palaeolithic winter. What remains is the Praeleaitz I choice for the digital ritual property of the strongest ray of solstice sunlight available to enter the given darkness of the cave. This inference is reinforced by the availability of Arno I which is also aligned on winter solstice sunrise but did not have a vulva-shaped entrance.

In summary we can see from the wide choice of caves available to local foragers in the River Deba valley Praileaitz I was not chosen as a site for calendar calculation, nor as a refuge from the anxieties of a Palaeolithic winter, nor the astronomical observation of solstice sunsets, nor winter solstice sunrises that drove their choice of a place of ritual, but as a dark cave for lunar rituals combined with the digital summer solstice penetration of sunlight through the vulva-shaped entrance.

6. INTERPRETING SKYSCAPE ARCHAEOLOGY

Variation in cave locations, internal dimensions and material culture provide another resource that might specify the reasons for the choice and organisation of Praileaitz I as a place of ritual. Praileaitz I is located 55 metres up a very steep hillside above the River Deba. If convenience of access in this steep topography was a consideration, then on the other side of the River Deba from Praileaitz I was Langatxo cave just 5 metres above the river level. While Langatxo was not aligned on any solstice and looks to the west, Iruroin XV had an alignment on winter solstice sunrise and is lower than Praileaitz I at 44 metres. Clearly convenient access alone was not a criterion and some degree of remoteness may have been a requirement for this ritual. Like Parpalló cave Praileaitz I's first inner chamber was two metre high in the centre and roughly 6.5 metres in diameter. It could conveniently hold a small group of people. A paved vestibule (Peñalver, 2014, 73) suggests it did.

The type of material culture and its sequence from the entrance and into the inner reaches of the cave carries high symbolic value. The site report suggests two attributions according to the shape of two of the stone pendants. The lozenge-shaped largest pendant is seen as signifying a vulva (Peñalver et al., 2006, 113). Other scholars have made the same attribution to the use of the lozenge motif in early cultures (Gopher et al., 2011; Barber, 1991; Smith, 1965). If we apply this genital attribution by shape to those stones that by silhouette are pillar shaped then by extension we see these as signifying a phallus and silhouettes intermediate between these two would be gender ambiguous (Sims, 2015). The variation in shape of the 26 stone pendants can be seen to reference female, male *and* the possible performative achievement of ambiguous gender. Second, the pendant accompanying the lozenge-shaped pendant is suggested to be an imitation of an atrophied *Cervus elaphus* (red deer) canine (Peñalver et al., 2006, 116). Other scholars have shown that these teeth, which need to be extracted from an opened upper mandible, were commonly used especially in the Iberian Magdalenian (Mussi, 2002, 268). If we see these pendants as representations of gender and of species, and of metamorphoses between genders and species, then they fit the predictions suggested by Power (1997) that the achievement of ritual sanctity requires a performative appropriation of ambiguity between both gender and species in blood rituals. The skyscape archaeology of Praileaitz I supports this prediction.

First into, or last out of, the cave we see two phallic shaped pendants combined with one gender ambiguous pendant (Figure 2). Facing them is the vestibule seat with to its right three *Rupicapra rupicapra* (wild mountain goat) teeth with the remnants of red ochre on one and a red ochre crayon. There follow along the same orientation as the three pendants and three goat teeth two more pendants - the lozenge-shaped and atrophied red deer tooth pendants. The line that all three groups fall upon coincides with the same line of summer solstice sunrise entering the cave. This ray of sunlight continues on to enter the right hand flank of the low entrance into the first inner cave chamber which contains the remaining 20 pendants. This movement through the cave entrance has taken us through its 5 metre high facsimile of a vulva. The three wild mountain goat teeth are placed in the vestibule after the two phallic and one gender ambiguous pendants. While the symbolism of vulva and blood (red ochre crayon) may signify either menstruation and/or childbirth, three additional significations are suggested by this choice and arrangement of materials. First, the symbolism of cutting (teeth, red ochre on teeth) suggests *vagina dentata* - a symbol universal to patriarchal cultures (Walker, 1983, 1034-7). Second, the wild mountain goat changes from a winter coat of blackish brown to red in the summer. Third, female *Rupicapra rupicapra* are gregarious and philopatric and exhibit strong coalitions. They herd in groups of on average over 20 adult females and kids, and females will care for orphaned kids (Gunderson, 2003). If males were appropriating these significations of female blood coalitions, as suggested by the change in pendant shape from phallic to vulva symbolism, then the presence of a red ochre crayon adds to the further possibility of the male appropriation of synchronous menstrual blood symbolism - gender and species ambiguity being another universal component of male initiation ceremonies (Gregor & Tuzin, 2001).

The further componential analysis of the material culture items found in Praileaitz I lends weight to this perspective. Peñalver (2014, 189) shows that the antler baton, found next to the crayon, has perforations and a worked shape that so weakens its integrity by length that it could not possibly have been used to throw or straighten spears (contra Underwood, 1965). However the baton's finely worked point has been ground to create a piercing and gouging instrument (Figure 3). Whoever sat in the vestibule seat faced those who entered. Equipped with red ochre crayon and a cutting baton such a person has the cave position and equipment to perform as a specialist in and instigator of a blood ritual. This baton comes from a *Cervus elaphus* (red deer) that loses its grey-brown coat of winter and like the wild mountain goat also grows a red coat during the summer. It is interesting that they did not choose antler for this tool from the readily available reindeer. Red deer are herd animals and their calves are born in June, signifying a time of collective birth. The redundant multi-media symbolic overlays of cave entrance, pendants, teeth, crayon, red ochre, baton and coalition behaviour and seasonal properties of wild mountain goat and red deer all mutually reinforce and focus a signification that strongly suggests a male initiation blood ritual imitating and possibly confiscating the properties of a menarche ritual. This inference is consistent with the evidence of explicit genital male representations in portable art from the European Upper Palaeolithic, and especially Magadalenian, which display phallic scarification, piercing and tattoos (Angulo et al., 2011). Instead of a simple fertility ritual the properties we have found for Praileaitz I, all aligned on the ray of light of summer solstice sunrise entering the cave, point towards a male initiation ritual that formally shares the structure of a lunar menarche ritual.

Symbolism continues deeper into the cave. For those entering the cave from the outside the low height of the vestibule ceiling to the left and the paving layout moving to the right sends entrants in an anti-clockwise direction through the vestibule and towards the first inner chamber in a line parallel to the line of the ray of sunlight at summer solstice sunrise. The low crawling height entrance requires descending onto all fours to gain access, infantilizing participants. In this darkening space are 20 pendants arranged in an arc, the middle three of which are all arguably signifiers of gender ambiguity - circular, oval and pear-shaped pendants (Figure 2). Crawling similarly into the next darker chamber reveals a central natural feature of a block of stalagmites and stalactites. Continuing further into the cave through another low entrance the now totally dark narrowing tunnel first ascends and then descends steeply to a fourth entranceway. By a sharp left turn into a tunnel just one metre wide it turns to the right and brings us to that part of the 1.5 metre wide passage which has parietal red ochre dots on walls which also evoke both uterine and vulva shapes. The small dimensions of this tunnel would not allow the presence of a group of people but just one or two individuals at most. According to Reznikoff (2014) there is a greater that an 80% chance that this position in the cave complex marked with red ochre dots will have maximum resonance properties. Returning to the first inner chamber and resuming the anti-clockwise circuit the pendant layout ends with the vulva-shaped and red deer tooth pendants bathed in summer solstice sunrise and perhaps signifying the completion of a gender and species metamorphosing initiation ritual.

In summary these twenty or so properties do not suggest a simple fertility ritual but a demanding and isolating gender and species ambiguous seclusion ritual by blood-letting. This is a universal characteristic of male initiation rituals (Whitehouse 1996).

7. CONCLUSION

From the 22 caves in the lower Deba Valley in the Basque country we have found that the Praileaitz I cave possesses a distinctive suite of characteristics that specifies aspects of its ritual purpose during the Magdalenian.

The 29.5 pendants, notation on some of these pendants and anti-clockwise route of vestibule pavement and pendant layout which extends into the first inner chamber all represent aspects of the lunar synodic cycle. These properties add credibility to Marshack's claims and weaken the claims of those who deny 'Palaeolithic astronomy'. The alignment of the Praileaitz I cave entrance on summer solstice sunrise, compared to locally available caves with every other alternative solstice alignment, indicates that this was a choice of cave entrance alignment made by local hunter-gatherers. The cave's entrance alignment suggests the displacement of lunar rituals conducted there from a monthly cycle onto an annualised ritual timed to synchronise with the summer solstice. Together these two astronomical properties of this cave's rituals represent a solarised lunar (synodic) ritual. This indicates a departure from Marshack's emphasis on lunar notational systems for discrete months, as at Parpalló cave five millennia earlier, and instead displays a substitution of an abstracted ritual month that can be mobilised at any moment. As such they suggest a confirmation of Marshack's thesis with the qualification of attenuating rituals from a discrete monthly lunar timescale to an annual solarised timescale.

The cave's vulva-shaped entrance, uterine and vulva characteristics of the inner chambers and tunnels, the red ochre crayon and the red ochre 'art' all suggest a female blood dimension to the ritual conducted in the cave. However six other caves also have vulva-shaped entrances, indicating that these were not sufficient characteristics for conducting ritual here. The artefacts of goat and deer teeth, sharpened baton, red ochre and the summer season redness of wild mountain goat and red deer support additional dimensions to this gender dimension. The overlapping and mutually reinforcing signifiers of cutting-blood-summer all suggest a blood ritual by piercing. The stone pendants are shaped to represent both gender and species identity and metamorphosis, suggesting a male appropriation of astronomically timed blood flow. The secluded location of Praileaitz I 55 metres above the valley floor, the inaccessibility of the cave's inner chambers, the complete darkness of the deep tunnel with red ochre 'art' and the extreme claustrophobia and sense of isolation engendered in the route to and dimensions of the red-dotted tunnel all suggest a fear-inducing experience. While this ritual may have been timed to begin and/or end with the start of the brightest longest days, once entering the cave it was digitally contrasted with the journey into a long dark cave complex that simulated the female sexual anatomy alongside symbolic artefacts for conducting a simulated lunar-scheduled blood ritual. These 34 or so enumerated properties all add to the single interdisciplinary model of a solarised lunar male initiation blood ritual.

ACKNOWLEDGEMENTS

We thank the anonymous reviewers for their constructive comments.

REFERENCES

Angulo, J. Garcia-Diez, M. and Martinez, M. (2011) Phallic decoration in palaeolithic art: genital scarification, piercing and tattoos. *Journal of Urology*, 186:6, 2498-503.

Archaeological Resumé (1995) *Munibe Anthropologia Arkeologia*. Supplemento 10. San Sebastián, Donostia

Barber, E. (1991) *Prehistoric Textiles*. Princeton, University Press.

d'Errico, F. (2009) The Oldest Representation of Childbirth. In P. Bahn (Ed.), *An Enquiring Mind.* Oxford, Oxbow, 99-110.
Esteban, C. and Tortosa, J. (2001) The winter sun in a Palaeolithic Cave: La Cova del Parpalló. In C. Ruggles, F. Prendergast and T. Ray (Eds.) *Astronomy, Cosmology and Landscape,* Bognor Regis: Ocarina Books, 8-14.
Farrer, C. (1998) *Living Life's Circle: Mescalero Apache Cosmovision.* Albuquerque, University of New Mexico.
Gaffney, V. et al. (2013) Time and a Place: A luni-solar 'time reckoner' from the 8th millennium BC Scotland. *Internet Archaeology* 34. http://dx.doi.org/10.11141/ia.34.1
Gopher, A. and Orelle, E. (2011) An Alternative Interpretation for the Material Imagery of the Yarmukian, a Neolithic Culture of the Sixth MilleBC in the Southern Levant. *Cambridge Archaelogical Journal* 6:2, 255-79.
Gregor, T. and Tuzin, D. (2001) *Gender in Amazonia and Melanesia.* London, University of California Press.
Gunderson, D. (2003) "Rupicapra rupicapra" (On-line). *Animal Diversity Web,* Accessed January 21, 2016 at http://animaldiversity.org/accounts/Rupicapra_rupicapra/
Hayden, B. and Villeneuve, S. (2011) Astronomy in the Upper Palaeolithic? *Cambridge Archaeological Journal* 21:3, 331-55.
HeyWhatsThat (2015) http://www.heywhatsthat.com/.
Marshack, A. (1972) *The Roots of Civilization.* London, Weidenfeld and Nicolson.
Mauss, M. (1969) *The Gift: Forms and Functions of the Exchange in Archaic Societies.* London: Routledge and Kegan Paul.
Mussi, M. (2002) *Earliest Italy: An Overview of the Italian Palaeolithic and Mesolithic.* London: Kluwer Academic/Plenum Publishers.
Peñalver, X., San Jose, S., J. Mujika and X. Otero (2006) *Praileaitz I.* San Sebastián, Donostia.
Peñalver, X. (2014) *La Cueva de Praileaitz I.* San Sebastián, Donostia.
Power, C. (1997) The Woman with the Zebra's Penis: Gender, Mutability and Performance. *Journal of the Royal Anthropological Institute,* 3:3, 537-560.
Rappaport, R. (1979) The obvious aspects of ritual. *Ecology, Meaning, and Religion,* California, North Atlantic Books, 173-221.
Rappenglück, M. (2015) Possible Astronomical Depictions in Franco-Cantabrian Palaeolithic Rock Art. In C. Ruggles (Ed.), *Handbook of Archaeoastronomy and Ethnoastronomy,* New York, Springer, 1205-12.
Reznikoff, I. (2014) On the Sound Related to Painted Caves and Rocks. Ikäheimo J., Salmi A., Äikäs (Eds.) *Sounds Like Theory,* Finland, Monographs of the Archaeological Society of Finland, 101-9.
Sims, L. (2006) The solarisation of the moon: manipulated knowledge at Stonehenge. *Cambridge Archaeological Journal* 16:2, 191-207.
Sims, L. (2015) Out of Africa: the solarisation of the moon. In F. Pimenta, N. Ribeiro, F. Silva, N. Campion, A. Joaquinito and L. Tirapicos (Eds.) *Stars and Stones: Voyages in Archaeoastronomy and Cultural Astronomy,* 200-7.
Smith, I. (1965) *Windmill Hill and Avebury.* Oxford, Clarendon Press.
Strauss, L. and Clark, G. (1986) *La Riera Cave: Stone Age Hunter-Gatherer Adaptations in Northern Spain.* Tempe, Arizona State University Anthropological Research Papers, 36.
Underwood, L. (1965) Le baton de commandement. *Man* 65, 140-4.
Villaverde, V. (1994) *Arte Paleolıtico de la Cova del Parpallo* Valencia. Servei d'Investigacio Prehistorica.
Walker, B. (1983) *The Woman's Encyclopedia of Myths and Secrets.* San Francisco, Harper.
Watts, I. (2005) 'Time, too, grows on the moon': some evidence for Knight's theory of a human universal. In W. James and D. Mills (Eds.), *The Qualities of Time: Anthropological Approaches,* 95-118.
Wengrow, D. and Graeber, D. (2015) Farewell to the 'childhood of man': ritual, seasonality, and the origins of inequality. *Journal of the Royal Anthropological Institute* 21:3, 597-619.
Whitehouse, H. (1996) Rites of terror: emotion, metaphor and memory in Melanesian initiation cults. *Journal of the Royal Anthropological Institute,* 2:4, 703-15.
Zeusse, E. (1979) *Ritual Cosmos: the sanctification of life in African religions.* Ohio, University Press.

3

SHRINES OF RAM-HEADED DIVINITIES AND CANOPUS: SKYSCAPING AT HERAKLEOPOLIS MAGNA

Juan Antonio Belmonte[1], M. Carmen Pérez Die[2] and Lucía Díaz-Iglesias Llanos[3]

[1] *Instituto de Astrofísica de Canarias and Universidad de La Laguna, Tenerife, Spain*
[2]*Museo Arqueológico Nacional, Madrid, Spain*
[3]*Centro de Ciencias Humanas y Sociales, CSIC, Madrid, Spain*

Corresponding author: Juan A. Belmonte (jba@iac.es)

ABSTRACT

In the first decade of the 21st century, the Egyptian-Spanish Mission on Archaeoastronomy of ancient Egypt performed a detailed statistical analysis of the orientation patterns of the temples of Pharaonic Egypt, resulting in most interesting outcomes such as the justified proposal of seven families of astronomical orientations (Belmonte, Shaltout and Fekri, 2009). A preliminary analysis of what we may dub as cosmic landscapes in certain Egyptian locations, such as the Giza pyramids or Karnak, followed up (Belmonte, 2012: 215-250). On the other hand, the Spanish Archaeological Mission at Herakleopolis Magna has been excavating for several decades in one of the most important ancient sites of Middle Egypt, making extraordinary archaeological discoveries in what once was the capital of Egypt during the 9th and 10th Dynasties and again of a chiefdom for a short period during the Lybian epoch (Pérez Die, 2009). It was hence decided that a new complete survey of the site should be made with an astronomy and landscape perspective in mind. This paper presents the result of such a survey where the relationship between land- and skyscape at the main monuments of the city is put in the spotlight. One important outcome has been the possible orientation to Canopus – the second brightest star of Egyptian skies – of the main temple of the city, the one devoted to the patron divinity of Herakelopolis, the ram-headed god Heryshef. Epigraphy is integrated in the analysis and possible mythological relationships are explored, including the connections with other ram-headed deities of the Egyptian pantheon, such as Amun-Re or Banebdjedet. Interestingly, the temples of these divinities at Thebes and Mendes also show alignments that could be related to Canopus, offering a new challenge in the relationship between skyscaping and religion in the civilization of Pharaonic Egypt.

KEYWORDS: Herakleopolis, Heryshef, landscape, skyscape, Canopus, ram-headed deities, Thebes, Mendes

1. INTRODUCTION: ASTRONOMY & LANDSCAPE IN HERAKLEOPOLIS MAGNA

This paper represents a true interdisciplinary effort which is the result of a collaborative work between three members of the Spanish Archaeological Mission at Herakleopolis Magna: an astronomer specialized in ancient Egyptian astronomy, the Egyptologist who possibly best knows the site under discussion and Director of the Excavation, and an epigraphist who has specialized in ancient Egyptian religious texts (notably those those built around Herakleopolis Magna, which where the backbone of her PhD dissertation, Díaz-Iglesias, 2012).

The ancient Egyptian city of Henennensu, current Ehnasya el Medina in Beni Suef province, was the capital of the 20th Upper Egyptian nome called *Naret Khentet* (i.e. the Upper Naret tree). The town is located very close to the right bank of the Bahr Yusuf, an arm of the Nile that waters the Fayum Oasis. This province was one of the main centres in the history of Pharaonic and Coptic Egypt. Its origins go back to the first dynasties (as written sources demonstrate) but, for the moment, the archaeological evidence dates back only to the end of the Old Kingdom.

The Greeks called it Herakleopolis Magna, identifying the main ram-headed god of the city, Heryshef, with the Hellenic Herakles. In this paper, the main results of an archaeoastronomical study (see Belmonte, Pérez Die and Díaz-Iglesias, 2015)[1] carried out at Ehnasya in ancient religious and funerary structures are presented. This was made within the auspices of the Spanish Archaeological Mission (Pérez Die, 2012). We will also try to contextualize these results within the framework of our knowledge about other ram-headed divinities, such as Amun or Banebdjedet, and their temples in other sites of Egypt.

The most significant results focus on the orientation and location of the main temples of the city (see Figure 1), notably the temple of Heryshef.

This sacred structure, in addition to its undeniable relationship with the in-wall necropoleis of the First and Third Intermediate Periods (in the following FIP and TIP, respectively; Pérez Die, 2010), would have been astronomically orientated (azimuth of c. 201°) to the setting of Canopus, the second brightest star of Egyptian skies, at the time of the first monumental building of the temple during the Middle Kingdom (c. 1850 B.C.). This orientation could be perhaps related to the stellar epithets of the tutelary deity of the province, the ram-god Heryshef.

Such stellar epithets (Díaz-Iglesias, 2014: 472-473) placed Heryshef as the leader of the cosmic sphere or of some of its elements (see Figure 2). Along with the more usual "King of the Sky" (*Nesu pet*) and "Lord of the Sky" (*Neb pet*), written sources at Herakleopolis, and even at Thebes, mention the exceptional description of Heryshef as "Pillar of the stars" (*Iunu sebau*) and "Ruler of the stars" (*Heka sebau*), originally assigned (at least the first one) to the deceased monarch in his ascent to the sky and then, within the divine pantheon, only to the Herakleopolitan ram-headed divinity. Other texts included the statement "he [who] made his place in the sky as the Lone Star" (see Fig. 2), linking an Osirianised Heryshef (he is often found in texts and reliefs in close connection with an Osiris of Naref, 'the One who is at the front of Naret') with a singular star. This Lone Star has seldom been identified with Venus as Morning Star or Capella (Krauss, 1997) but Canopus seems also a reasonable candidate. All these epithets show a trend towards the stellar and solar transfiguration of Heryshef.

In the celestial sphere, it is important to take into account both the stars and the big luminaries, the Sun and the Moon, and their relationship to the calendar and festivals as possible frames of reference. The documented presence of stellar titles among the epithets of Heryshef justifies it (see Belmonte, Pérez Die and Díaz-Iglesias, 2015). In addition, the relevance of the stars is well characterized in the case of the bright Sirius (a reference for another family of temple orientations and herald of the inundation). It was personalized in the goddess Sopdet, one of the manifestations of Isis (and of Hathor, Heryshef's partner).[2] Within this context we may put other important constellations such as Meskhetyu, widely analyzed in previous studies as the dominant in the northern skies where the Egyptians recognized a most relevant part of the seat of immortality (Belmonte, Molinero Polo and Miranda, 2009).

Heryshef was the patron god of Herakleopolis Magna. Already witnessed by epigraphic sources from the early dynasties, his spheres of action are deployed into four spheres: water and purification, royal connections, reverential fear, and external manifestation or capacity of action (in Egyptian called "ba", the same word used to denote rams). These spheres of action related him to defining attributes of ram-headed divinities. His stellar and royal facets are highlighted (Díaz-Iglesias, 2014: 467-475) as well as his relationship to Osiris (as his "ba" or external manifestation), and perhaps to Amun.

---

[1] The rough data of the archaeoastronomical fieldwork is presented in Table 1 in this publication.

[2] A Queen Shepensopdet has been reported at Heracleopolis in the excavations of the TIP Necropolis (Pérez Die 2010: 443).

These defining factors of Heryshef seem to be linked to the astronomical orientation of his sanctuary in Ehnasya. However, this case is not unique, since other parallels point in the same direction. For example, the unfinished and aniconic temple of Qsar As-Sagha, located north of the Fayum Lake some 60 Km to the northwest of Herakleopolis as the crow flies was certainly built in the Middle Kingdom similarly to the first confirmed monumental complex of Heryshef. It has an orientation that would be compatible to the rising of Canopus in the appropriate period. More examples will follow. But let us come back to Herakleopolis since this is not all what the city has to offer.

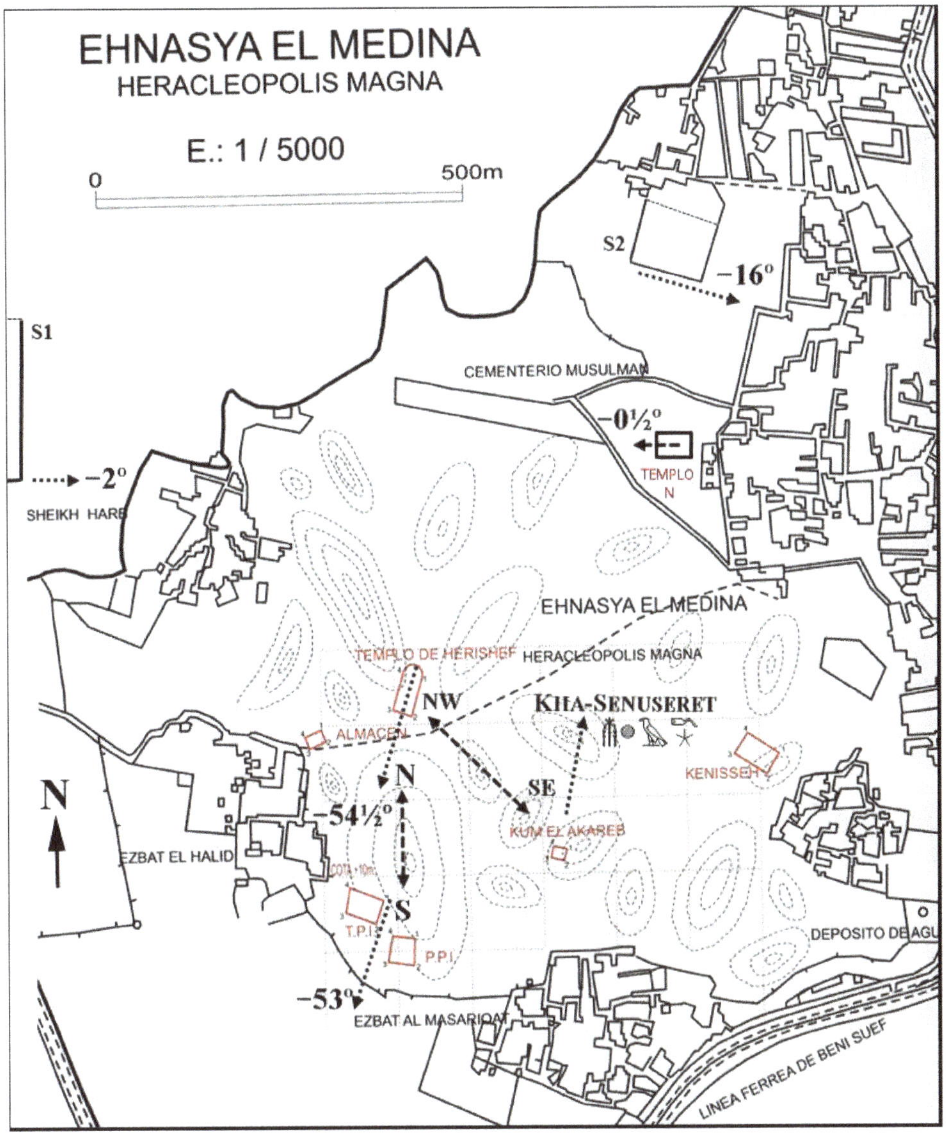

Figure 1. Plan of Herakelopolis Magna within the context of the modern village of Ehnasya el Medina, including the main monuments of the city. The most significant topographical and astronomical (represented by the value of the declination) alignments are plotted. The temple of Heryshef could have originally been orientated to the star Canopus as the chapel of the Third Intermediate Period necropolis (T.P.I. in the plot) twelve centuries later. The sanctuary at Kum el Akareb may have shown a double alignment to the "Imperishable" stars of the northern skies (notably those of Meskhetyu) which were orbiting above the pyramid complex of the Middle Kingdom Pharaoh Senuseret II in Lahun. The recently excavated – by an Egyptian team – "North Temple" is Equinoctially orientated, precisely. S1 and S2 stand for the two hypothetical saltpans that could be the modern counterparts of the classical Natron and Maat Lakes related by many ancient source to the city of Herakleopolis Magna . The expansion of the Muslim cemetery is challenging the northern sector of the site at the northern limit of the Spanish concession. See the text for further discussion. Figure elaborated by the authors upon a site plan of the excavation area by courtesy of the Archive of the Spanish Archaeological Mission at Herakleopolis Magna.

The site included two splendid ancient necropoleis of the FIP and TIP (periods when Herakleopolis rose to preminence) excavated by the Spanish Mission (see Fig. 1). The former was located in a north-south (cardinal) relationship to the main temple, a fact which might be significant (the "imperishable" stars would orbit above the temple). The latter was built on the axis of the temple and nearly with the same pattern of alignments. Interestingly, the chapel on the burial precinct (from out looking inside) has an orientation that would fit the setting of Canopus c. 800 B.C., adjusting the diachronic variation of declination of the star due to the precession of the equinoxes. In the vicinity of the necropolis of the FIP, in a place called Kum el Akareb (see Fig. 1), there are remains of another building that must almost certainly have been an imposing temple. Nowadays, only part of the portico is well preserved consisting of several cyclopean rectangular blocks, some columns and an ashlar with the name of the king Neferusobek, a woman who was the last sovereign of the glorious 12th Dynasty. Three colossi of Ramses II, possibly reutilized, were found in three of the building corners. Two of them, exposed today in the garden of the Cairo Museum, have been recently tentatively identified as portraits of the Middle Kingdom king Amenemhat IV, direct predecessor of Neferusobek and perhaps her brother. The third one, found in wild excavations during the 2011 "revolution", still unpublished and now in the Service of Antiquities Magazine at Ehnasya has a cartouche of Senuseret (III).

Figure 2. Temple, iconography and "celestial" epithets of Heryshef. Adapted aerial image of Heryshef temple showing where the data were collected and golden statue of the deity with his name in hieroglyphs, meaning "The One who is upon his Lake". The five celestial epithets from top to bottom are: "King of the Sky", "Lord of the Sky", "Sovereign of the Stars", "Pillar of the Stars", and "He who is stablished in the Sky as the Lone Star". Diagram of the authors based on images by courtesy of José Javier Martínez, Boston MFA, and the Spanish Archaeological Mission at Herakleopolis Magna.

The topographical orientation of this large sacred structure to *Kha-Senuseret*, the pyramid complex of Senuseret II at Lahun, more than ten kilometres to the north, is very appealing and could reinforce the idea of an original construction of this building in the Middle Kingdom. An astronomical orientation could also be suggested since the "imperishable" constellation of Meskhetyu would have had its easternmost position in the sky above Senuseret II complex c. 2000 B.C., precisely.

An additional monument, recently excavated by a team of Egyptian archaeologists of the Department of Antiquities under the lead of Ahmed Galal, is a spectacular sanctuary located close to (and nearly surrounded by) the expanding Muslim cemetery of Ehnasya. The construction reuses an enormous amount of blocks of the Middle and New Kingdoms presumably from other temples, possibly including the one of Heryshef and perhaps of the still unidentified temple of Hathor, his consort and partner. However, several additional blocks show the typical *opus quadratum* of Roman buildings. Hence, this third temple could have been built in the Roman or Byzantine periods.

The temple shows a very marked equinoctial orientation, which is exceptional in Egypt later than the pyramid age (when it was the dominant one) but is frequent in Hellenistic buildings and Christian churches (González-García and Belmonte, 2014 and 2015). This poses an interesting question mark on the kind and purpose of this building. Unfortunately, the outcomes and archaeological discoveries of this excavation have not been published yet, remain completely unknown and cannot be used to extract further conclusions.

Therefore, the archaeological site of Herakleopolis Magna includes a series of temples and sanctuaries of different epochs, in a peculiar environment, likely to be interpreted within a general context of landscape archaeology, where the archaeoastronomical analysis of the skyscape plays a most relevant role.

As a corollary, we will enter now in a more controversial issue. When analysing the fieldwork data of Ehnasya we used modern satellite images of the area comparing them with old reports and maps of earlier researchers on the site (e.g. Naville, 1894; Wilcken, 1903). There were two rectangular enclosures that strongly called our attention, located to the west and the north of the Spanish concession (but unfortunately out of it). These are marked as S1 and S2 in Figure 1. According to early reports these would be two saltpans of unknown origin, called Ma'm al Kôm el Assâra and Mellaha, respectively, but that were in use in the 19th century. Their orientation (nearly cardinal in one case) and location (in close apparent connection to the axes of the North temple, see Belmonte, Pérez Die and Díaz-Iglesias, 2015) was appealing.

In the religious written sources related to Herakleopolis (from the Coffin Texts onwards, Díaz-Iglesias, 2014) there are several references to a couple of sacred, perhaps seasonal, lakes which received the names (with different variants) of Lake Natron and Lake Ma'a(t). The former could perhaps be connected with an endorreic area in the contour of Herakleopolis (sometimes confused with the lake produced at high water within the temple of Heryshef, see Fig. 2) where different salt deposits ("natron") could be collected in low waters. The latter had a name that could be tentatively translated as "Upper Lake" or "Lake that is in the height" (nothing to do with its nearly homophonous word *ma'at*, i.e. the cosmic order) which suggests that it was located in an upper or superior area which might only be filled up during the inundation.

Interestingly, structure S2 major axis could be assigned a declination of c. −16° (see Fig. 1) which fits, within the errors, to the declination of Sirius/Sopdet, the harbinger of the inundation, for a long part of Egyptian history. Could structures S1 (cardinally orientated) and S2 (orientated to Sopdet) be modern adaptations or rearrangements of the lakes of ancient sources? These have till now escaped any correct archeological identification. Hence, it would be fascinating if this were the case. Indeed a more detailed analysis of both structures ought to be performed in future seasons that may hopefully shed some light over this important, controversial but also very suggestive issue.

2. RAN-HEADED DEITIES, TEMPLES AND SKYSCAPING

During the six campaigns of the Egyptian-Spanish Mission of Archaeoastronomy of ancient Egypt a total of some 350 temples were measured. This permitted a pioneer statistical study of Egyptian temple alignment patterns which discovered a total of seven families of astronomical orientations (Belmonte, Shaltout and Fekri, 2009). One of these families (numbered V) was assigned to Canopus. Nearly twenty temples across Egyptian geography could be related to it. Interestingly, several temples associated with ram-headed divinities - who were also somehow related to Osiris - were among them.

In this sense, some inscriptions recording stellar epithets of Heryshef have been found in the Amun-Re religious complex at Karnak (Thebes) in two reliefs dating to the reign of Ramses III. Amun-Re was normally represented in human form, although his relationship with the ram is well-known. Suffice it to remember that his bark and the sphinx avenues were always decorated with ram-headed images and statues. The relationship between the nocturnal aspect of Re (who is depicted as a ram-headed creature) and Osiris is also a well stablished fact as the famous relief of Nefertari's tomb – and in other royal tombs in the Valley of the Kings – demonstrates. In the same line of argument, numerous structures at Karnak had their axes presumably orientated to the setting of Canopus. These include the VII and VIII pylons creating a secondary axis of the complex (Gabolde, 1999) that was perpendicular to the main axis of the temple which was solstitially orientated (Hawkins, 1973; see Figure 3). The temple of Ramses III and the TIP chapel of Osiris Heqadjet (although with their gates open to the north) could have had their axes orientated to the setting of Canopus as well (Belmonte, Shaltout and Fekri, 2009: Appendix II). This series of facts would reinforce our hypothesis.

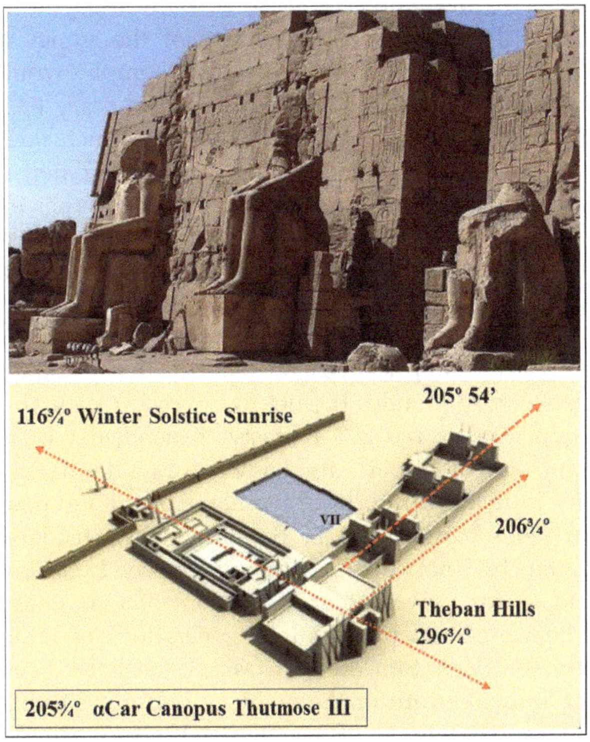

Figure 3. The VIIth pylon of the temple of Amun-Re at Karnak, the first to be erected in the processional way to the temple of Mut during the reign of Thutmose III and Hatshepsut. This pylon, and the VIIIth one, defined a secondary axis of the temple that was not exactly perpendicular to the main axis which was orientated to sunrise at the winter solstice. This axis, however, was aligned, either by chance or probably by design, to the setting of Canopus during the reign of Thutmose III. See the text for further discussion. Diagram by the authors

It is worth noticing that the temple of the Oracle at the Oasis of Siwa, where Amun was conceived and represented as a ram-headed god (Fakhry, 2004), could also be oriented to Canopus (Shaltout, Belmonte and Fekri, 2007). The Oasis of Siwa is one of the most important areas of Egypt outside the Nile Valley and, from our personal point of view, is by far the most fascinating of all the oases of the Western Desert. After the foundation of Cyrene by Dorian colonizers in 631 B.C., the kings of Egypt became interested in the oasis which was administered as a sort of vice-kingdom until Roman times.

The Oasis of Siwa is very rich in archaeological sites of Egyptian typology, with the remains of at least 12 temples of which the temple of the Oracle of Amun at Aghurmi is by far the best preserved and the most interesting. This temple has a splendid location in a low rocky outcrop above the palm-tree sea of the oasis with open views in all directions. It was built during the reign of Amasis (570-526 B.C.) perhaps over an earlier building, but the foundations of the temple have never been excavated in depth. The southeast horizon as seen from the sancta sanctorum of the temple is dominated by the hills of Djebel Takrur, where the quarry of the stones to build the temple was located. This orography would have been dominated in antiquity (6th century B.C.) by the appearance of Canopus in the winter skies. The temples of Khamisa and Zeitun, both dated in late Ptolemaic or Roman times, would be other examples of presumable orientation to the second brightest star in the Siwan skies (Shaltout, Belmonte and Fekri, 2007).

Heryshef had a possible Lower Egyptian counterpart in the ram-headed god Banebdjedet of the ancient city of Mendes (Djedet in Egyptian), capital of Egypt during the 29th Dynasty. Banebdjedet ("The Ram Lord of Djedet", also known as Mendes like the city in the Greek form of the name of the Egyptian king Nesobanebdjedet, i.e. Smendes) was considered as the "ba" (note the homophony between *ba* for ram and *ba* for power of manifestation) of four deities who were known to be four generations of father and son (Geb, Shu, Re and Osiris), the latter being the most relevant of the quartet (Redford, 2010). A substantial temple of the god existed at Mendes whose most representative structures were four gigantic, monolithic naoi dedicated to each of these divinities of which only one survives today (see Figure 4).

Although the temple of Banebdjedet built by the kings of the 29th Dynasty was open to the north, there are archaeological evidences that this was not the case in earlier periods and that the temple - or parts of it – could be facing south in previous periods of Egyptian history. Hence, with an azimuth of 22° (202° if south, Shaltout, Belmonte and Fekri, 2007; see Fig. 4), it would be nearly parallel to the temple of Heryshef at Herakleopolis Magna and, taking into account the different latitude, and possibly an alternative epoch (c. 600 B.C. or later), the temple axis could also have been orientated to the setting of Canopus. This similarity could indeed reinforce the association of Heryshef's temple at Herakleopolis, and even of the ram-headed god himself, with this bright star.

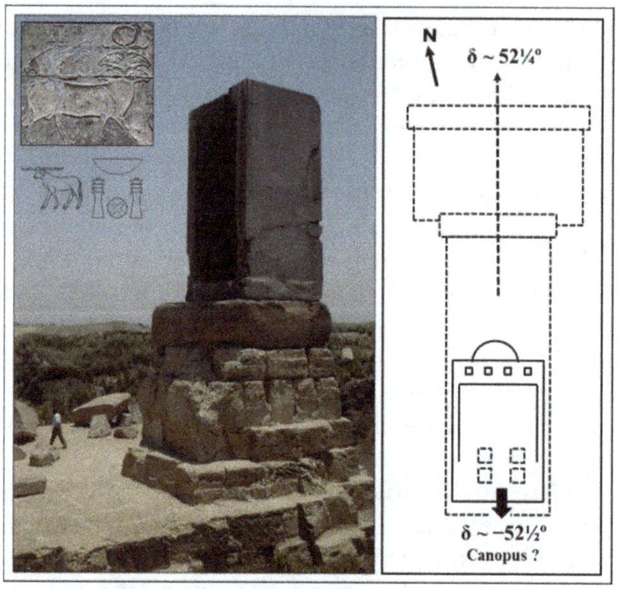

Figure 4. The only surviving exemplar of the four huge naoi once erected at the sancta sanctorum of the temple of the four ram-headed god Banebdjedet (image and glyph name within the diagram) at Mendes (ancient Egyptian Djedet), capital of Egypt during the 29th Dynasty. The temple plan shows that it is today orientated north. However, archaeological excavations on the site show that it might have originally been orientated towards south and hence to the setting of the star Canopus. See the text for further explanations. Diagram of the authors.

However, it is worth mentioning that neither of the two best known temples of another ram-headed deity, the "creator" god Khnum, at Elephantine and Esna, which are orientated closer to southeast, had any alignment compatible with the Canopus family (Belmonte, Shaltout and Fekri, 2009: Appendix II). A detailed comparison between the mythology associated with this divinity and his ram-headed counterparts could be interesting to stablish similarities and discrepancies which may shed some light on this particular issue.

On the contrary, a very interesting case of the Canopus Family would be that of the Isis temple complex at Philae. The main axis of the temple of Isis on the site was diverted to a declination of −53½° and hence could have been orientated to the setting of Canopus. We should not forget that Isis was the consort of Osiris. In this role, it might have been significant to orientate her temple to a star which could have been associated with ram-deities, who themselves were so closely related to Osiris.

3. CONCLUSIONS

It is not self-evident whether the ancient Egyptians recognized Canopus or not. They probably did so; being the second brightest star in their skies, its presence would not have passed unnoticed, giving way to mythological associations also reflected in architectural constructions. However, we have been unable to undoubtedly identify its name, astronomical correlations or possible religious connections (Lull and Belmonte, 2009). There are, however, a couple of interesting exceptions. On the one hand, the very late reference by M. Capella, who called it *Ptolemaeus* in honour of King Ptolemy Lagos (Allen, 1963). It should be remembered that kingship is one of the main attributes of Heryshef.

On the other hand, a possible relation between Canopus and Osiris (and hence to the mythology of Isis) can be inferred in Plutarch [*The mysteries of Isis and Osiris*, lxvi], although it is not clear if the text refers to the star itself or to the pilot of the vessel Argo. However, Argo Navis, clearly the constellation now (where Canopus is located at its helm), is related to the boat of Osiris within the same paragraph. The traditional Coptic name of the star may also relate it to sailing. Besides, this correlation could also be inferred from a controversial passage of the *Book of Day and Night* mentioning Osiris, who is "behind" Sah (Piankoff, 1942) and hence of a celestial body which would be located behind this ancient Egyptian constellation which is often identified as Orion or parts of it (Lull and Belmonte, 2009). So the star Canopus may be related in some way or the other to the Osirian Triad and hence to the deities who were somehow connected to this charismatic god as a celestial manifestation of them.

This might be one explanation for the stellar attributes of the god Heryshef (see Fig. 2), his relationship with Osiris, and the orientation of his temple at Herakleopolis. The same might apply for other ram-headed divinities, such as Amun-Re or Banebdjedet who were somehow linked to Osiris and Re as well.

The land- and skyscape of Herakleopolis have proven to be deeply interconnected. The orientation of the main axis of the Temple of Heryshef allows confirmation of its membership to the Canopus family of orientations together with its visual connection with the TIP necropolis. The orientation of the chapel at this site seems to suggest a diachronic variability in the alignments that could perhaps be related to the precession of the equinoxes. The orientation of the building known as Kum el-Akareb to the funerary complex of Senuseret II is very appealing and suggests an original construction of this shrine in the Middle Kingdom. This will need to be further explored by the members of the Spanish Mission in highly desirable future excavations if socio-political circumstances allow it.

Finally, the equinoctial orientation of the North "Temple", possibly of the Roman period, suggests a totally different context (a sanctuary of the Egyptian divinities or perhaps a Christian church) that will also need to be explained once the results of the excavations by the local Egyptian team of archaeologists have been published, contrasted and evaluated. The expectances for future research in Herakleopolis Magna are hence very promising.

ACKNOWLEDGEMENTS

The authors wish to thank the other members of the Spanish Archaeological Mission in Herakleopolis Magna, especially those of the 2013 campaign: Jose Ramón Pérez-Accino, Gema Garrido, Antonio Guio, Antonio Gómez, Bettina Bader, Mª Antonia Moreno, Mª Fernanda Pascual and José Javier Martínez, for their human quality and know-how, which made the field on the terrain infinitely more pleasant and productive. Our gratitude should also go to the inspectors of the Service of Antiquities of Beni Suef who have accompanied us during this campaign for their invaluable support and help. This work has been funded through Grants for Archaeological Projects Abroad awarded annually by the Spanish Ministry of Education, Culture and Sport (General Directorate of Fine Arts and Cultural Assets, Libraries and Archives) and the projects AYA2015-66787"Orientatio ad Sidera IV" of the Spanish MINECO, and P/309307 "Arqueoastronomía" of the IAC.

REFERENCES

Allen, R.H. (1963) *Star names, their lore and meaning*, New York, Dover Inc. Pub..

Belmonte, J.A. (2012) *Pirámides, templos y estrellas: astronomía y arqueología en el Egipto antiguo*, Barcelona, Crítica.

Belmonte, J.A., Molinero Polo, M.A. and Miranda, N. (2009) Unveiling Shesat: new insights into the stretching of the cord ceremony. In *In Search of Cosmic Order: selected essays on Egyptian Archaeoastronomy*, J.A. Belmonte and M. Shaltout (eds.), Cairo, Supreme Council of Antiquities Press, 195-212.

Belmonte, J.A., Shaltout, M. and Fekri, M. (2009) Astronomy, landscape and symbolism: a study of the orientation of ancient Egyptian temples. In *In Search of Cosmic Order: selected essays on Egyptian Archaeoastronomy*, J.A. Belmonte and M. Shaltout (eds.), Cairo, Supreme Council of Antiquities Press, 213-285.

Belmonte, J.A., Pérez Die M.C. and Díaz-Iglesias, L. (2015) Astronomía y paisaje en Heracleopolis Magna: un estudio de los templos. In *Ex Aegypto lux et sapientia, Homenatge al profesor Josep Padró Parcerisa*, N.

Castellano, M.Mascort, C. Piedrafita and J. Vivó (eds.), *Nova Studia Aegyptiaca* IX, Barcelona, 107-123.

Díaz-Iglesias Llanos, L. E. (2012) *La proyección mitológica de Heracleópolis Magna: continuidades, transformaciones y contextualización de los mitemas heracleopolitanos en fuentes funerarias y cultuales*. Ph.D Thesis, Universidad de La Laguna.

Díaz-Iglesias Llanos, L. (2014) *El ciclo mítico de Heracleópolis Magna: continuidad y reelaboración a partir de las fuentes funerarias y cultuales. Aula Aegyptiaca Studia* VII, Barcelona. 1 vol + 1 CD.

Fakhry, A. (2004) *Siwa Oasis* 5th Ed., Cairo, American University of Cairo Press.

Gabolde, L. (1999) Canope et les orientations nord-sud de Karnak étables par Thoutmosis III, *Revue d'Égyptologie*, Vol. 50, 278-282.

González García, A.C. y Belmonte J.A. (2014) Sacred architecture orientation across the Mediterranean: a comparative statistical analysis, *Mediterranean Archaeology and Archaeometry*, Vol. 14, 95-113.

González García, A.C. y Belmonte J.A. (2015) The Orientation of Pre-Romanesque Churches in the Iberian Peninsula *Nexus Network Journal*, Vol. 17, 353-377.

Hawkins, G.S. (1973) *Beyond Stonehenge*, New York, Routledge.

Krauss, R. (1997) *Astronomische Konzepte und Jenseitsvorstellungen in den Pyramidentexten*, Ägyptologische Abhandlung Band 59, Wiesbaden.

Lull, J. and Belmonte, J.A. (2009) The constellations of ancient Egypt. In *In Search of Cosmic Order: selected essays on Egyptian Archaeoastronomy*, J.A. Belmonte and M. Shaltout (eds.), Cairo, Supreme Council of Antiquities Press, 155-195.

Naville, E. (1894) *Ahnas el Medineh (Heracleopolis Magna): with chapters on the Mendes, the Nome of Thoth, and Leontopolis*. Memoir of the Egypt Exploration Fund 11, London.

Pérez Die, M.C. (2009) Excavaciones de la Misión Arqueológica Española en Heracleopolis Magna (Ehnasya el Medina), 54-66. In *120 years of Spanish archaeology in Egypt*, M.C. pérez Die and W. Al-Sadiik (eds.), Madrid, Ministerio de Cultura, 54-65.

Pérez Die, M.C. (ed.) (2010) *Heracleópolis Magna (Ehnasya el Medina, Egipto). La necrópolis "real" del Tercer Periodo Intermedio y su reutilización*, Madrid, Ministerio de Cultura, 2 vols. + 1 CD.

Pérez Die, M.C. (2012) El proyecto de Investigación "Heracleópolis Magna" (Ehnasya el Medina). Trabajos 2008-2009. In *Novos Trabalhos de Egiptologia Ibérica, IV Congreso Ibérico de Egiptología*, L.M. De Araujo and J. Das Candeias Sales (eds.), Lisboa, Universidade de Lisboa, 923-935.

Piankoff, A. (1942) *Le livre du jour et de la nuit*, Cairo, IFAO.

Redford, D.B. (2010) *City of the Ram-Man: the story of ancient Mendes*, Princeton, Princeton University press.

Shalthout M., Belmonte J.A. and Fekri, M. (2007) On the orientation of ancient Egyptian temples: (3) key points at Lower Egypt and Siwa Oasis, *Journal for the History of Astronomy*, Vol. 38, 141-160 (PI) and 413-442 (PII).

Wilcken, U. (1903) Die berliner Papyrusgrabungen in Herakleopolis Magna im Winter 1898/9, *Archiv für Papyrusforschung*, Vol. 3, 294-336.

4

ROMANS IN THE NEAR EAST: THE ORIENTATION OF ROMAN SETTLEMENTS IN PRESENT-DAY JORDAN

A. Rodríguez-Antón[1,2], J.A. Belmonte[1,2] and A.C. González-García[3]

[1] *Instituto de Astrofísica de Canarias, 38200, La Laguna, Spain*
[2] *Universidad de La Laguna, 38200, La Laguna, Spain*
[3] *Heritage Sciences Institute, Incipit-CSIC, Santiago de Compostela, 15704, Spain*

Corresponding author: A. Rodríguez-Antón (ara@iac.es)

ABSTRACT

An essential difference between the western and eastern provinces of the Roman Empire is the fact that sophisticated urban cultures had developed in Asia Minor and the Levant centuries before the Romans arrived. Underlying the Hellenized, and later Roman, veneer was a myriad of older local traditions and languages, which had an immense impact upon Roman religious tradition through elements such as the introduction of new religious practices. Following the path of previous studies, in this article we try to discern how Roman culture was inherited and adapted to the heterogeneous Eastern traditions and how it could be reflected in the architecture and urban layout, mainly in what concerns to the orientation of the urban structures. Considering ancient writings, such as those of Higynius Gromaticus (*Constitutio*, I), the orientation of these features could follow the position of certain celestial bodies, mainly the sun, which would imply a careful observation of the sky. Developing the lines of previous studies on the orientation of Roman settlements in the western part of the Empire (González-García et al., 2014 & Rodríguez-Antón et al., 2016), a number of Roman cities and military settlements in modern-day Jordan, Syria and Palestine are analysed here. Through this approach, we try to obtain a first insight into whether their orientations looked towards astronomical positions and wether there existed common patterns comparing with those sites previously measured in Hispania or Britannia. This would help us to obtain a wider vision of Roman ritual practices, cosmovisions and how Roman culture could have evolved, spread and became assimilated through lands and time.

KEYWORDS: Orientation of Roman settlements; Roman urbanism; ancient Jordan; Eastern Roman Empire; Limes Arabicus; Decapolis.

1. INTRODUCTION

By the time of the Roman arrival in the Near East in the 60s BC, the region had already hosted sophisticated urban cultures for centuries. The Hellenistic presence made Greek a common language but there were also a myriad of older local traditions that turned this region into a mosaic of different zones with different histories. From the creation of the province of Syria in 64 BC by General Pompey, to the victory of the Islamic forces in the Battle of Yarmuk in the 7th century (Kennedy, 2004), the Roman Empire in the East underwent a clear transformation. The Roman identity of the Republic and early Empire was successively merging with the local ones, creating new mentalities and religious practices.

These interactions with indigenous cultures and the non-static character of the 'Romanism' affected the urban style of the eastern Roman cities. Unlike other sites previously studied in the western Roman provinces (González-García et al., 2014), here there already existed an old urban tradition by the time of the Roman arrival. So, instead of funding new cities *ex-novo*, the Romans limited themselves to overlaying distinctive features on pre-existing settlements. These new elements might be representative public buildings such as theatres. But, most relevant for us is the introduction of one or more thoroughfares (Ball, 2000). These have been identified with processional ways in some cases, as the *cardus* of Gerasa (Ball, 2000), or regarded as a dominating feature in the cityscape, which could have reflected ancient astrological beliefs (Rabbabeh, 2014). Another significant difference is the almost complete absence of traditional *fora* in the eastern Roman provincial cities, being the heart of a Roman city in the West.

Examples of how Romans adapted their urban tradition to the eastern one can be observed in the Decapolis cities, which have been historically regarded as a sort of Graeco-Roman city-states and, more recently, as a region in northern Jordan, southern Syria and Palestine (Khouri, 1986). Although the definition of Decapolis is still vague, ancient writers such as Pliny[1], who listed the ten cities that conform the entity, mentioned it. We have introduced in our sample the eight cities that have been currently identified from that list (see Table I). We should not ignore the cohabitation of Graeco-Macedonians, Arabs (Nabateans), Jews and Romans, in order to identify whether this fact could affect the orientation patterns.

Owing to its bordering position within the Roman Empire (more concretely its south easternmost boundary), there is also a great number of military settlements in the area studied. For that reason, it needed to be protected from external pressures and a number of military infrastructures and garrisons were required. This situation arose mainly as a result of the stability policies of the 2nd century AD. This complex of military infrastructures formed the so-called *limes arabicus* (e.g. Parker, 1992 and Bowersock, 1976) and were interlinked through communication roads, such as the *Via Nova Traiana*, which spanned from Bosra to the Red Sea at Aila.

Based on all these facts, and following the line of previous archaeoastronomical studies on Roman urbanism (e.g. González-García et al., 2014) we would like to learn how the Romans adapted their traditions in this region. With this information, we seek to compare the results of the present work with those obtained in the western provinces of the Roman Empire. Our aim throughout this study is to discern how Roman practices evolved over time and discover whether a connexion between astronomical observations and urban layout existed, as can be inferred from ancient treatises[2]. All these aspects would furthermore contribute to better understand how the local customs of the indigenous peoples of all the lands of the Empire would affect the way they assimilated Roman practices, that is to say, how so-called Romanization worked in the different territories of the Empire and, particularly in this case, in frontier regions.

2. METHODOLOGY AND DATA SAMPLE

Since architecture was a key component in the consolidation of Roman power over the settled lands, we have analysed the orientation of a number of Roman cities and military settlements in modern-day Jordan, Syria, and Palestine.

The sample consists of the measurements of the azimuth and the altitude of the horizon of the main urban features of 13 Roman cities, as well as the main axes of 17 forts and fortresses built and occupied in Roman times (see Fig.1 and 2). In the case of the cities, those features are their principal streets, except for Philadelphia (present-day Amman) where we have considered the sides of the forum. Eight of these cities were originally part of the Decapolis, according to Pliny's list and were thus of Hellenistic origin. Although Roman cities in the East do not strictly fit the commonly assumed orthogonal layout (Castagnoli, 1971), they usually contain typical main streets: *cardus* and *decumanus*, running north-south

[1] Pliny. Naturalis Historia 5.16.74. He mentioned Damascus, Philadelphia, Raphana, Scythopolis, Gadara, Hippos, Dium, Pella, Gerasa and Canatha.

[2] Frontinus De Agrimensura, 27 and Hyginus Gromaticus. Constitutio, 1.

and east-west, respectively. In those cases in which both types of streets are not clearly identified, we have considered the direction of a presumably important street, which would follow the leading orientation of the city.

In the case of the military settlements, there exist various architectonic styles in this region that differ from others found in the western provinces. Nevertheless, the playing card-shape design, common in Britain, is also present in the East to a lesser degree, as in Qasr el-Azraq surrounding the later fortification (Kennedy, 2004). There is also a lack of temporary camps, which are numerous in Britain. Additionally, there are several cases of re-occupation and re-adaptation of pre-existing settlements, sometimes without excessive modifications. This makes significantly complicated to specify which structures are properly Roman, or what was their period of construction (Parenti and Gilento, 2012). This is a problem in which archaeoastronomy could provide further information to that extracted from other archaeological works.

The data for eight cities and three military settlements were obtained in situ during a fieldwork campaign in Jordan performed in 2011 by members of our group. The instruments used were two tandems with a compass (error 0¼°) and a clinometer (error 0½°), and a GPS. The remaining site's azimuths have been measured with Google Earth, and the angular altitudes of the horizon by a digital reconstruction of the terrain, *HeyWhatsThat* (http://www.heywhatsthat.com). To determine the estimated error of these sites we have compared the data acquired in situ with the same measurements obtained by Google Earth and *HeyWhatsThat*. Based on these calculations, we have considered an average error of 1° for the azimuths and 0½° for the altitudes of the horizon. The azimuths measured with the compass have been corrected for magnetic declination with the calculator of the US National Geophysical Data Center (http://www.ngdc.noaa.gov/geomag/declination.shtml).

Assuming the existence of two axes, even in the cases in which one of them cannot be identified, we have considered four perpendicular orientations per site. We have divided the horizon into four azimuthal portions 90° wide so that, for a single site, each of its perpendicular azimuths matches one of those sectors. By this division, we consider a *decumanus* to be the streets within 45° to 135°, and 225° to 315°. The *cardus* would fall on the remaining azimuthal sectors, towards the north and south. On this basis, we classify the streets and avoid two perpendicular azimuths falling in the same division.

In addition, the '*decumanus* sectors' comprise the lunisolar range for Jordanian latitudes, that is, the azimuth of the sun at the solstices and the moon at its major standstills for an average latitude of the sample (Φ≈31°).

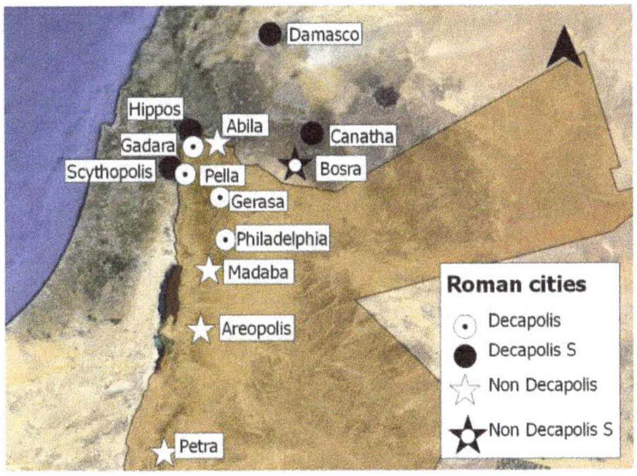

Figure 1. Roman cities of the sample.

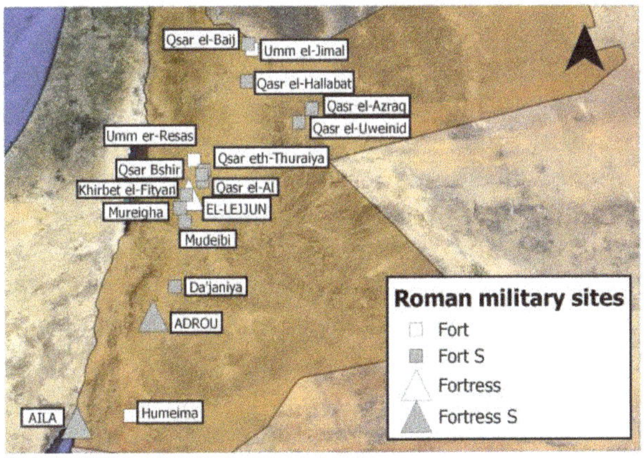

Figure 2: Roman military settlements of the sample.

In order to link properly these orientations with astronomical positions we have computed the declination for all the orientations. Knowledge of this astronomical quantity enables us to see how local topography affects the rising and setting positions of celestial bodies. Moreover, declination is independent of the geographical location. The estimated error for the declination of the data obtained in situ is 0¾°, while that for the places measured through satellite images and *HeyWhat'sThat?* is 1½°. This translates into an error of approximately 2 and 4 days, respectively.

All the data are given in Table I and Table II. Both tables provide a set of four perpendicular azimuths per site, the angular altitude of the horizon in those directions and the corresponding declination per sector. We have also introduced Petra, which is mostly Nabatean but contains Roman features, such as its colonnaded street.

2.1 Orientation of the sample

The azimuths of 13 cities are represented in an orientation diagram, where the four sectors and the solar and lunar extremes are indicated (see top of Fig. 3). It can be seen from the diagram, all the cities are within the lunisolar azimuth range. From these, all except Madaba fall within the solar range. The azimuths of this city fall on the southern and northern major lunar standstills towards the east (where the altitude of the horizon is 0°) and west, respectively. The non-Decapolis cities are concentrated around the cardinal points, and around the winter solstice towards the east, and the summer solstice towards the west. Two further azimuth groups may also be considered, one almost cardinal which comprises Damasco, Gadara, Canatha (three Decapolis cities) and Petra, and another around 114° composed of Areopolis (present-day Rabba), Hippos and Gerasa (present-day Jerash).

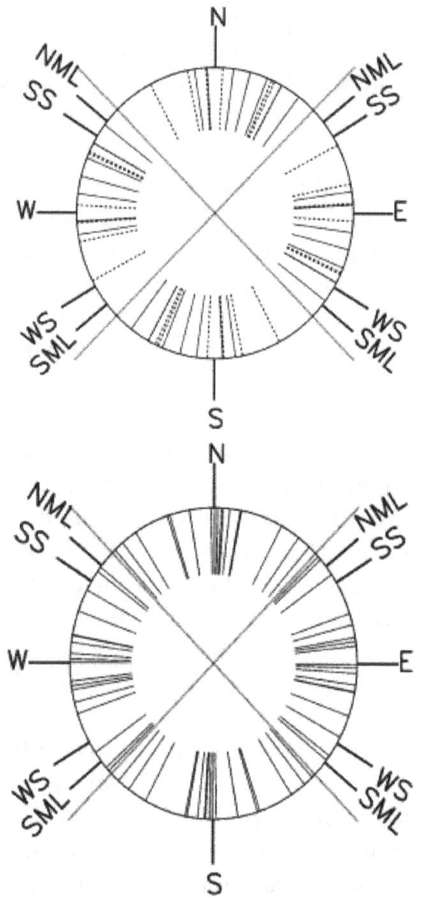

Figure 3: Orientation diagram of the Roman cities (top) and military settlements (bottom). Each short line gives the azimuth of one of their main axes. There are four directions per site, one per sector. The cities of the Decapolis are represented in dotted lines. The long dashed lines give the separation into four azimuthal sectors. SS and WS stand for sunrise and sunset at summer and winter solstice, and NML and SML stand for the northern and southern major lunar standstills, respectively, for an average latitude of the sample.

Regarding to the declinations (see Table I), at first sight we do not find that the cities follow any common trend but we can highlight some striking cases. Assuming declination values of ±24° for the sun at the solstices, there are three solstitially-orientated cities towards the sunset: Scythopolis, Gerasa and Pella. The first two look towards summer solstice sunset while the third one looks towards winter solstice sunset. We may also consider equinoctial orientations in the cities of the cardinal azimuth group, according to the estimated error in declination. For the second azimuth group (114° azimuth) we observe similar declinations towards the east for the three cities, and towards the west in Hippos and *Areopolis*. These last concur with sunset at the beginning of August, being the month of Augustus.

Concerning the military settlements, we count on the measurements of 17 structures, these being auxiliary forts and three legionary fortresses, but there are in total 18 sets of data due to the irregular shape of the *castellum* of Umm el-Jimal (Table II). Three of the sites were measured in situ: Umm el-Jimal, Umm er-Resas and Humeima. We estimate that our sample encompasses, at least, 40% of the measurable military sites in the region. Although we are conscious that further sites should be measured and that there are limitations on the data obtained with Google Earth, it is possible to make a first approach to the orientation patterns. In Figure 3 (bottom) the azimuthal distribution can be appreciated, with an accumulation around the meridian line. Four of the sites that share that orientation are relatively close.

The declination has also been computed and represented in two histograms (Fig. 4) towards east and west. We chose a band pass of 2½°, considering an average between the systematic errors from the data obtained by fieldwork and satellite images. We have overwritten the declination distributions obtained from Roman military sites in *Britannia* (Rodríguez-Antón et al., 2016). The distributions obtained in the present study are broad with two main peaks that are more clearly defined in the eastern distribution. They correspond to the beginning of March and mid-October (δ≈-7°). The distribution is less defined towards the west, where such peaks blend. The maximum here would correspond to the beginning of April (δ≈4°) sunset. In both directions there are also smaller peaks beyond the solar range.

However, we should not lose sight of the fact that the sample managed in this study is not large. The histograms just show a first approach to the declination distribution, but the real data are more scattered with few sites looking towards these 'relevant' directions exactly.

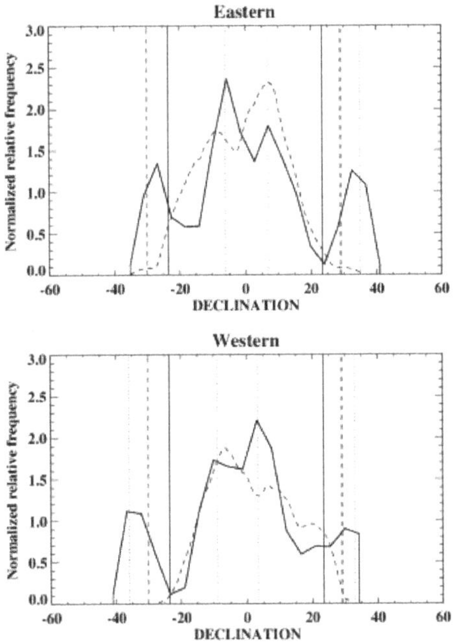

Figure 4: Declination histograms of Roman military sites in Jordan and Britain towards east (top) and west (bottom). The distributions indicated by a solid line are from Jordan and those distributions drawn with a dashed line are from Britain. Vertical solid lines indicate the extreme declinations of the sun, vertical dashed lines indicate the extreme lunar declinations, and vertical dotted lines indicate the solar declination for particular dates (see text for details).

3. DISCUSSION

3.1. *Orientation of Roman cities*

We have seen in the previous section that most of the orientations of the Roman cities in this area are within the solar range. But, getting into the results, we can extract some remarkable cases. Regarding the solstitial orientations, these would agree with similar results previously obtained in western regions of the Empire (Magli, 2008; González-García et al., 2014). Moreover, this pattern is also present in Nabatean monuments (Belmonte et al., 2013). According to the estimated error in declination, cities from the cardinal azimuth group mentioned in the previous section could be regarded as equinoctial. This orientation was also obtained in Greek (Boutsikas, 2009) and Nabatean temples (Belmonte et al., 2013), comprising both cultures the main cultural substratum that Romans found in this area. In fact, the orientation of the *cardus* of Petra, a city occupied by the Romans but of highly Nabatean nature, is almost equinoctial towards the east ($\delta=-1\frac{1}{4}^{\circ}$) (Belmonte et al., 2013). This orientation could indicate that Romans did not transform the entire city plan in accordance with their precepts. Other evidence of the permanence of previous traditions would lie in the fact that some Semitic names, such as Gadara or Gerasa, were preserved. It seems likely that the local languages were the most spoken ones, remaining Latin and Greek been used as purely for administrative purposes.

Although urban development and population reached a maximum in Roman times, we should not forget that all the cities here studied had exist before the arrival of the Romans. They were simply transformed, to a greater or lesser degree, during that period. With this fact in mind, we find particularly interesting the orientation of Hippos, *Philadelphia* and *Areopolis* (present-day Amman and Rabba, respectively). These last two were the Iron Age cities of *Rabbath-Ammon* and *Rabbath-Moab,* respectively. The southern declination values of these three cities are almost orientated towards the rising point of Canopus ($\delta=-52\frac{3}{4}^{\circ}$). This star, Arabic *Suhail,* has been a well-known celestial body for different cultures in the area since ancient times. We can appreciate its presence in previous architecture, for example, in the orientation of Nabatean buildings (Belmonte et al., 2013), as well as in the pre-islamic temple of *Ka'aba* in Mekka (Hawkings and King, 1982).

3.2. *Orientation of Roman military settlements*

Regarding to the azimuths of the military settlements, we found an accumulation around the meridian line. Four of those settlements in relatively close proximity. Nevertheless, rather than fully Roman structures some of them are considered by archaeologists to be previously existing sites re-occupied by the Roman army. That is the case for Qasr el-Al. There, a great deal of Iron Age and Nabatean pottery were found, and its structure is more likely to be from one of those periods (Parenti and Gilento, 2012).

The results obtained in the declination histograms seem remarkable since those orientations could be related to solar positions in March. During that month a number of festivities were traditionally celebrated in honour of Mars, a Roman god related to war. This was so at least in earlier Roman times, as mentioned in the Ovid's *Fasti* (*Fasti*, Book 3) and in the *Feriale Duranum*[3] (Espinosa et al., *at this voulme*).

The declination histograms of Jordan military sites (Fig. 4) seem to show that a March orientation is more likely to be towards the sunrise. The main peak towards the west is around the beginning of April, although there is also a secondary accumulation

[3] Kal(endis) Ṃ[a]ṛtis ob c̣[e]r[imo]niạ[s natalicias Martis Patris Victoris (Papiry.info: Transcriptions of Feriale Duranum)

around the beginning of March. Compared to what was obtained in *Britannia*, the resulting distributions are similar but like a mirror image in each sector. That is, the peaks towards the east in Jordan match better with those towards the west in Britain, and vice versa. In the case of Britannia, we suggested that the direction of the beginning of March towards the west could be due to the high number of temporary camps in the sample (Rodríguez-Antón et al., 2016 *in press*). In that case, the hypothesis was that they were built when troops arrived at a new place at the end of the journey. So, in the case that they looked for solar orientations, they would take that of the sunset to establish the axes prior to starting the construction of the camp. This is not the case in the settlements in Jordan, where warfare activity and strategy were different, thereby resulting in the vast majority of the defensive settlements being made permanent.

3.3 Comparing samples

We have conducted a Kolmogorov-Smirnov test in order to broaden our understanding of the results and check whether the declination distributions for Britain and Jordan are drawn from the same parent population. This test checks whether the null hypothesis, that both samples are drawn from the same parent population, can be rejected. We obtained a probability of 0.84, thus we cannot reject that this condition is fulfilled. This does not confirm the null hypothesis either.

In addition, although orientations such as those considered equinoctial are not common in the western provinces of the Roman Empire, we obtained a number of solstitially-orientated cities in this study. These would be in line with what was extracted in Hispania (González-García et al., 2014) or Italy (Magli, 2008).

4. CONCLUSIONS

From a first overview of the results, we found that all the cities and most of the military settlements are orientated within the lunisolar azimuth range. Moreover, considering to the corresponding astronomical declinations, some of the resultant orientations are in accordance with what is observed in the western provinces of the Roman Empire as well as in some pre-existing native sites.

One remarkable distinction between this area of the Roman Empire and other provinces in the West is the different degree of urban development prior to the arrival of the Romans (Laurence et al., 2011). Whereas in some western areas urbanization really started after the Roman conquest, this was not the case in the Greek-speaking East. A perfect example are the Decapolis cities, of a presumably Hellenistic origin, as well as Petra and Bosra, both originally Nabatean cities. This fact could explain, for instance, the observation of almost equinoctial orientations in Damasco, Canatha, Gadara and Petra, which are non-Roman in origin. This orientation is practically non-existent in Hispania and Italy, and could result from Nabatean or Greek influence.

Taking all the above into account we may conclude that, in contrast to the general urbanization process in the West, here the previous cultural substratum played a major role in the cities layout. Even when Romans settled those cities and introduced their own distinctive features, they would not change the entire layout. This does not mean that they had nothing to do in the reorganization of the urban space but, in several cases, they probably limited themselves to adapting what they found to the Roman style.

In relation to the military settlements, we cannot reject the hypothesis that the declination distributions in Jordan and Britain are drawn from the same parent population. Furthermore, the vast majority of the elements of both samples are completely Roman buildings constructed with a common aim: military activity. The fact that in both cases the distributions show a tendency to cluster around dates in March is, at the very least, noteworthy. This fact constitutes added plausibility to the consideration that there could be traces of intentionality underlying the orientation of those sites, and, what is more, that the orientation could be relate to the position of the sun on those days that corresponded to the Roman army's religious feasts.

Considering the above and despite the potential role of the indigenous cultures, we cannot dismiss a possible continuity in the orientation patterns in both of these widely separated Roman territories that might have been due to the prevalence of a more ancient Roman practice still in use during the period in which all those settlements were erected.

All these considerations might have resulted from an attempt to maintain, in a greater or lesser degree, an Imperial identity in order to keep united all the distant pieces that formed the Roman puzzle.

Table I: Orientation for the 12 Roman cities in Jordan, Syria and Palestine. The places measured by Google Earth images are indicated by an asterisk. Each site is characterized by four azimuths (A1 to A4) and four altitudes of the horizon (h1 to h4), which means one per azimuthal sector. Also indicated is the latitude of the place (Φ) and the declinations computed in each azimuthal sector (δe, etc). Angular altitudes marked with asterisks were measured with a digital reconstruction of the horizon (http://www.heywhatsthat.com/), and declinations with a question mark have been calculated for a flat horizon (h=0°) because it was not possible to obtain a horizon profile in those directions.

Site	A1(°)	h1(°)	A2(°)	h2(ª)	A3(°)	h3(°)	A4(°)	h4(°)	Φ(°)	δe(°)	δw(°)	δn(°)	δs(°)
Damasco* (D)	356½	3*	86½	0*	176½	0½*	266½	2.3*	33.50	2.6	-1.8	59.1	-56.3
Canatha* (D)	3	0½*	93	3½*	183	2.6*	273	-0½*	32.70	-0.7	1.9	57.2	-54.8
Abila	351¼	--	81¼	-0.5	171¼	--	261¼	--	32.70	6.7	-7.7?	55.7?	-56.8?
Hippos* (D)	24	3¾*	114	--	204	0¼*	294	0.45*	32.70	-18.4?	20.0	53.2	-50.5
Gadara (D)	356	-0½	86	0*	176	--	266	-0½	32.60	3.1	-4.0	56.0	-57.7
Bosra*	14½	0½*	104½	--	194½	--	284½	--	32.52	-12.5?	11.8?	54.6	-55.2?
Scythopolis* (D)	28½	--	118½	--	208½	--	298½	--	32.50	-24.1?	23.4?	47.4?	-48.3?
Pella (D)	332	0.8*	62	16½	152	--	242	1	32.45	32.1	-23.0	48.4	-48.6?
Gerasa (D)	25½	2½	115½	2½	205½	2½	295½	4	32.30	-20.0	23.5	51.7	-47.8
Philadelphia (D)	348¼	13	78¼	1	168¼	5½	258¼	4.8*	31.90	10.3	-7.4	68.2	-51.1
Madaba	37	--	127	0	217	--	307	--	31.70	-31.1	30.4?	42.4?	-43.2?
Areopolis	22	0.85*	112	0*	202	0.85*	292	0*	31.20	-19.0	18.4	52.8	-52.1
Petra	7½	--	97½	10	187½	--	277½	7.5	30.32	-1.4	10.2	58.3?	-59.4?

Table II. Orientation for the 17 Roman military settlements in Jordan. The name given is the modern one. The places measured by Google Earth images are denoted with an asterisk. Each site is characterised by four azimuths (A1 to A4) and four angular altitudes of the horizon (h1 to h4), which means one per azimuth sector. Also indicated is the latitude of the place (Φ) and the declinations computed in each azimuth sector (δe, etc). Altitudes marked with asterisks were measured with a digital reconstruction of the horizon (http://www.heywhatsthat.com/), and declinations with a question mark have been calculated for a flat horizon (h=0°) because it was not possible to obtain a horizon profile in those directions.

Site	A1(°)	h1(°)	A2(°)	h2(°)	A3(°)	h3(°)	A4(°)	h4(°)	Φ(°)	δe(°)	δw(°)	δn(°)	δs(°)
Qasr el-Baij*	28	0¼*	118	0¼*	208	0*	298	0*	32.37	-23½	23.0	48.1	-48¼
Umm el-Jimal	3¾ 11	0 0½*	100¼ 109½	0.8* 0.7*	183.75 191¼	0*	280¼ 289½	0¼*	32.33	-8.6 -16.3	8.5 3.2	56.9 56.0	-58.0 -56.5
Qasr el-Hallabat*	327	0¼*	55	0½*	147	0*	235	1½*	32.09	29.0	-28.6	45	-45.7
Qasr el-Azraq*	36	0¼*	126	0*	216	0.2*	306	--	31.88	-30.3	-43.6?	42.2	29.9
Umm er-Resas	6¼	-0½	93	-0½*	186¼	-1*	273¼	-0½*	31.50	-2¼	2.6	57.0	.59.9
Qasr eth-Thuraiya*	0¼	0*	90¼	0*	180¼	0¼*	270¼	0¼*	31.50	-0.7	0½	58.0	-58.8
Qasr Bshir*	320½	0½*	47	0½*	140½	0½*	227	0½*	31.34	35.6	-35.6	41.2	-41.2
Khirbet el-Fityan*	1½	--	91½	0*	181½	0.7*	271½	1*	31.24	-1.6	1.6	58.1	-58.2
Lejjun	341	3	71	0½	161	11	251	3½	31.24	16.2	-14.6	55.9	-53.6
Mureigha*	350	0*	80	0*	170	1¼*	260	1½*	31.14	8.2	-8.1	56.8	-56.6
Mudeibi*	350	1¼*	81½	0¼*	170	0½*	261½	1½*	31.04	7.1	-6.7	58.1	-57.7
Da'janiya*	316½	1½*	46½	0*	136½	-0½*	226½	0.8*	30.50	36.0	-36.1	39.4	-39.4
Udruh*	358¾	--	83¼	-0½*	178¾	0*	263¼	1¾*	30.33	5.1	-5.1	59.1?	-60.2
Humeima	10½	-0½	100½	1	190½	0	280½	-1½	29.50	-8.8	7.9	57.8	-59.3
Qasr el-Al*	3	0*	97¾	0.1*	183	0*	277¾	--	31.50	-6.8	6.3?	57.9	-58.9
Qasr el Uweinid*	342	--	75¼	--	162	0*	255¼	0¼*	31.78	12.2?	-12.6	53.4?	-54.4
Aila*	41½	2½*	128	8*	221½	0¾*	308	2½*	29.53	-27.6	33.67	42.1	-40.5

ACKNOWLEDGEMENTS

This work was partially financed under the framework of the project P31079 'Arqueoastronomía' of the IAC, AYA2011-26759 "*Orientatio ad Sidera* III" of the Spanish MINECO.

REFERENCES

Ball, W. (2000) *Rome in the East. The transformation of an empire*, Routledge, London

Beard, M., North, J. and Price, S. (1998) *Religions of Rome. Volume I. A History*, Cambridge University Press, Cambridge

Belmonte, J.A., González-García, A.C. and Polcaro, A. (2013) Light and Shadows over Petra: astronomy and landscape in Nabatean lands, *Nexus Network Journal*, Vol. 15, 487-501.

Boutsikas, E. (2009) Placing Greek Temples: An Archaeoastronomical Study of the Orientation of Ancient Greek Religious Structures, *Archaeoastronomy*, Vol. 21, 14-19.

Bowersock, G.W. (1983) *Roman Arabia*, Harvard University Press, US

Castagnoli, F. (1971) *Orthogonal Town planning in antiquity*. MIT Press, Cambridge, MA, U.S.A.

Espinosa Espinosa, D., González-García, A.C. and García Quintela, M.V. (2016) On the Orientation of Roman Towns in the Rhine Area. *Mediterranean Archaeology and Archaeometry,*at this volume.

Fishwick, D. (1988) Dated inscriptions and the Feriale Duranum. *Syria. Art et Histoire*, Vol. 65(3), 349-361.

Gilliver, C.M. (1999) *The Roman art of war*, TEMPUS, Stroud, UK

González-García. A.C. and Magli, G. (2014) Roman city planning and spatial organization. In C.L.N. Ruggles (ed.) *Handbook of Archaeoastronomy and Ethnoastronomy: Building bridges Between Cultures*, Cambridge University Press, Cambridge, U.K., 374-381.

González-García, A.C., Rodríguez-Antón, A. and Belmonte, J.A. (2014) The orientation of Roman towns in Hispania: preliminary results, *Mediterranean Archaeology and Archaeometry*, Vol. 14 (3), 107-119.

Hawkins, G.S. and King, D.A. (1982) On the orientation of the Ka'ba, *Journal of the History of Astronomy*, no.13, 102-109.

Kaizer, T. ed. (2008) *The variety of Local Religious Life in the Near East in the Hellenistic and Roman Periods*, BRILL, Leiden

Kennedy, D. and Riley, D. (1990) *Rome's desert frontier from the air*, B.T. Bratsford Limited, London

Kennedy, D. (ed.) (1996) The Roman Army in the East, *Journal of Roman Archaeology Supplementary Series* no.18, Ann Arbor

Kennedy, D. (2000) The Frontier of Settlement in Roman Arabia: Gerasa to Umm el-Jimal...and beyond, *Mediterraneo Antico. Economie Societá Culture*, Vol.3 (2), Instituti Editoriali e Poligrafici Internazionali, 397-453.

Kennedy, D. (2004) *The Roman Army in Jordan,* The Council for British Research in the Levant, London

Kennedy, D. and Bewley, R. (2004) *Ancient Jordan from the Air*, The Council for British Research in the Levant, London

King, D. A.(1995) The orientation of medieval Islamic religious architecture and cities, *Journal of the History of Astronomy*, no. 16, 253-274.

Khouri, R. (1986) *Jerash. A frontier city of the Roman East*, Longman group LTD, London and New York

Laurence, R., Simon Esmonde, C. and Sears, G. (2011) *The city in the Roman West c.250 BC – c. AD 250*. Cambridge University Press, Cambridge, UK

Magli, G. (2008) On the orientation of Roman towns in Italy. *Oxford Journal of Archaeology*, Vol. 26(1), 103-108.

Millar, F. (1993) *The Roman Near East 31BC-AD 337*, Harvard University Press, US

Moschek, W. (2009) The Limes: Between open Frontier and Borderline, In Fryde, N. and Reitz, D. (ed). *Walls, Ramparts, and Lines of Demarcation. Selected Studies from Antiquity to Modern Times*, LIT Verlag, Münster

Nock, A.D. (1952) The Roman Army and the Roman Religious Year, *The Harvard Theological Review*, Vol. 45(4), 187-252.

Papyri.info, The *Feriale Duranum* translations

Parenti, R. and Gilento, P. (2012) Limes Arabicus and still-standing buildings. In Vannini, G. and Nuccioti, M. (ed.) *La Transgiordania nei secoli XII-XIII e le 'frontiere' del Mediterraneo medieval*. Archaeopress, Oxford, UK, 111-124.

Parker, S.T. (1987) *The Roman frontier in Central Jordan. Final report on the Limes Arabicus Project, 1980-1989*, Vol. I-II, British Archaeological Reports. International Series340 (i), Oxford

Rababeh, S. et al. (2014) Colonnaded streets within the Roman cityscape: a "spatial" perspective. *Journal of Architecture and Urbanism*, Routledge, Vol. 38 (4), 293-305.

Raja, R. (2012) *Urban Development and Regional Identity in the Eastern Roman Provinces 50 BC-AD 250. Aprhodisias, Ephesos, Athens, Gerasa.* Museum Tusculanum Press, University of Copenhagen

Rodríguez-Antón, A., González-García, A.C. and Belmonte, J.A. (2016) Orientation of Roman camps and forts in Britannia: reconsidering Alan Richardson's work, *Sophia Centre Press*, 135-144.

5

ON THE ORIENTATION OF TWO ROMAN TOWNS IN THE RHINE AREA

David Espinosa Espinosa[1], A. César González-García[2] and Marco V. García Quintela[1]

[1]*University of Santiago de Compostela*
[2]*Institute of Heritage Sciences, Incipit-CSIC*

Corresponding author: Second Author (a.cesar.gonzalez-garcia@incipit.csic.es)

ABSTRACT

The aim of the present paper is to extend the archeoastronomical study sample on the orientation of Roman cities to the analysis of a number of cases in the Rhine area. The starting point is a study of the orientation of *Augusta Treverorum* (present day Trier; Goethert, 2003). Goethert assumed that the orientation of the *decumanus maximus* was towards sunrise at the autumn equinox, on September 23rd as the *dies natalis* of the city. This event would deliberately coincide with the anniversary of the birth of Augustus, and would have determined the establishment and orientation of the new urban layout. However, our *in situ* measurements of the orientation of the urban network at several sites of the Roman town rule out this hypothesis. We find an orientation that is more in line with those documented for other Roman cities and camps elsewhere in the Roman provinces (González-García *et al.*, 2014; Rodríguez-Antón *et al.*, 2016). Moreover, measurements made in the *Lenus Mars* temple indicate a recurrent phenomenon of cultural hybridization. Here the temple, located outside the city walls on the west bank of the river Moselle, combines a possibly Celtic orientation with Roman symbolic beliefs. In reality, the alleged orientation towards the *dies natalis* of Augustus is verified for Cologne. There are a number of circumstances that make this choice logical for a city that was initially planned as the capital of the Augustan province of *Germania* and the seat of an *ara* of imperial worship.

KEYWORDS: Roman Towns. Orientation. Limes Germanicus. Augusta Treverorum. Celtic rites. Cologne.

1. INTRODUCTION

Roman political supremacy in the provinces was established at different levels, ranging from practical engineering through to the religious and symbolic realm. Recent publications on the orientation of Roman cities and military camps highlight this latter aspect by indicating the application of certain design criteria that coincide with sunrise at particular moments of the Roman religious calendar (Magli, 2008; González-García and Magli, 2014; González-García *et al.*, 2014; Bertarione and Magli, 2015). Despite this, the study of particular cities such as Lyon (García Quintela and González García, 2014; González García and García Quintela, 2014) or St. Bertrand de Comminges (González García et al., this issue) demonstrates the existence of syncretistic processes in the methods of city planning and development.

Here we present the results for two Roman cities in present-day Germany, one in the province of *Gallia Belgica* and the other in the Augustan province of *Germania*: *Augusta Treverorum* (Trier) and *Ara Ubiorum*, (Cologne). These results extend the sample of our measurements towards the Germanic *limes*, but also have sufficient entity to be considered the subject of specific studies.

2. *AUGUSTA TREVERORUM*

During Roman times, the city of Trier received the name of *Augusta Treverorum* (Mela, 3, 2, 20) (Fig. 1), a designation that specifically links the town with the political action of Augustus, who would have founded it to provide the *Treveri* with a new *caput civitatis* replacing the chief *oppidum* at Titelberg (Pétange, Luxembourg; Trunk, 2010). As we know from Pliny, the *civitas Treverorum* (defined as the legally constituted political community of the *Treveri*) enjoyed a peregrine administrative status of *civitas libera* until an unspecified time in the reign of Augustus, at which time it would have received the status of a Latin colony in the context of the creation of *Augusta* (Scheid, 2006).

Today there is no definitive agreement on the date of foundation of the city. Authors like Goethert (2003) maintain a late chronology based on the construction of a Roman bridge over the river Moselle, dated by dendro-chronology to the year 17/16 BC. Others like Morscheiser-Niebergall (2009) defend an early chronology based on the archaeological evidence supplied by the city, which seems to point to a period between the year 8/7 BC and the change of era. Regardless of this undoubtedly important fact, the city had to have been built and its governing bodies operating during the first decade of the first century AD, when a dedication to Gaius and Lucius Caesar was erected in AD 4 (Trunk, 2010).

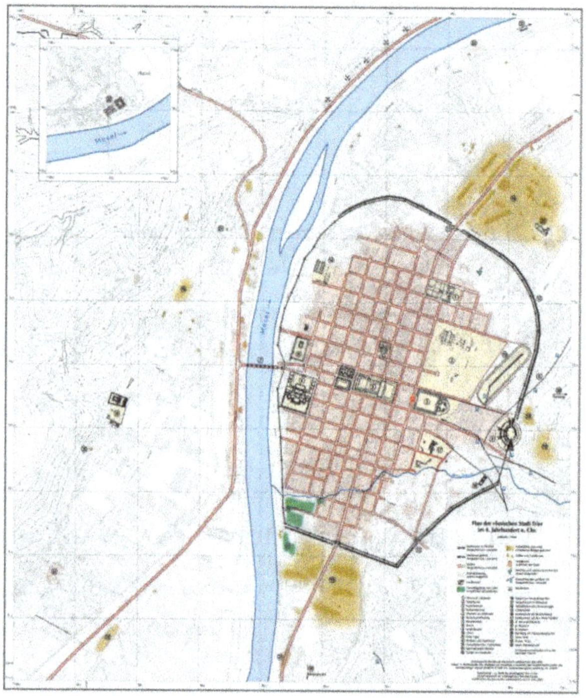

Figure 1. Plan of Augusta Treverorum from Rettet das archäologische Erbe in Trier (2005). The red dot indicates the point from where the picture in Fig. 3 was taken.

In the year 30/29 BC and prior to the foundation of *Augusta*, Rome built a military settlement on the nearby hill of Petrisberg (Trunk, 2010). It stands on a plateau on a hill on the east bank of the valley with a clear visual control over the alluvial plain and the ford of the Moselle. Archaeologists and historians have linked its construction to the uprising of the *Treveri* in 29 BC. (Cassius Dio, 51, 20, 5). Geophysical prospecting work and emergency excavations documented a fortified area of 50 ha with buildings made of mud and wood. Among them were the *Principia*, the *Praetorium* and some barracks aligned with the alleged *via Praetoria*. Its life would have been brief, and was possibly abandoned around the year 27 BC (Trunk, 2010).

2.1. *The Lenus Mars Temple Complex at Irminenwingert*

On the west bank of the Moselle, at the feet of Markusberg, are the remains of a religious complex outside of the city walls. This is a sanctuary dating from the time of Augustus and the second half of the first century AD. At a functional and architectural level, the shrine would have been organised around four places of worship: a properly delimited outdoor venue with individual chapels on its western flank that seem to have housed a cult of *Mars Iovantucarus* related to rites of passage from childhood to adolescence; a Gallo-Roman temple of classical tradition with a *temenos* consecrated to *Lenus Mars*; a processional entry to the temple complex with three exedras and altars offered to *Mars, Lenus Mars, (Mars) Intarabus* and *Ancamna*; and a monumental theatre (Scheid, 2006; Derks, 2008).

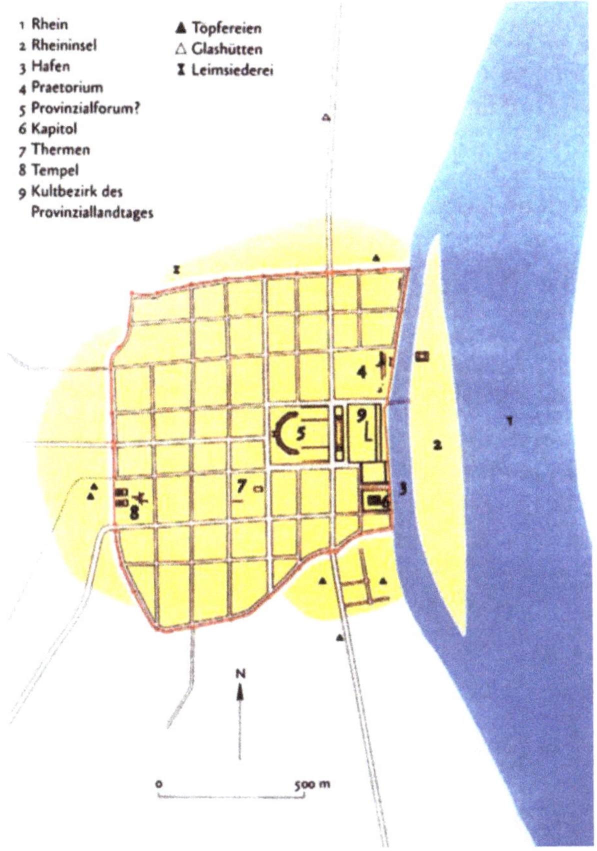

Fig. 2. Ara Ubiorum towards the end of the 1st Century A.D. Map from Eck, 2004.

3. *ARA UBIORUM*

In Roman times, the city of Cologne had two names as a result of its complex formation history: *Ara Ubiorum* (since its foundation in the time of Augustus to AD 50) and *Colonia Claudia Ara Agrippinensium* (from AD 50 onwards) (Bechert, 2013) (Fig. 2).

Both names refer to the same reality: the presence of an altar in Cologne. Its political and religious significance would have led to the incorporation of the term to the official name of the city as a sign of identity and distinction. Literary sources inform us about the existence of these *ara* from the time of Augustus (Eck, 2004). This is where the representatives met from the Germanic *populi* of the Augustan province of *Germania* (probably before the disaster of Varus in AD 9). As for the cult professed in it, there is agreement on the idea that this *ara,* by analogy with the one in *Lugdunum,* was consecrated to Rome and Augustus, especially considering a passage from Tacitus (*Annales* 1, 59, 1) where he refers to the "human" nature of the Roman gods worshiped there (*sacerdotium hominum*). This would have been led by a *sacerdos*, probably chosen among the assembled Germanic *populi*, similar to Lyon (Fishwick, 2002; Eck, 2004).

Regarding the date of its construction, there are two main opinions. On the one hand, Fishwick (2002) argues that the altar of Cologne was erected at the same time as the altar of Lugdunum, probably between 12 and 9 BC. He bases his argument on the evidence of an important religious policy on implementing the imperial cult in *Tres Galliae* and *Germania* promoted by Drusus. Among the results of this policy would be both altars, a temple to Augustus consecrated on 9 BC in the *civitas* of the Lingons (Cassiodorus, *Chronica.* 2, p. 135, 586-7; Fishwick 2002) and the widespread dissemination of the priesthood associated with the imperial cult. On the other hand, Eck (2004) and Bechert (2013) consider a slightly later date in line with the building work in the city from 7 BC.[1] This hypothesis clashes with Fishwick's proposal, who dissociates the erection of the *ara* from the urban construction work, a possibility (in the current state of research) that cannot be excluded. Thus, the existence of the altar would have determined the choice of the site for the foundation of the city, the official place name and the layout of the urban area.

Concerning the foundation date of the city, *communis opinio* considers that it was before the year AD 4/5, a *terminus ante quem* that has managed to be established by the dendrochronological dating of the "*Ubiermonument*", a square structure made of stones and wooden piles that may have been a defensive enclosure or a port infrastructure (Trier, 2014). The *terminus post quem,* however, has been fixed by two stratigraphic levels from the area of Cologne Cathedral and Great St. Martin Church, dated between 7 BC and AD 6/5-14 (Bechert, 2013). Eck (2004) thus proposes the building of the city at the end of Tiberius campaigns (9/8-7 BC), as part of an ambitious program of urban and municipal development launched by Augustus in *Tres Galliae* and *Germania* (Morscheiser-Niebergall, 2009).

As for the possible location of the altar, there are two opposing theories. The first situates the *ara* on the western flank of the *forum* in the heart of the city, perhaps as part of a semi-circular structure that enclosed the public square on that side (Bechert, 2013). The second, however, locates it at the eastern edge of the town, in a prominent location overlooking the Rhine, straddling the *Praetorium* and the *Capitolium,* and the straight *decumanus maximus*. In this area, according to Eck (2004), there are several peripheral structures bordering a vast terrace of 220 x 120 m^2. These structures encircled a building (in the opinion

[1] There are no clear material remains attributable to the altar, and so it is not possible to confirm those dates.

of Eck) reminiscent of the Augustan monument erected in *Nicopolis* to commemorate the victory at *Actium*.

4. ARCHAEOASTRONOMICAL MEASUREMENTS

The measurements were obtained with a Suunto Tandem including a precision compass and a clinometer, with an accuracy of ¼° in the determination of azimuth and ½° for the horizon height. The measurements were corrected from magnetic declination using the WMM magnetic declination model included in the NOAA webpage[2]. When the horizon was blocked due to the presence of buildings we used the digital terrain model available to reconstruct the horizon[3]. This, together with the errors due to the state of preservation of the remains, translates in a mean error of about ¾° in declination.

Table I. Buildings measured in Augusta Treverorum (Trier), Petrisberg (latitude 49° 45′ N) and Ara Ubiorum (Cologne) (Latitude 50° 56′N). We provide measurements towards E and W. The columns show the site or structure measured, the azimuth (A) and horizon altitude (h) and the calculated astronomical declination (δ). A * means that the horizon was calculated using a digital terrain model. The last column gives the possible astronomical event or date. For details see text.

Petrisberg	A	h	δ	Event
Urban layout (E)	131	1¾	−23¾	Winter Solstice
(W)	311	1¼	26	---
A. Treverorum	**A**	**h**	**δ**	**Event**
Urban Layout (E)	106	4½	−7	3rd Mar/11th Oct
(W)	286	5½	14¼	29th Apr/14th Aug
Lenus Mars (E)	113¼	1¼	−14	11th Feb/1st Nov
(W)	283¼	11½	23¾	Summer Solstice
Ara Ubiourm	**A**	**h**	**δ**	**Event**
Urban Layout (E)	90½	0½*	0.07	21st Mar/23rd Sep
(W)	270½	0½*	0¼	21st Mar/22nd Sep

The orientations for *Augusta Treverorum* and *Ara Ubiorum* were obtained by measuring several different structures in Trier and Cologne and verifying that the orientations were consistent (within the measurement errors). Moreover, in the case of the Roman camp at Petrisberg (Trier), we obtained measurements from detailed prospection maps kindly made available to us by the persons in charge for surveying.

The measurements are condensed and presented in Table I, assuming for each site a general east-west orientation. The general hypothesis considered here is that the orientation of the Roman city, generally speaking, could be connected with sunrise or sunset on specific dates. As a result, measurements towards the north and south were considered and obtained, but are not included in the Table for the sake of space.

It is interesting that the first of the three settlements studied here, the Roman camp of Petrisberg, is oriented towards sunrise at the winter solstice (the orientation towards sunset would not work due to the horizon height in that direction). In contrast, the urban layout of *Augusta Treverorum* seems oriented towards the sunrise at the beginning of March or the middle of October. This orientation was verified *in situ* on March 3rd 2015 (see Fig. 2). The orientation towards the west could also be relevant, as this may be linked with the start of May, one of the Celtic beginning of season dates discussed elsewhere (González-García *et al.* this volume). This orientation towards the beginning of season festivities appears more clearly in the main orientation of the *Lenus Mars* temple complex (Fig. 4). In the case of *Ara Ubiorum*, the eastern orientation of the city seems connected with the astronomical equinox.

Figure 3. Sunrise on March 3rd 2015 as seen from the Kaiserthermen in Trier. Sunrise on these dates happens on the hill where the Roman camp of Petrisberg was located. The sunlight is perfectly in line with the remains of the decumanus maximus. Photo: David Espinosa.

5. DISCUSSION

5.1. *The Orientation of the Roman camp at Petrisberg and Augusta Treverorum*

The Roman camp at Petrisberg has an orientation consistent with the sunrise at the winter solstice. This seems at odds with the patterns observed for Roman Camps in other areas of the Empire (like *Britannia*, Rodríguez-Antón *et al.*, 2016; or the *limes Arabicus*, Rodríguez-Antón this volume). However, it would be in agreement with other military camps and Roman towns studied in *Hispania* (González-García *et al.*, 2014) or Italy (Magli, 2008).

[2] http://www.ngdc.noaa.gov/geomag-web/

[3] http://www.heywhatsthat.com/

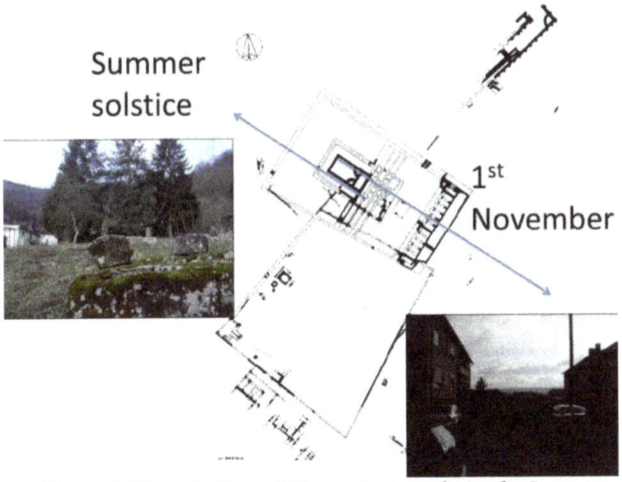

Figure 4. Orientation of the main temple in the Lenus Mars complex. The temple is facing the banks of the river Moselle. The back of the temple is towards the steep slopes of Markusberg.

A question of interest is related to the toponymical evolution of Petrisberg. As we know from the entry on Trier in the *Topographia Archiepiscopatuum Moguntiensis, Trevirensis et Coloniensis* from the *Topographia Germaniae* of Matthäus Merian (published between 1642 and ca. 1660), Petrisberg hill was known as *Mons Martis*, while the hill located on the left bank is mentioned as *Apollinis Mons*. This circumstance seems to abound in the importance and permanence of the cult of Mars among the *Treveri*. Moreover, the ancient tradition places the Monastery of San Martin (commissioned by Magneric, Bishop of Trier, in the sixth century in honour of St. Martin of Tours) on *Mons Martis*. This would have meant the Christianization of an ancient place name, Marsberg, transformed by metonymy in Martinsberg. This is one of the many possible methods of Christianization of the Landscape known throughout Europe, including some that have been studied from an archaeoastronomical perspective.

As regards the urban layout or *Augusta Treverorum*, the *decumanus maximus* and the *forum* are oriented towards sunrise on the first days of March and mid-October (see Fig. 3). Interestingly, this sunrise occurs on top of *Mons Martis*. Here we propose that these dates are related to the "war season", ritually defined in Roman traditions and perhaps also in Celtic culture (*vid. infra*). This orientation is at odds with that found in the small sample we still have for *Gallia* (see González-García *et al.*, this volume), but could be compatible with other Roman towns in *Hispania* (González-García *et al.*, 2014) and the military camps in *Britannia* (Rodríguez-Antón *et al.*, 2016). In fact, the orientation of cult structures towards the "war season" has also been indicated for *Hispania* in the pre-Roman or Romanized settlements of the NW (García Quintela *et al.*, 2014), and could also be found at the so-called Janus temple outside the city walls of Autun (García Quintela and González García, 2016), as well as in the orientation of the temple of Mercury at the summit of the Puy-de-Dôme (Latitude 45°46′; Azimuth 101¼°; altitude of the horizon 0¼°; declination −8°; dates: 1st March/14th October; measured by us *in situ*).

Therefore, the choice of these dates in the case of *Augusta Treverorum* does not seem arbitrary, and could have been motivated by the desire of the Roman authorities to signal the start and end of the "war season" and, according to this, the "government of Mars" in the civic life of the *Treveri*.

In general, the cult of Mars was highly important among the *Treveri* (in fact, there are five shrines consecrated to this god in association with native and Roman gods) with deeply martial values that permeated their activities. For instance, Caesar considered the *civitas Treverorum* to be the most powerful in Gaul for its cavalry (Caesar, *De Bello Gallico,* 5, 3, 1 and 4), stressing the good reputation enjoyed by their riders among the Gauls themselves (Caesar, *De Bello Gallico,* 2, 24, 4), a potential that would be used by Rome from the time of Caesar himself.

According to him, it was customary among the *Treveri* (*more Gallorum*) to organize a military assembly (*armatum concilium*) before setting off to war (Caesar, *De Bello Gallico* 5, 56, 1-2). In addition to the speech of the military commander, the last soldier who had joined the assembly was slaughtered after undergoing all kinds of torture. Once the war came to an end, the *Treveri* sacrificed the captured animals that had survived in honour of Mars, depositing the rest of the booty in a consecrated *tumulus* (Caesar, *De Bello Gallico,* 6, 17, 1-2). In the opinion of Roymans (2004), these testimonies would demonstrate the existence of a deeply martial ideology among the Gauls, a circumstance that Rome would use and promote by large-scale recruitment of auxiliary troops (preferably from the *Treveri* and Batavian). This ideology, according to Roymans, was conveyed through the establishment of certain myths and cults like that of Mars among the *Treveri,* and Hercules among the Batavians.

In the Roman sphere, the importance of March and October in military and liturgical calendars is well known. Both months were the beginning and end of the "war season" (as defined by Dumézil, 1987), a period subject to a series of legal and religious rules that would have taken shape in a series of rites and ceremonies of consecration and desecration of the army. March, as we know from Ovid (*Fasti* 2, 857-861; 3, 85-88), was the month consecrated to *Mars Gradivus*, "the god marching to war". On the 1st of this month the *ceremonias natalicias* of *Mars Pater* were held, i.e. the *dies natalis* of Mars in his role as father of Romulus, founder of Rome. A series of celebrations and rituals began on this day with the army as one of its protagonists: the second *Equirria* (March 14), the *Sacrum Mamurio* or *Mamuralia* (14/15 March), the *Agonium Martiale* (17 March), the *Quinquatria* (March 19) and the first *Tubilustrium* (March 23) (Scullard, 1981). As for October, this month marked when military operations came to an end, when series of rituals took place dedicated to Mars and aimed at the *lustratio* of army and weapons: the October *Equus* (October 15) and *Armilustrium* (October 19) (Scullard 1981). Furthermore, from 19 B.C. onwards, the *Augustalia* were held in honour of Augustus every October 12 (*Res Gestae Divi Augusti* 11).

In view of the above, the orientation of the urban grid of *Augusta Treverorum* to the sun rising on Petrisberg (a hill probably consecrated to Mars) during the first days of March and mid-October would not have been accidental. Instead, it seems to be the result of a deliberate decision by the Roman and *Treveri* authorities to match the orthogonal layout of the city with the beginning and end of the "war season" according to religious, military and astronomical criteria. A very different matter is to propose that one of these two moments was the *dies natalis* of *Augusta Treverorum*, a difficult issue to resolve in the current state of the sources.

In this sense, Goethert (2003) raised the possibility that the *dies natalis* of the city was the *dies natalis* of Augustus, that is, September 23rd. This would have determined the orientation of the urban grid to the dawn of that day. This approach, developed theoretically without measurements of any kind, was rejected by Morscheiser-Niebergall, who argued that the orientation of *Augusta Treverorum* followed only geomorphological criteria (Morscheiser-Niebergall, 2009). While the application of these criteria should not be ruled out, measurements taken *in situ* on certain Roman public buildings revealed that cosmological principles had decisively influenced the orientation of the city that would not have been determined (as proposed by Goethert) by the birthday of Augustus, but by the opening and closing ceremonies of the "war season" consecrated to Mars.

The choice of Mars (and, thereby, the choice of dates to which the urban fabric was oriented), in addition to responding to religious reasons related to military activities, may have been prompted by three main reasons. First, the existence since Pre-Roman times of a profoundly martial ideology with which the *Treveri* identified, and which recognized by the rest of the Gauls and even Rome itself. Secondly, the desire of the *Treveri* authorities to legitimize their position through the public acceptance of political and military values supported by Augustus and personified by Mars as *Pater* and *Ultor*, that is, as founder of the *nomen Romanum* as father of Romulus (Velleius Paterculus 2, 131), and as tutelary deity (with *Venus Genetrix* and *divus Iulius*) of the gens *Iulia* (Ovid, *Ars Amatoria*, 1, 203-204; *Tristia*, 2, 295-296; *Fasti*, 5, 549-598; Suetonius, *Caesar*, 6, 1). And thirdly, the interest in making Mars the main god of the colony due to his apotropaic and protective qualities as *custos civitatis* or *custos urbis* against any external threat (Scheid, 1992).

It is not surprising that three centuries later, when Constantine stayed in Trier, Mars (with helmet, spear, *chlamys* and/or shield) was chosen as the main allegorical figure for the bronze coinage issued in *Augusta Treverorum* with the inscription MARTI PATRI CONSERVATORI or MARTI PATRI PROPUGNATORI (e.g., *Roman Imperial Coinage*, VI, 725, 862-864, 896-897).

5.1.1. The *Lenus Mars* Temple Complex

The Roman orientation of the urban grid of *Augusta Treverorum* contrasts sharply with the syncretic character of the orientation of the temple consecrated to *Lenus Mars* on the western outskirts of the city. The main cult building was oriented towards dawn on November 1st and towards the sunset at the summer solstice. The choice of these dates, which belong to different religious and symbolic spheres, could be an indication, as the name of *Lenus Mars* suggests, of a process that was agreed and sanctioned by the Roman authorities for cultural survival, tolerance and syncretism, which would have allowed the *Treveri* authorities to establish a bond of kinship with Rome while safeguarding their local identities.

This Celtic orientation towards the sunrise on November 1st (the beginning of the season and of the whole year) would be in line with several other dates already studied by our research group in different parts of Gaul (Lyon: García Quintela and González-García 2014; Bibracte-Autun: García Quintela and González-García 2016; see also González-García *et al.* this volume) and Spain.

Moreover, the extramural location of the sanctuary would be analogous to that of the so-called Janus temple in *Augustodunum* (Autun, France). However, this temple is facing the mountains where Bibracte is located, and is orientated towards sunset on the days of the above mentioned "war season". Also, the orientation of *Augustodunum* changes from beginning of season to solstitial, but the orientation of some of the most remarkable elements in Bibracte is towards the beginning of season dates. We can see a possible parallelism emerging, where an analogous process of syncretism is present between the Roman and Aeduan population in *Augustodunum* to the situation in *Augusta Treverorum*.

5.2. *The orientation of Ara Ubiorum*

In the case of *Ara Ubiorum*, its equinoctial orientation seems to be in line with a number of other Roman settlements in Gallia, such as *Vesunna* (Perigueux), *Ruscino* (Château-Roussillon, Perpignan) and *Augustonemeton* (Clermont-Ferrand) (González-García *et al.*, this volume). In the case of *Ara Ubiorum*, the choice of an orientation towards the sunrise on September 23rd (*ante diem VIIII Kalendas Octobres*) may have been motivated by the desire of the Roman and Ubian authorities to coincide the *dies natalis* of this city (conceived as capital of the Augustan province of *Germania*) with the birthday of Augustus, a circumstance that has been suggested (for example) in the Augustan *municipium* of *Segobriga* in Spain (Abascal and Almagro-Gorbea, 2012).

Indeed, September 23rd was an important feast day of the Roman liturgical calendar, possibly established as *dies ferialis* by Augustus himself. This appears in the *fasti publici* (as in the case of the *Fasti Pinciani, Fasti Palatii Urbinatis, Acta Fratrum Arvalium, Fasti Maffeiani, Fasti Vallensis, Fasti Pighiani* and *Fasti Philocali*) and the *ferialia* (such as the *Feriale Cumanum* and *Feriale Duranum*).

In addition, a series of texts and epigraphs document the celebration of ceremonies and games in honour of Augustus, as well as offerings, during this date both in the eastern and western provinces. In this respect, a decree passed between 27 and 11 BC in the Greek city of Mytilene reports on the granting of "divine honours" (*isotheoi timai*) to the emperor, the construction of a temple, the creation of a cult identifying Augustus with *Zeus Olympios* and the monthly and yearly celebrations held in his honour during the *dies natalis* (*Inscriptiones Graecae ad Res Romanas pertinentes,* IV, 39). On the other hand, around 19 BC Athens approved a decree on the celebration of the birth of Augustus, establishing similar sacrifices for the emperor to those made during the festival of Apollo (*Supplementum Epigraphicum Graecum*, XVII, 34). In connection with these celebrations, but from Philadelphia (Lidia), an epigraph dated AD 40 shows that a local family paid the expenses as a result of Augustus' *dies natalis* celebration (*Inscriptiones Graecae ad Res Romanas pertinentes*, IV, 1615). However, this religious behaviour was not confined to the eastern territories of the Empire (which were familiar with the cult of the Hellenistic rulers), but was widespread in Italy and in other provinces of the West. According to Dio Cassius (54, 26, 2), in 13 BC the *praetor Iullus* organized *ludi circenses* in Rome for the birthday of Augustus. Two decades later, according to this author, Tiberius and Germanicus planned a horse race in AD 11 at the end of their campaign in *Germania* to commemorate the *dies natalis Augusti* (Cassius Dio, 56, 25, 2-3). The feast, after the death of Augustus in AD 14, had to be celebrated in Rome with games similar to those established in honour of *Mars Ultor* (*ludi Martiales*) (Cassius Dio, 56, 46, 4).

Another, possibly complementary type of celebration to this was to hold *immolationes* and *supplicationes* to Augustus in his different aspects (*genius Augusti, numen Augusti* and *divus Augustus*) on 23 and 24 September. This practice is attested in *Narbo Martius*, where an epigraph commemorates the consecration of the *ara numinis Augusti* in AD 11 and prescribes the individual sacrifice of a victim by three *equites Romani* and three freedmen on September 23. Furthermore, they had to pay for incense and wine on September 24th for the *supplicationes* that the *coloni et incolae* had to carry out to the *numen Augusti* (*Corpus Inscriptionum Latinarum*, XII, 4333). A similar procedure is attested in *Forum Clodii* in AD 18, where two victims should be sacrificed on the *ara numinis Augusti* for 23 and 24 September (*Corpus Inscriptionum Latinarum*, XI, 3303).

Based on the amount of testimonies, the political and religious significance of September 23 as *dies natalis Augusti* is quite clear. This importance is further emphasised if we consider that in 9 BC, the Greek cities of the Roman province of *Asia* set this day as the beginning of the year and the "beginning of all good things."

Therefore, we believe that obtaining September 23 on the basis of the measurements and calculations is not the result of chance, but is the consequence of a planned intention to coincide the orthogonal urban layout of *Ara Ubiorum* with dawn on Augustus' birthday, a fact that provides a lasting testimony of *pietas* and *fides,* with the altar used as a central place of worship and meeting point for all of the Germanic *populi* on this day until the Varian disaster. It would therefore be consistent with this interpretation that the *ara* were part of the eastern façade of the city in an open area facing towards *Germania* and in line with the *forum*. The fact that the first rays of the sun every September 23rd aligned with the religious and political areas of *Ara Ubiorum* would have given the whole complex a theatrical effect full of symbolism, creating a prodigious vision attributable to the manifestation of the *numen* of Augustus.

ACKNOWLEDGEMENTS

We are grateful to Prof. Markus Trunk for his cooperation and helpful explanations on the Petrisberg site. ACGG is a Ramon y Cajal fellow of the Spanish MINECO.

REFERENCES

Abascal, J. M. and Almagro-Gorbea, M. (2012) Segobriga, la ciudad hispano-romana del sur de la Celtiberia. G. Carrasco Serrano (coord.), *La ciudad romana en Castilla-La Mancha*, Cuenca, Ediciones de la Universidad de Castilla-La Mancha: 287-370.

Bechert, T. (2013) Ara Ubiorum. Zum Namen des frühkaiserzeitlichen Köln und zum Standort des Kaiseraltars. *Carnuntum Jahrbuch* 2012, 9-16.

Bertarione, S.V. and Magli, G. (2015) Aosta Under Augustus' sign: the role of Astronomy in the foundation of Augusta Praetoria Salassorum. *Cambridge Archaeological Journal*, Vol. 25:1, 1–15.

Derks, T. (2008) Les rites de passage et leur manifestation matérielle dans les sanctuaires des Trévires. D. Castella and M.-F. Meylan (dirs.), *Topographie sacrée et rituels. Le cas d'Aventicum, capitale des Helvètes*, Basel, Archéologie Suisse: 191-204.

Dumézil G., 1987, *La Religion romaine archaïque,* deuxième éd., Paris, Payot.

Eck, W. (2004) *Geschichte der Stadt Köln. Köln in römischer Zeit. Geschichte einer Städt im Rahmen des Imperium Romanum.* Cologne, Greven Verlag.

Fishwick, D. (2002) *The Imperial Cult in the Latin West. Studies in the Ruler Cult of the Western Provinces of the Roman Empire.* Vol. III: *Provincial Cult.* Part 1: *Institution and Evolution.* Leiden-Boston-Colonia, Brill.

García Quintela, M. V. and González-García, A. C. (2014) Le 1er août à Lugdunum sous l'empire romain: bilans et nouvelles perspectives. *Revue Archéologique de l'Est*, Vol. 63, 157-177.

García Quintela, M.V. and González-García (2016) De Bibracte à Augustodunum: observations archéoastronomiques. *Revue Archaeologique de l'Est.* Vol. 65, 283-297.

García Quintela, M.V., González-García, A.C. and Seoane-Veiga Y. (2014) The Iron Age Saunas of NW Iberian Peninsula: An Archaeoastronomical Perspective. *Mediterranean Archaeology and Archaeometry*, Vol. 14(3), 133-141.

González-García, A.C. and García Quintela, M.V. (2014) The 1st of August at Lugdunum: Astronomy and Imperial Cult in Gallia. *Mediterranean Archaeology and Archaeometry,* Vol. 14(3), 83-91.

González-García. A.C. and Magli, G. (2014) Roman city planning and spatial organization. C.L.N. Ruggles (ed.) *Handbook of Archaeoastronomy and Ethnoastronomy: Building bridges Between Cultures*, Cambridge, U.K., Cambridge University Press: 374-381.

González-García, A.C., Rodríguez-Antón, A. and Belmonte, J.A. (2014) The orientation of Roman towns in Hispania: preliminary results. *Mediterranean Archaeology and Archaeometry*, Vol. 14 (3), 107-119.

Goethert, P.-K. (2003) Untersuchungen zum Gründungsschema des Stadtplanes der Colonia Augusta Treverorum. Die Geburt der Stadt an der Mosel. *Archäologisches Korrespondenzblatt*, Vol. 33, 239-258.

Magli, G. (2008) On the orientation of Roman towns in Italy. *Oxford Journal of Archaeology*, Vol. 27(1), 63–71.

Morscheiser-Niebergall, J. (2009) *Die Anfänge Triers im Kontext augusteischer Urbanisierungspolitik nördlich der Alpen.* Wiesbaden, Harrassowitz Verlag.

Roymans, N. (2004) *Ethnic Identity and Imperial Power: The Batavians in the Early Roman Empire*. Amsterdam, Amsterdam University Press.

Scheid, J. (1992) Épigraphie et sanctuaires guérisseurs en Gaule. *Mélanges de l'Ecole française de Rome. Antiquité* 104/1 : 25-40.

Scheid, J. (2006) Réflexions sur le Mars trévire. V. Brouquier-Reddé, E. Bertrand, M.-B. Chardenoux, K. Gruel and M.-C. L'Huillier (dir.), *Mars en Occident. Actes du colloque international 'Autour d'Allones (Sarthe), les sanctuaires de Mars en Occident' (Le Mans, 4-5-6 juin 2003)*, Rennes, P.U.R.: 35-44.

Scullard, H.H. (1981) *Festivals and Ceremonies of the Roman Republic.* London, Thames & Hudson.

Trier, M. (2014) Das *oppidum Ubiorum* – Köln in der frühen Kaiserzeit. M. Trier and F. Naumann-Steckner (eds.), *14 AD: Römische Herrschaft am Rhein*, Cologne, Wienand: 46-55.

Trunk, M. (2010) Augusta Treverorum. Débuts et développement d'une métropole. R. González Villaescusa (ed.), *Simulacra Romae II. Rome, les capitals de province (capita provinciarum) et la création d'un espace commun européen.* Une aproche archéologique, Reims, Société archéologique champenoise: 195-200.

6

LOOKING THROUGH A TELESCOPE WITH AN OBSIDIAN MIRROR: COULD SPECIALISTS OF ANCIENT CULTURES HAVE BEEN ABLE TO VIEW THE NIGHT SKY USING SUCH AN INSTRUMENT?

Josef Vit[1] & Michael A. Rappenglück[2]

[1]*Independent Scholar, Oberbettingen, Germany*
[2]*Adult Education Centre and Observatory, Gilching, Germany*

Corresponding author: Michael A. Rappenglück (mr@infis.org)

ABSTRACT

Convex cut mirrors manufactured from the volcanic glass obsidian had been known since Neolithic time (7400/7100 to about 6200 BCE) in Çatalhöyük, Turkey. A Herschelian type telescope made with an obsidian mirror (∅ 12 cm) allowed to see the Moon's craters sharp and distinct, the phases of Venus as well as the discs of Venus and Jupiter. The moons of Jupiter however are not visible due to the low degree of reflection of the used obsidian mirror. A specimen with a much better reflection or a bigger one would result in improved views of celestial objects. The paper reports on the making of the telescope and its potential application. Moreover, as a general basis, the study addresses the prehistory and symbolism of mirrors, with special focus on a possible assignment for skywatching.

KEYWORDS: Obsidian mirrors, prehistory of reflecting telescopes, symbolism of mirrors, Çatalhöyük, Neolithic optics

1. INTRODUCTION

In the 17th c. different types of reflecting telescopes had been developed and sometimes practical used (Wilson 2007, 1-15): ca. 1616 by Niccolò Zucchi (1586-1670), using a Herschelian design (see below); before 1626 by Cesare Caravaggi (died before 1626); 1632 by Bonaventura Francesco Cavalieri (1598-1647); and 1636 by Marin Mersenne (1588-1648) (Wilson 2007). The most advanced constructions had been developed 1663 by James Gregory (1638-1675) - the Gregorian, 1668-1672 by Sir Isaac Newton (1643-1727) - the Newtonian, and 1672 by Laurent Cassegrain (ca. 1629-1693) - the Cassegrain. The brothers John Hadley (1682-1744), Henry Hadley (1697 - 1771) and George Hadley (1685-1768), in 1721 constructed an improved reflecting telescope type 1721 using a parabolic mirror as primary.

Some (Atalay 2006; Buys 2007; Atalay and Wamsley 2009, 163-164) consider Leonardo da Vinci (1452-1519) to be the first describing the theory and the use of concave mirrors for stargazing purposes (Leonardo da Vinci, Codex Arundul, 1513). It is discussed that he might have constructed and used a reflecting telescope of the Herschelian type.

The history of polished concave and convex mirrors however is much older, being backtracked fifth or even ninth millennia (Albenda 1985; Vedder and Swogger 2005; Badisches Landesmuseum Karlsruhe 2007; Anderson 2007; Enoch 2006, 2007).

That raises the question of whether a few specialists of archaic cultures in principle would have been able to construct and manufacture a kind of reflecting telescope albeit without having an ocular, which they could have used for valid observations of the night sky phenomena. This paper reports on a kind of experimental archaeoastronomy, done by Josef Vit (Oberbettingen, Germany), who grinded the obsidian mirror and used it as a Herschelian type telescope. Michael A. Rappenglück added research on the cultural background of mirrors, in particular of specimens, which might have been used for sky watching, all over the world and through epochs back to the Neolithic time (ca. 7500 BCE; Enoch 2006, 775-776).

2. THE EARLIEST MIRRORS

Earliest mirrors had been made of different matter (Nordenskiöld 1926; Albenda 1985, 2; Taube 1992a, 169, 1992b, 33; Schechner 2005, 141-145; Saunders, 2003, 19; Calvo 2003, 5, 10-11; Matsumoto 2013, 103): Obsidian, jet, copper, gold, silver, bronze, selenite, slate, haematite, magnetite, anthracite, pyrite, and later glass. Hitherto obsidian is the oldest known material used for producing mirrors, as it is attested by specimens from the 6th millennium BC in Anatolia, Syro-Mesopotamia, and the Levant (Albenda 1985; Vedder and Swogger 2005; Enoch 2006, 2007; Badisches Landesmuseum Karlsruhe 2007).

Usually the mirrors are hand-sized (9-16 cm) and have a round or oval shape (Nordenskiöld 1926; Albenda 1985; Enoch 2006, 775, 777-778; Matsumotu 2013, 103). Their handles are long. The artisans apparently designed most of the mirrors for feeling good in the hand. Nevertheless, there are also bigger specimens, ∅ up to 30 cm, and smaller ones, ∅ down to 4 cm (Matsumotu 2013, 103; Enoch 2006, 778).

The curvatures of earliest mirrors typically are flat, but a few had been grinded convex or concave (Nordenskiöld 1926; Albenda 1985; Enoch 2006): At Çatalhöyük, Turkey, 7,400/7,100-6,200 BC, the hitherto found obsidian mirrors are convex (Vedder and Swogger 2005; Enoch 2006, 775-776, 778; Badisches Landesmuseum Karlsruhe 2007). In ancient Egypt, starting with ca. 2,900 BC, mirrors, made of different matter, have a convex or concave curvature (Enoch 2006, 776). Since approximately the first millennium BC, the Olmec (Mesoamerica) manufactured concave and some convex mirrors (Drucker, Heizer and Squier 1959, 176-184 and plate 43a; Lunazzi 1995, 1996; Enoch 2006, 778; Mexicon 2011, 36-37). Surprisingly a few are slightly shaped parabolic (Gullberg 1959, 280-283, and pl. 62). They have diameters up to 10 cm and focal lengths between 5 and 80 cm (Enoch 2006, 778). The mirrors are well polished. They still have excellent optical imaging and enlarging abilities.

Ancient people had been well aware of different types for mounting and holding a mirror: cells, cords, handles or tangs made of different materials (Albenda 1985; Juliano 1985, 36; Lerner 1996, 11; Enoch 2006, 775-776, 778, and Table 1/2).

Since at least the ninth millennium BCE advanced and more than sufficient technologies had been available for shaping (pecking and flaking), grinding and polishing surfaces (Enoch 2006, 780-781; Astruc et al. 2011, 3415-3424, 3418, fig. 5), which could be used for producing excellent mirrors. There are evidences for the usage of certain kinds of mechanical devices supporting the manufacturing process of very smooth and well-shaped surfaces (Astruc et al. 2011, 3421-3422).

Finally it is noteworthy that Mesoamerican people produced liquid mirrors using water and in a few cases probably mercury (Matsumoto 2013, 105).

3. SYMBOLISM AND USAGE

Studying cultures across the world and through time gives evidence for multiple profane and sacral use of flat and curved mirrors (Lerner 1996, 11; Moyer 2012). They helped providing self-perception (Juliano 1985, 37; Lerner 1996, 11, 31; Pendergras

2003; Mexicon 2011, 36-37; Moyer 2012; Matsumotu 2013, 104-105): They served as toiletry for personal preparation of the body. They permitted self-observation putting on cloths. They had been used as fashion accessories. People often associated the mirror, just as e.g. the spindle and the distaff, with woman's activity (Lerner 1996, 31). Mirrors figured into entertainment (Lerner 1996, 11). Moreover they had been a practical but also a ritual aid for the illumination of dark interiors (Russell 1988, 672 and fn. 45) by directing sunlight into them. People used them for medical purposes (Lunazzi 1996, 412).

Moreover mirrors were important figuratively: In general they symbolized the possibility of self-awareness and self-contemplation embodied by the power of reflection (Lunazzi 1995; Moyer 2012; Matsumotu 2013, 104-105). People believed that mirrors, especially if they are curved, allow seeing deeper into the everyday world. Moreover they open insight and give access into the normally hidden otherworld (Coqueugniot 1998, 358; Cauvin 1998; Moyer 2012). They reveal benevolent or malevolent power beings disguised in the shape of a human being or other creatures. Hence in the opinion of people mirrors served for getting knowledge.

The Latin word for 'mirror' is *speculum* (Pokorny 1959, 984). The word is rooted in the Indo-European *spek̂-* meaning "to watch". Derivations refer to the activities of observation, exploration, spying, and scrying (Pokorny 1959, 984). Similar the Yucatec (Maya language) word for (obsidian) mirror means "considering, contemplating, foretelling" (Taube 1992b, 34; Matsumotu 2013, 113-114). The Warau in South America used one term for denoting a shadow, the image in a mirror and the spiritual being of a decedent (Goeje 1930, 62 [415]). These linguistic examples are complement to the traditions of past and present cultures according to those mirrors served for catoptromantic and scrying purposes, using reflecting surfaces, e.g. water, crystals, crystal balls, and other material (Taube 1992b, 33-34; Lunazzi 1995; Chevalier and Gheerbrant 1996, 657-661; Saunders 2001, 224; Maillet 2004, 49-55; Schechner 2005, 141; Blainey 2007; Moyer 2012; Matsumoto 2013, 105).

According to archaic conceptions mirrors allowed spiritual specialists gazing into the divine real. They worked as 'openings', which enabled communication and encounter with 'other worlds', wherein deities, ancestors, certain powers or other entities exist. They specifically played an important role during the passage between life and death, and for the renewal of life (Kano 1979, 9-10; Moyer 2012; Matsumoto 2013, 104-105, 109-110). They therefore were important in rituals dealing with those topics, e.g. funerary customs (Lerner 1996, 11, 31; Moyer 2012). Because mirrors offered access to other realms of the world they had been associated figuratively to doorways, caves, wells and pools of water (e.g. Mesoamerican and other cultures) (Moyer 2012). Mirrors, functioning as 'gates', were protected and used by supernatural beings (Waterbury 1952, 117-136; Bulling 1960; Strätz 1996; Moyer 2012). These ideas often were related to archaic cosmologies.

Moreover ancient people considered mirrors to be a kind of 'eyes' and interrelated them with vision and contemplation (Schele and Miller 1986; Lunazzi 1996, 416; Blainey 2007; Matsumotu 2013, 104, 112-113). Cosmovisions in particular are often linked with both 'tools' for 'viewing': the eye and the mirror. That explains why some animal species (often carnivores), who have a superior night vision, caused by a layer of tissue behind the eye's retina (tapetum lucidum; Ollivier 2004), which reflects the visible light back through it whereby increasing the light for the receptors, had been correlated to shamans or rulers: Like for example a felidae, who uses the 'eye shine' ('mirror like eyes') in the night, shamans and rulers probably wanted to see into 'other worlds' (Matsumotu 2013, 104).

The above-mentioned ideas led to the view that mirrors have magical, apotropaic, protective, destructive, political and divine power (e.g. Juliano 1985, 36; Witzel 2005; Blainey 2007; Moyer 2012; Matsumotu 2013, 106-107).

4. SKYGAZING AND ANCIENT MIRRORS

Ancient Traditions passed down the concept of the mirror as a tool for viewing far away objects (Laufer 1928).

Mirrors in the underworld are dark (Roe 1982, 135). They are complemented by bright mirrors, e. g. embodied by the sun, who reflected and focused the light onto the Earth (Roe 1982, 228). Obsidian 'dark' mirrors are thought to be opposed to 'bright' pyrite ones (Taube 1992b, 33; Cauvin 1998). Mirrors collect cosmic energy (Saunders 2003, 25).

All over the world archaic cultures associated mirrors with fire and sun (Kano 1979, 9-10; Carlson 1981, 128; McCrickard 1990, 64, 65, 67, 69, 91, 97, 102, 122, 124, 125, 126, 170, 194, 210, 214, 225, 242, 251; Taube 1992a, 192-195; Milbrath 1999, 92, 88-89, 102, 226; Matsumotu 2013, 104). The mirror symbolized the Japanese Sun-goddess Amaterasu-Omi-Kami. It was an imperial implement in Shinto shrines (Witzel 2005, 15, 16, 19, 21, 24, 26).

The Maya Moon Goddess Ixchel is giving birth to a rabbit, a symbol of the moon, onto (or in front of) a mirror situated directly where the cosmic World Tree grows (Miller and Taube 1993: 127. 142: Schele and Miller 1986: 143-144).

Amerindians considered the Moon as a mirror and the stars as tiny mirrors (Roe 1982, 249; Milbrath 1999, 133 Fig. 410, 198, 283; Matsumoto 2013, 112). In ancient China the 'Broken Mirror', was related to the Moon's phases, except Full Moon, which was regarded as a whole mirror (Gheerbrant 1996, 658).

The Aztec and Toltec god Tezcatlipoca ('smoking mirror') is equipped with a round obsidian mirror and linked to the Big Dipper asterism and lightning (Lehmann-Nitsche 1938, 73-76, 79; Taube 1992b, 34, 75; Saunders 2001). Mirrors coming from ancient China show animals of the cosmic cardinal quarters, Sun, Moon, the five planets, the Big Dipper asterism and Polaris, the zodiacal signs and the stems (Schafer 1978-79; Juliano 1985, 36; Von Falkenhausen 1989, 222-223). They have apotropaic functions collecting light from the heaven, e.g. rays from the stars (Schafer 1978-79, 58).

In ancient China during the New Fire ceremony in spring fire was kindled with the aid of a mirror (Frazer 1919, 166). In the Roman Empire the Vestals produced sacred fire collecting the sun rays by a mirror (Frazer 1919, 168, fn. 347). Mirrors may have served for fire making in Olmec rituals, too (Lunazzi 1995; but Mexicon 2011, 36-37). In the Late Han-Dynasty (23/25–220 AD), during a special ceremony the "Director of Sun Fire" a concave Bronze mirror facing the Sun at noon used to focus solar fire (Seidel and Kalinowski 1982; Needham 1996, 87). In addition it was used to collect dew, being exposed to the Full Moon's light. Mirrors were associated to fire, Sun, water and Moon (Moyer 2012, 285, 291, 292).

According to the traditions of shamans in the Altai region Sun and Moon act as the two heavenly mirrors, displaying events on Earth (Holmberg 1927: 419). From both the images are reflected again into the magic mirrors, which a shaman uses. They are accessories of his costume (Holmberg 1964, 419, 437). These magical mirrors symbolize Sun and Moon (Rappenglück 2016, 22). Often ancient cultures considered the Moon to be a mirror reflecting the Sun's light (Chevalier and Gheerbrant 1996, 658; Moyer 2012, 304). They sometimes are decorated with the animals of the zodiac (Holmberg 1964, 437). Orphism (6th to 5th c. BCE) handed down a very similar idea about the Sun mirroring the primeval light and the Moon reflecting the Sun's rays (Eisler 1910, 656-657, 687). A comparable concept is delivered by Pythagoreanism, 6th c. BCE (Eisler 1910, 687, 724 fn. 4). Later Empedocles (c. 490 - c. 430 BCE) regarded the celestial luminaries as reflections of the primeval cosmic light (Eisler 1910, 692). In the Old Testament, Job 37.18 (English Standard Version), the complete sky was considered to be a cast metal mirror. Even today Nahua Natives regard the dome of the sky as a living (crystal) mirror filled with the sparkle of the sun and stars (Saunders 2003, 20).

The Polynesians had concave polished mirrors (Kelley and Milone 2005, 346). It remains unclear whether they used them for sky watching. The Kogi people in South America used obsidian mirrors for magical purposes related to the sun, shining vertical through a hole in the temple (Reichel-Dolmatoff 1982, 177-178; Kelley and Milone 2005, 433).

Modern indigenous Maya people still use liquid water surfaces for indirect watching reflections of eclipses (Matsumotu 2013, 105). Some researchers assume that ancient Maya observed these and perhaps also other astronomical phenomena alike (Milbrath, 1999: 27; Matsumotu 2013, 105). Moreover some scientists speculated that the Maya applied obsidian mirrors for astronomical observations possibly in relation to shamanistic practices (Schagunn 1975; Milbrath 1999, 27; Kelley and Milone 2005, 353, 354 fig. 12.2, 433; Blainey 2007, 26; Matsumotu 2013, 105). The Olmec, too, seemed to have used mirrors for astronomical purposes (Lunazzi 1995).

It is not clear what the few representations of persons (rulers, shamans) viewing into a mirror or into a ceramic bowl, which sometimes is equipped with a mirror placed at the bottom, really show (Kelley and Milone 2005, 353, 354 fig. 12.2, 433; Matsumoto 2013, 105-106). Is an act of self-reflection, of meditating, or of sky gazing (Kelley and Milone 2005, 353, 354 fig. 12.2, 433) depicted?

5. OLDEST OBSIDIAN MIRRORS

During Mesolithic and Neolithic time Aşıklıhöyük (ϕ: 38° 20′ 57″ N, λ: 34° 13′ 47″ E, 1119.5 m.a.s.l.) and Çatalhöyük (ϕ: 37° 40′ 3″ N, λ: 32° 49′ 42″ E; 903 m.a.s.l.), both in Anatolia (Turkey), were important centres of obsidian trade close to the volcanic glass resources at the volcano of Hasan Dağ (Turkey).

A highly polished obsidian bracelet from Aşiklı Höyük, Central Anatolia 8,300-7,500 cal. BC. (Astruc et al. 2011, 3415-3424, 3418, fig. 5), reveals that the artisans in those days had been able to search and select obsidian of high quality as well as applying an excellent drilling and abrasive technology. It seems that certain complex polishing techniques for obtaining a surface quality down to nanometres had been at their disposal. That precision is in the same scale as needed by telescope lenses today.

From Wadi Rabah Culture, Kabri's Area A (Israel) comes an obsidian mirror, which is dated to ca. 6th/5th millennium BCE (Ben-Tor 1992, 39; Gopher, Marder, and Barkai 2011). The source for the raw material was ancient Anatolia. Because of having a handle, that beautiful worked specimen probably served for cosmetic purposes.

Some centuries older and much more impressive are the convex spherical obsidian mirrors, perfectly smoothed, from Çatalhöyük, Turkey, 6,520-6,220 cal. BC (Museum of Anatolian Civilizations, Ankara, Turkey; Enoch 2006, 775-776).

6. ÇATALHÖYÜK, 6,520-6,220 CAL BC

Eight totally preserved mirrors made of obsidian had been dug up in Çatalhöyük, Turkey. A few more were found broken (Vetter 2005; Badisches Landesmuseum Karlsruhe 2007, #336; Haddow 2014, 39, 217-218). Some of them still show perfect optical imaging quality (Vedder 2001; Enoch 2006, 775-776, and Figure 2). The smallest mirror (Ø 4 cm, H: 2 cm; found in space 226) has not been finished.

Most of them had been found in graves of levels VI-IV, c. 6,520-6,220 cal BC (Mellaart 1967, 79, 80 table 13, 156, 208, 209, and plate XII; Badisches Landesmuseum Karlsruhe 2007, #336). Several burials, mostly of women, yielded circular, hand-sized objects, about 9 cm in diameter, made of highly polished obsidian (Enoch 2006, 775-776). Some of the obsidian still show an excellent image (Hodder 2012, 9). The finest of these was set into lime plaster. Archaeologists left open the question whether these mirrors had been used for ritual or for another purpose. Some but not all obsidian 'mirrors' seemed to be related with skulls, azurite and ochre, beads, and hackberries. A solely relation of the mirrors to female gender is not allocable.

A kind of serial production is evident, showing that the craftsmen of that time controlled the technology (Badisches Landesmuseum Karlsruhe 2007, #336; Astruc 2011). Experiments showed: First the surface for the mirror was squared. Then followed repeated working steps of cutting, grinding, and polishing using raw and fine whetstones, sand, clay, and water (Badisches Landesmuseum Karlsruhe 2007, #336; Vit, pers. communication). The production of one mirror took about 7-8 hours (Badisches Landesmuseum Karlsruhe 2007, #336; Astruc 2011, 3421-3422).

A photograph (Fig. 1), taken from the object during the exhibition showed the reflection of a fluorescent lamp, which allowed to determine the corresponding concave mirror's focal length. That served as a template for making an obsidian mirror telescope. The surface of the mirror is polished spherical and is slightly convex. The backside is conical and only a bit carved.

Answering the question whether an obsidian mirror telescope could work, required an experimental assembly, which for a 12 cm obsidian mirror was done by Josef Vit (see below).

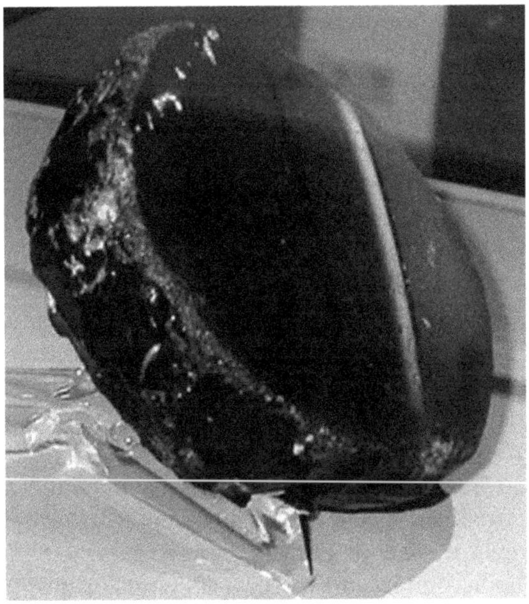

Figure 1: 8,500 years old example of a spherical convex obsidian mirror (Ø: 9.5 cm, H: 5.1 cm); exhibition "Anatolian Monuments", Karlsruhe, Germany, 2007. Photo: Josef vit.

7. MAKING THE OBSIDIAN MIRROR

Obsidian (Ericson et al. 1975) is a deep black or blackish green translucent glass (70–75% SiO_2, MgO, Fe_3O_4) of volcanic origin. Its properties are similar to quartz, but it is softer, having Mohs values of 5.0-6.0. Obsidian is an amorphous material with a density of 2.3-2.6 g/cm^3. As a special characteristic obsidian shows a conchoidal surface fracture.

Though its reflectance (R%) is only 4-5%, which is quite low compared with other material (e.g. silver [Ag]: 93.3%; copper [Cu]: 64.4%; pyrite [FeS_2]: 51.7%; hematite [Fe_2O_3]: 26.4-29.9%; magnetite [$Fe^{2+}(Fe^{3+})_2O_4$]: 21%; (Craig and Vaughan 1994, Table A2.1, 405-410; Lunazzi 1995, 4), it can serve as material for the production of flat or curved mirrors (Lunazzi 1995;). Obsidian mirrors can be etched and polished up to the limit allowed by properties of the material.

Sphericity results from the grinding process, which also gives a pair of concave-convex surfaces. First experiments (Vedder 2001; Vedder and Swogger, 2005) lead to a largest mirror surface of ca. 4 by 6 cm. Grinding and polishing was done by granite, sand, sandstone, fine-grained stone, and clay. The mirrors were slightly convex and had a good imaging.

Resuming Vedder's construction of the past, Vit started making a mirror of obsidian. Raw obsidian material coming from different source worldwide was carefully proven with the unaided eye concerning purity, homogeneity, frictions, and cracks. Then a piece of Mexican obsidian was cut for preparing a round disk. Subsequent the specimen was grinded and polished. It turned out that grinding an obsidian mirror results more quickly in a deep curvature than grinding quartz glass, because of a lesser Mohs value (obsidian: 5.5-6.0 / quartz: 7). Finally the proof of quality was done using a modern method: a Bath-Interferometer (Bath 1973).

The theoretical base for making the mirror followed Bath (2007), though ancient experts doubtless will have applied the trial and error method. The computations depend on standard values for near and far point, visual acuity, exit pupil, reflectivity of obsidian, myopia, and some other parameters etc. The finished mirror is characterized as follows: ∅ = 12 cm, focal length f = 213.5 cm, ratio = 1:17.8, theoretical limiting magnitude: 9.0 mag, magnification: 20x, true field of view: 3.4°. The degree of reflection however was about 3%, less than the expected 5%.

The secondary mirror in a Newtonian design, if manufactured as an obsidian reflector, too, would consume another 5% of the incoming light, resulting in only 0.25% degree of reflection for the complete telescope. That would make the instrument useless. With regard to the used material obsidian and the modern variants of telescope designs, only one telescope design seems to be sufficient: *the Herschelian type,* named after William Herschel (1738-1822), who applied that type of construction for his reflecting telescopes (Hoskin 2011). For reducing errors caused by the Herschelian design - observing off the optical axis - the mirror must have a long focal ratio.

Figure 2: The Herschelian telescope with the obsidian mirror, ready for observing; Full Moon observed with the obsidian mirror. Still image taken from a film. Photos: Josef Vit.

The 12 cm Herschelian reflector (f = 213.5 cm, based on a mirror made of obsidian, was set in a wooden mounting, which allowed aligning and a bit reposition of certain astronomical targets by the observer or / and an assistant. The abilities for carpetering and woodturning of such a bracket had been excellent at the time and place of Çatalhöyük (Mellaart 1967, 213-215, and Fig. 55; Asouti 2013).

The Herschelian telescope with its obsidian mirror shows a brilliant, sharp and detailed image of the Moon (Figure 3), with lunar maria, craters, lunar rays, and mountains. The observer can recognize the phases of Venus, the disk of Jupiter, and also some other celestial objects (e.g. Open Clusters, Orion Nebula). The Galilean moons of Jupiter or the Orion Nebula (M42) however are not visible due to the low degree of reflection of the used obsidian mirror. A specimen with a much better reflection or a bigger mirror (e.g. ∅ = 16 cm or 19 cm) for sure would result in improved views and an extension of the range of celestial objects: If the obsidian has 5% reflectivity and the diameter of the obsidian mirror is 19 cm, the limiting magnitude would reach about 10 mag (Bath 2007).

8. CONCLUSION

An obsidian mirror could have been both a catoptromantic and a proto-scientific tool, comparable to the ancient Chinese geomantic compass (Needham 1996: 254-273). There is some evidence for sky gazing and astronomical referenced time-reckoning at Çatalhöyük (Matossian 1980): People observed the annual cycles of bees and barley. Counting days they correlated them with the star phases of the Pleiades and perhaps asterisms in the area of today's Bull and Ram, which they recognized there. Whether the interest in sky gazing and the experimentally verified ability of producing high quality obsidian mirrors actually resulted in the development and use of a Herschelian type obsidian based reflector is not proven. At present there is no evidence for or against that the society at the time of Çatalhöyük would have required an astronomical telescope. This article however demonstrates that the technical abilities of constructing a Herschelian type obsidian basically have been existing in the Neolithic epoch.

REFERENCES

Albenda, P. (1985) Mirrors in the ancient Near East. *Notes in the History of Art* 4 (2/3), 2-9.

Asouti, E. (2013) Woodland vegetation, firewood management and woodcrafts at Neolithic Çatalhöyük. In: Hodder, I. (ed.), *Humans and landscapes of Çatalhöyük: reports from the 2000-2008 seasons.* Cotsen Institute of Archaeology, UCLA, Los Angeles, 129-162.

Astruc, L. et al. (2011) Multi-scale tribological analysis of the technique of manufacture of an obsidian bracelet from Aşikli Höyük (Aceramic Neolithic, Central Anatolia). *Journal of Archaeological Science* 38, 3415-3424.

Atalay, B. (2006) *Math and the Mona Lisa.* New York, Harper Collins Books.

Atalay, B. and Wamsley, K. (2009) *Leonardo's Universe. The Renaissance World of Leonardo da Vinci.* Washington, D.C., National Geographic Books.

Badisches Landesmuseum Karlsruhe (2007) *Vor 12000 Jahren in Anatolien. Die ältesten Monumente der Menschheit.* Stuttgart, Konrad Theiss.

Bath, K.-L. (1973) A Simple Interferometer for Testing Astronomical Optics. *Sterne und Weltraum* 6 (1), 177-180.

Bath, K.-L. (2007) *Mitteilungsblatt der Sternfreunde Breisgau e.V.* http://www.sternfreunde-breisgau.de/steinzeitspiegel.html

Ben-Tor, A. (1992) *The Archaeology of Ancient Israel.* New Haven, Yale University Press and Tel Aviv, Open University of Israel.

Bulling, A. (1960) The Decoration of Mirrors of the Han Period. *Artibus Asiae. Supplementum* 20.

Buys, A. (2007). *Astronomical Technology and the Exploration of the Universe.* http://www.up.ac.za/media/shared/Legacy/UserFiles/astro-technology-lecture.pdf [accessed 20/02/2016].

Calvo, M. L. and Enoch, J. M. (2005) Introduction to the History of lenses and Visual Corrections: A Reference to Spain and the Spanish Colonies in the New World (XV-XVI C.). *Revista Cubana de Física* 22 (1), 3-12.

Calvo, M. L. and Enoch, J. M. (2007) Ancient Peruvian Optics with emphasis on Chavín and Moche culture. *Atti della Fondazione Giorgio Ronchi,* Year LXII.

Carlson, J. B. (1981) Olmec Concave Iron-Ore Mirrors: The Aesthetics of a Lithic Technology and the Lord of the Mirror. *The Olmec and their Neighbours* (E. P. Benson, ed.). 117–148. Dumbarton Oaks, Washington, D.C.

Cauvin, J. (1998) La Signification Symbolique de L'obsidienne. *L'obsidienne au Proche et Moyen-Orient: Du Volcan à l'Outil* (M.-C. Cauvin et al.), 379-382. BAR International Series 728.

Chevalier, J. and Gheerbrant, A. (1996) *Dictionary of Symbols, 2dn edition.* London, Penguin Books.

Coqueugniot, É. (1998) L'obsidienne en Méditerranée Orientale aux Époques Post-Néolithiques. *L'Obsidienne au Proche et Moyen Orient: Du Volcan l'Outil* (Cauvin, M.-C. et al.), 351-361. BAR International Series 728.

Craig, J. R. and Vaughan, D. J. (1994) *Ore Microscopy and Ore Petrography, 2nd edition.* New York, John Wiley & Sons, INC.

Eisler, R. (1910) *Himmelsmantel und Weltenzelt. Religionsgeschichtliche Untersuchungen zur Urgeschichte des Antiken Weltbildes.* 2 Bde. München, C. H. Beck'sche Verlagsbuchhandlung Oskar Beck.

Enoch, J. M. (2007) Archaeological optics: the very first known mirrors and lenses. *Journal of Modern Optics* 54 (9), 1221–1239, 2007.

Enoch, J. M. (2006) History of Mirrors Dating Back 8000 Years. *Optometry and Vision Science* 83 (10), 775–781.

Enoch, J. M. and Calvo, M. L. (2007) Chavín de Huántar and Moche Cultures: Capability for lens optics. *Atti della Fondazione Giorgio Ronchi, Year LXII.*

Ericson, J.E., Makishima, A., Mackenzie, J.D., and Berger, R. (1975) Chemical and physical properties of obsidian: a naturally occurring glass. *Journal of Non-Crystalline Solids* 17 (1), 129-142.

Von Falkenhausen, L. (1989) KOSHI SHUNJŪ. *Early China* 14, 213-226.

Frazer, J. G. (1919) *The Golden Bough: A Study in Magic and Religion. Vol. X. of XII. Part VII: Balder the Beautiful. The Fire-Festivals of Europe and the Doctrine of the External Soul. 3rd edition.* New York and London, MacMillan and Co.

Goeje, C. H. de (1930) The inner structure of the Warau language of Guiana. *Journal de la Société des Américanistes* 22 (1), 33-72.

Gopher, A., Marder, O., Barkai, R. (2011) An Obsidian Industry from Neolithic Hagoshrim, Upper Galilee. In: Healey, E., Campbell, S., Maeda, O. (eds.). *The State of the Stone: Terminologies, Continuities and Contexts in Near Eastern Lithics.* SENEPSE 13, 395-402. Berlin, ex oriente.

Gullberg, J.E. (1959) Technical Notes on Concave Mirrors. *Smithsonian Institution Bureau of American Ethnology, Bulletin* 170, 280-283, and pl. 62.

Haddow, S. D. (2014) Çatalhöyük 2014 Archive Report by members of the Çatalhöyük Research Project. http://www.catalhoyuk.com/sites/default/files/media/pdf/Archive_Report_2014.pdf [accessed 16(01/2016].

Healey, E. (2013) Exotic, aesthetic and powerful? The non-tool use of obsidian in the later Neolithic of the Near East. In: *Interpreting the Late Neolithic of Upper Mesopotamia* (Nieuwenhuyse, O. P., Bernbeck, R., Akkermans, P. M. M. G. and Rogash, J., Eds.), 251-266. Brepols, Turnhout.

Heizer, R.F. and Gullberg, J.E. (1981) Concave Mirrors from the Site of La Venta, Tabasco: Their Occurrence, Mineralogy, Optical Description, and Function. *The Olmec and Their Neighbours* (E.P. Benson ed.), 109-116. Dumbarton Oaks, Washington.

Hodder, I. (2012) Çatalhöyük is added to UNESCO World Heritage List. *Stanford Archaeology Center Newsletter* 2012, 9.

Hoskin, M. (2011) William Herschel and Herschelian Reflectors. *History of Science* 49, 115-120.

Holmberg (-Harva), U. (1964), *Finno-Ugric, Siberian* (= *The Mythology of All Races IV*). New York, Cooper Square Publishers (Reprint of Boston, Archaeological Institute of America, 1927).

Juliano, A. (1985) Possible Origins of the Chinese Mirror. *Notes in the History of Art* 4 (2/3), 36-45.

Kano, Ch. (1979) The Origins of the Chavín Culture. *Studies in Pre-Columbian Art and Archaeology* 22, 1-3, 5-87

Kelley, D. H. and Milone, E. F. (2005) *Exploring Ancient Skies. An Encyclopedic Survey of Archaeoastronomy.* New York, Springer.

King, H. C. and Spencer Jones, H. (2003) *The History of the Telescope.* Dover, Dover Publication.

Laufer, B. (1928) The Prehistory of Television. *The Scientific Monthly* 27 (5), 455-459.

Lehmann-Nitsche, R. (1938) Tezcatlipoca und Quetzalcouatl. Ihre ursprüngliche Sternnatur. *Zeitschrift für Ethnologie* 70 (1/2), 67-82.

Lerner, Judith (1996) Horizontal-Handled Mirrors: East and West. *Metropolitan Museum Journal* 31, 11-40.

Lunazzi, J. J. (1995) On the quality of the Olmec mirrors and its utilization II. *Reunión Iberoamericana de Óptica, Guanajuato - GTO - Mexico, 18-22 Sept.* SPIE V 2730, 2-7.

Lunazzi, J. J. (1996) Olmec Mirrors: an Example of archaeological American mirrors. *Trends in Optics* (Anna Consortini, ed.), Vol. 3, 411-421. International Commission for Optics (ICO), Ac. Press.

Lunazzi, J. J. (2011) An Olmec Concave Mirror from El Salvador. *Mexicon* 33 (2), 36-37.

Maillet, A. (2004) *The Claude Glass: Use and Meaning of the Black Mirror in Western Art.* New York, Zone Books.

Matossian, M. K. (1980) Symbols of Seasons and the Passage of Time: Barley and Bees in the New Stone Age. *Griffith Observer* 44 (11), 9-17.

Mexicon (2011). An Olmec Concave Mirror from El Salvador. *Mexicon* 33 (2), 36-37.

Russell, J. M. (1988) The Program of the Palace of Assurnasirpal II at Nimrud: Issues in the Research and Presentation of Assyrian Art. *American Journal of Archaeology* 102 (4), 655-715.

McCrickard, Janet (1990) *Eclipse of the Sun. An investigation into Sun and Moon myths.* Glastonbury: Gothic Image Publications.

Matsumotu, M. (2013). Reflection as Transformation: Mirror-Image Structure on Maya Monumental Texts as Visual Metaphor for Ritual Participation. *Estudios de Cultura Maya* (41), 93–128.

Mellaart, J. (1967) *Çatal Hüyük. A Neolithic Town in Anatolia.* New York, McGraw-Hill Book Company.

Moyer, A. (2012). *Deep Reflection: An Archaeological Analysis of Mirrors in Iron Age Eurasia.* University of Minnesota. Ph.D. Thesis.

Needham, J. (1996) *Science and Civilisation in China*, Vol. IV, 1. Cambridge, Cambridge University Press.

Seidel, A. and Kalinowski, M. (1982) 'Tokens of Immortality in Han Graves'. *Numen* 29/1, 97-98.

Nordenskiöld, E. (1926) Miroirs convexes et concaves en Amérique. *Journal de la Société des Américanistes de Paris* 18 (1), 103-110.

Ollivier, F. J. et al. (2004) Comparative morphology of the tapetum lucidum (among selected species). *Veterinary Ophthalmology* 7 (1), 11–22.

Pokorny, J. (1959) *Indogermanisches etymologisches Wörterbuch. 3 Vol.* Bern/München, Francke.

Pendergras, M. (2003) *Mirror, Mirror: A History of The Human Love Affair With Reflection.* New York, Basic Books.

Rappenglück, M. (2016) Astral high-fashion clothing: relations between costumes and astronomy. In: Michael A. Rappenglück, Barbara Rappenglück, Nicholas Campion, Fabio Silva (eds.). Astronomy and Power: How Worlds Are Structured. Proceedings of the SEAC 2010 Conference, 19-28. BAR International Series 2794.

Reichel-Dolmatoff, G. 1982. Astronomical models of social behaviour among some Indians of Columbia. *Ethnoastronomy and Archaeoastronomy in the American Tropics* (A.F. Aveni and G. Urton, eds.), Annals of the New York Academy of Sciences 385, 165-181.

Roe, P. G. (1982) *The Cosmic Zygote Cosmology in the Amazon Basin.* New Brunswick, NJ, Rutgers University Press.

Saunders, N. J. (1988) Chatoyer: Anthropological reflections on Archaeological Mirrors. *Recent Studies in Pre-Columbian Archaeology* 1 (N. J. Saunders and O. de Montmollin, eds.), 1–40. BAR International Series 421, Oxford.

Saunders, N. J. (2001) A Dark Light: Reflections on Obsidian in Mesoamerica. *World Archaeology* 33 (2), 220–236.

Saunders, N. J. (2003) 'Catching the light': Technologies of power and enchantment in Pre-Columbian goldworking. In J. Quilter and J. W. Hoopes (eds.), *Gold and Power in Ancient Costa Rica, Panama, and Colombia*, 15–47. Washington D.C.: Dumbarton Oaks.

Schafer, E. H. (1978–79) A T'ang Taoist Mirror. *Early China*, 4, 56-59.

Schagunn, J. O. (1975) La Venta Mirrors' Possible Use as Astronomical Instruments. *Balance y perspectiva de la antropología de Mesoamérica y del centro de México,* Arqueologia I, XIII mesa redonda, 293-298. Sociedad Mexicana de Antropología, México.

Schechner, S. J. (2005) Between Knowing and Doing: Mirrors and Their Imperfections in the Renaissance. *Early Science and Medicine* 10 (2), 137-162.

Schele, L. and Miller, J. H. (1983) The Mirror, the Rabbit and the Bundle: Accession Expressions from the Classic Maya Inscriptions. *Studies in Pre-Columbian Arts and Archaeology* 25. Dumbarton Oaks, Washington, D .C.

Schele, L. and Blainey, M. G. (2007) *Surfaces and Beyond the Political, Ideological, and Economic Significance of Ancient Maya Ancient Maya Iron-Ore Mirrors.* Peterborough, Ontario, Canada, Trent University. MA Thesis.

Strätz, V. (1996) Materialien zu Tierkreisen in China. *Monumenta Serica* 44, 213-265.

Taube, K. A. (1992a) The Iconography of Mirrors at Teotihuacan. *Art, Ideology, and the City of Teotihuacan* (J. C. Berlo, ed.), 169–204. Dumbarton Oaks, Washington, D.C.

Taube, K. A. (1992b) The Major Gods of Ancient Yucatan. *Studies in Pre-Columbian Art and Archaeology* 32, i-v, vii-viii, 1-160.

Taube, K. A. (2011) An Olmec Concave Mirror from El Salvador. *Mexicon* 33 (2), 36-37.

Vedder, J. F. (2001) Grinding it Out. Archaeological Institute of America, archive.archaeology.org/online/news/mirrors.html [accessed 05-01-2016].

Vedder, J. F. and Swogger, J. G. H. (2005) The Obsidian Mirrors of Çatalhöyük 597. *Changing materialities at Çatalhöyük: reports from the 1995–99 seasons* (Hodder, Ian, ed.). British Institute at Ankara, BIAA Monograph 39, 597-619.

Waterbury, F. (1952) Bird-Deities in China *Artibus Asiae. Supplementum* 10, 1-139, 141-165, 167-181, 183, 185-191.

Wilson, R. N. (2007) *Reflecting Telescope Optics 1 Basic Design Theory and its Historical Development. 2nd edition.* Berlin, Heidelberg, Springer.

Witzel, M. (2005) Vala and Iwato, The Myth of the Hidden Sun, in India, Japan, and beyond in the case of Amaterasu - in China are intermediaries between the round mirror (Heaven) and the square mirror (Earth). *Electronic Journal of Vedic Studies* (EJVS) 12-1, 1-69.

7

ARCHAEOASTRONOMY IN BRONZE AGE SITES OF LA MANCHA (SPAIN)

César Esteban*,[1,2] and Luis Benítez-de-Lugo-Enrich[3,4]

[1] *Instituto de Astrofísica de Canarias, La Laguna, Tenerife, Spain*
[2] *Departamento de Astrofísica, Universidad de La Laguna, La Laguna, Tenerife, Spain*
[3] *Departamento de Prehistoria y Arqueología. Universidad Nacional de Educación a Distancia, Valdepeñas, Ciudad Real, Spain*
[4] *Departamento de Prehistoria y Arqueología. Universidad Autónoma de Madrid, Madrid, Spain*

Corresponding author: César Esteban (cel@iac.es)

ABSTRACT

We present archaeoastronomical results of an interdisciplinary project to study Bronze Age sites of the so-called *Cultura de las Motillas* in the Spanish region of La Mancha. We find that winter solstice sunrise was of special importance in the funerary – and perhaps religious – practises of these peoples. The impressive megalithic monumental complex of Castillejo del Bonete shows a remarkable marker of the winter solstice sunrise on the most peculiar mountain of its horizon as well as alignments with this and other singular solar events as equinox and summer solstice. There seems to be a correspondence between the orientations of some of the main architectural structures of the monument and the general arrangement of two of the main galleries of a natural cave that lies just beneath it. Markers and orientations to the winter solstice sunrise seem to be present in other nearby contemporary sites such as the necropolis of Cerro Ortega and the *motilla* of El Azuer. Castillejo del Bonete stands as the first evidence of a clear solar marker in a megalithic site of the Iberian Peninsula, indicating that the precise location of the monument was carefully chosen. The alignments defined by several of its structures further reinforce its astronomical symbolism.

KEYWORDS: Archaeoastronomy, Bronze Age. Late Prehistory, Iberian Peninsula, Monumental Barrow, Motilla, La Mancha, Winter Solstice.

1. INTRODUCTION. *CULTURA DE LAS MOTILLAS*

La Mancha is a natural and historical region that occupies a large fraction of the Spanish autonomous community of Castilla-La Mancha, located close to the centre of the Iberian Peninsula. Although our knowledge about the Late Prehistory of La Mancha is rather reduced - mainly because of the paucity of studies - the megalithic phenomenon has been very rarely reported in this region (e.g. Benítez de Lugo Enrich *et al.* 2014a). We are involved in an interdisciplinary project to study the archaeology, hydrogeology, palaeobotany and archaeoastronomy of the culture that developed the first attested settlements in the La Mancha - the so-called *Cultura de las Motillas* - during the Bronze Age (III-II millennia BCE, Mejías Moreno *et al.* 2015). Unlike the previous Neolithic and Chalcolithic periods, the research on this culture has been certainly remarkable over the past four decades (e.g. Molina *et al.* 2005; Aranda *et al.* 2008; Benítez de Lugo Enrich *et al.* 2014b). During the Bronze Age, La Mancha suffered a crisis of extreme aridity due to an abrupt climate change, known in the scientific world as 4200 cal. BP Event (2350-1850 BCE, Magny *et al.* 2009), which affected other major cultures from Mesopotamia to China. The peoples of La Mancha built large fortified settlements around deep wells - called *motillas* - in order to get control of the underground water resources. Certainly the *motillas* can be considered one of the oldest systems of relatively large exploitation of underground water in Europe. Despite the relative large amount of information we have about the *motillas*, very little is known about the funerary practises of these peoples because few excavations have been carried out in the scarce necropolises dated of this period.

In this paper we present the results of the first archaeoastronomical study ever made in archaeological sites belonging to the *Cultura de las Motillas*. These sites are three necropolises - Castillejo del Bonete, Cerro Ortega and Bocapucheros - and one *motilla* - Motilla de El Azuer, all of them located in the province of Ciudad Real

2. METHODOLOGY

The fieldwork was carried out in September 2014. The instruments used were a precision compass, a handheld clinometer, a theodolite, a global positioning device (GPS) and a digital camera. The methodology used in the data collection is described in detail in Esteban and Moret (2006), and summarized briefly below.

The GPS was used to obtain the geographic coordinates of the sites, as well as for timing the measurements of the position of the solar disc to correct for the zero level of the horizontal angles provided by the theodolite. This was made centering the solar disc in the reticle of the viewfinder of the theodolite several times - usually three - during each visit to the site. We also used this instrument for measuring the azimuth and height of the topographic features on the horizon around the sites, which accuracy is 0.05° in both cases. Height measurements were corrected for the atmospheric refraction effect near the horizon. The orientation of the man-made stone structures was measured with a precision compass, which provides an uncertainty of about 1°. The magnetic declination was determined on each site by comparing the horizontal angle given by the compass with the azimuth provided by the theodolite for different topographic features.

3. THE MONUMENTAL COMPLEX OF CASTILLEJO DEL BONETE

This impressive monument dates from the last quarter of the third millennium BC (Benitez de Lugo Enrich *et al.* 2014a). It comprises a monumentalized cave with a series of ground-level structures such as megalithic corridors and tumuli associated with funerary remains and votive deposits. It consists of several tumuli - the largest one of about 20 m of diameter - with radial corridors, some of them connecting tumuli. It is located in Terrinches (in the province of Ciudad Real) on the top of a hill with a dramatic view of the southern half of the horizon. High peaks of the mountain ranges of Alcaraz and Segura can be seen to the southeast from the site. After measuring the azimuth and height of different topographic features of the horizon, we check their possible relationship with rising or settings of the most important celestial bodies, finding an exceptional event: the sunrise at winter solstice takes place over the centre of the top of the most striking mountain visible from the site: Peña del Cambrón. This mountain is 1550 m high, located about 30 km away and has a curious rectangular profile, with a flat top and very steep edges - the width of the flat summit is 1.8° (see Figure 1). We observed the sunrise from the site on December 22, 2014 - winter solstice - taking the photograph shown in Figure 1. In this figure, we also compare the position of the solar disc at present-day winter solstice sunrise (with a declination of δ = -23.45°) and the expected position at 2000 BCE (δ = -23.9°), as it would de observed by the builders of Castillejo del Bonete.

Calculations made for 2000 BCE indicate that the time interval elapsed since the southern edge of the sun touches the northern edge of the Peña del Cambrón until the winter solstice is about 18 days. That is, the Sun travels for 18 days the flat summit of the mountain, moving slowlier southward as it

Figure 1. Winter solstice sunrise of 2014 (December 22nd) from the megalithic monumental complex of Castillejo del Bonete. The solar disc appears on the central part of Peña del Cambrón (photo: Oppida). Inset: Detail of the profile of Peña del Cambrón. The dashed circle indicates the size and position of the rising sun during the present-day winter solstice. The solid circle shows that position around 2000 BC, the approximate date of construction of the monument. The grey dotted arrow represents the path of the south edge of the solar disk. Note that the sun touched the right edge of the mountain at winter solstice. North is to the left.

Table I. Orientations defined in structures of the monumental complex of Castillejo del Bonete.

Structure	Azimuth (°)	Height (°)	Declination (°)	Astronomical Event
Corridor no. 1	235±3	−0.3±0.1	−27±3	Winter solstice sunset
Enclosure no. 4	61±2	1.0±0.1	+23±2	Summer solstice sunrise
Corridor A	269±2	1.0±0.1	−0±2	Equinox sunset
Corridor B	127±3	−0.1±0.2	−28±3	Winter solstice sunrise?
Corridor no. 2	335±2	1.0±0.1	+46±2	-

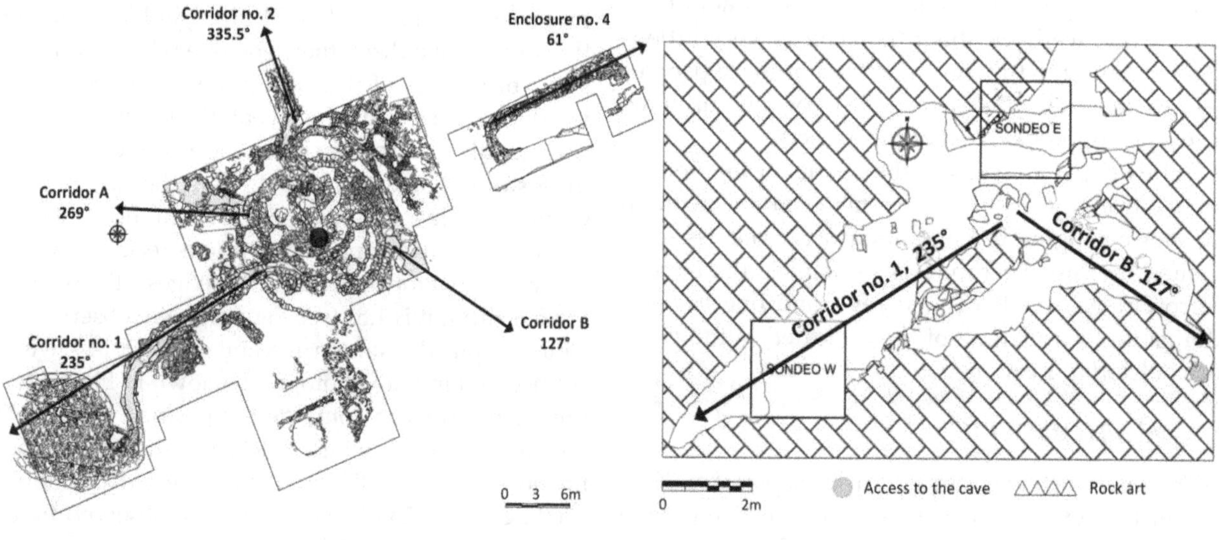

Figure 2. Left: Plan of the megalithic monumental complex of Castillejo del Bonete including the azimuth of the orientations of some of its structures. The black circle at the centre indicates the access to the cave. Right: Plan of the cave located under the tomb. Two of the galleries show the same general orientations than corridors B and no. 1, related to the sunrise and sunset at the winter solstice, respectively. North is up and east to the right. The grey circle at the right edge of the plan of the cave indicates the position of the access to the cave (plans by Enrique Mata and Norberto Palomares).

approaches the winter solstice. During the last days before solstice, the sun almost stops moving only 0.1° in that period touching the southern edge of the mountain just at the solstice (see Figure 1). Afterthat moment, the sun reverses the sense of its movement on the summit, taking another 18 days to abandon completely the flat summit of the mountain. We believe that an observer experienced in counting the days elapsed between the successive passes of the sun between the edges of Peña del Cambrón might have been able to estimate the exact day of the winter solstice. Therefore, this striking coincidence may have been used as a calendrical marker of the winter solstice and/or an indicator of a funerary ritual in that moment of the year.

Besides the analysis of the horizon that surrounds Castillejo del Bonete, we also took bearings of the longest and more linear corridors and tumuli walls, in the sense from the centre out of the main tumulus. The results – included in Table 1 and graphically shown in Figure 2 – are of great interest and suggest an astronomical planning of the monument.

- Corridor no. 1, the longest excavated corridor, has a long rectilinear initial structure that curves linking with tumulus no. 2. The straight part of the corridor seems to point – with an uncertainty of 2°–3° – to the point the horizon where the winter solstice sunset occurs (see Figure 3).
- The north and south walls of Enclosure no. 4 are parallel and oriented similarly to Corridor no. 1 but about 6° apart. The north wall – the best-preserved one – faces the summer solstice sunrise with great precision (uncertainty between 1°–2°).
- The walls of Corridor A are not parallel, the south wall – the longer and best preserved one – is highly rectilinear and is oriented in the east-west line (within an uncertainty of between 1°–2°). Therefore, it could be related to the sunset at the equinoxes or dates close to them.
- Corridor B is rather badly preserved, but its few standing elements are oriented towards the Peña del Cambrón, the point on the horizon where the winter solstice takes place.
- Corridor no. 2 is oriented northerly, about 25° apart from true north (uncertainty between 1°–2°). Changing the sense of the bearing, from the north edge to the corridor to the centre of the main tumulus, we find a curious alignment with a peaked mountain located close to the south: El Yelmo, one of the highest in the mountain range of Segura. It is interesting to comment – but in the field of speculation – that the rising of the stars Rigil Kentaurus (α Cen) and Acrux (α Cru) occurred around that mountain at the time of construction of the monument (2000 BCE). These are the main stars of the constellations of Centaurus and Southern Cross and are among the brightest ones in the southern hemisphere. Today they are no longer visible from the latitude of Castillejo del Bonete due to the precession of the equinoxes.

Figure 3. Photo taken from the initial part of Corridor no. 1 and looking to the southwest horizon. The circle represents the position and size of the solar disc at the winter solstice sunset of 2000 BCE. The end of the corridor is no longer aligned with this astronomical event, twisting to bind to the nearby tumulus no. 2.

4. NECROPOLIS OF CERRO ORTEGA

This Bronze Age necropolis – discovered by chance in 1997 (Barrio Aldea and Maquedano Carrasco 2000) – is located at the bottom of a rocky promontory called Cerro Ortega, in Villanueva de la Fuente (Ciudad Real), only about 11 km apart from Castillejo del Bonete. The tombs are artificial cavities carved on the surface of the wall of the promontory. In the absence of measurable structures except the very collapsed openings of the tombs, we simply determined the general orientation of the rock wall and performed the analysis of the horizon.

The perpendicular to the rock wall just at the point where the tombs are placed is facing an azimuth range comprised between 125° and 128° (Figure 4) that projected to the local horizon and considering its height, includes the point where the winter solstice sunrise occurred at 2000 BCE. This sunrise coincides with the most distant peak of the mountain range of Sierra del Relumbrar that can be seen from the necropolis. As we can see in Figure 4, the rock wall varies its orientation along its entire length, but the tombs are excavated precisely in the zone where the entrance of the graves are roughly oriented towards the winter solstice sunrise. We believe that, although this result by itself might be considered casual, the coincidence with the main astronomical target found in Castillejo del Bonete, suggests that perhaps both necropolis share the same tradition, perhaps the same element in the funereal ritual related to the solar cycle.

5. *MOTILLA* OF EL AZUER

This is one of the most representative *motillas* and the most extensively excavated one. We included it in our study in order to explore the archaeoastronomical potential of this kind of sites. It is a fortified mound built between 2200 and 1500 BCE and it has been recently restored (see Figure 4; Benitez de Lugo, 2010). We could not obtain bearings of their structures because all of them present curvilinear geometry without apparent symmetry axes that may define directions. The site is located in an extended plain, with an excellent view of the distant horizon, which is rather poor in remarkable topographic features. The mountains that are seen from the site are not very high and are mostly located, to the southeast – mountain ranges of Sierra de Alhambra and del Cristo – except a small group to the northwest – Sierra de Malagón. The only result of astronomical interest that we have found in this *motilla* is that the winter solstice sunrise occurs on one of the highest peaks of Sierra de Alhambra, located at about 37 km apart (Figure 4). As in the case of Cerro Ortega, the astronomical relationship found in *motilla* of El Azuer is not striking and could also be the result of chance but seems significant because it is consistent with the findings of the other two sites studied in this paper.

6. DISCUSSION AND CONCLUSIONS

The results of our study at Castillejo del Bonete indicate the high archaeoastronomical interest of this megalithic monumental complex. It presents a striking marker of the winter solstice sunrise on the most conspicuous topographical feature of the horizon, Peña del Cambron. Moreover, the astronomical relevance of the site is reinforced by the orientation of different corridors and enclosures of the complex. It is particularly significant that corridors no. 1 and B - as well as two of the main galleries of the cave extending just below the monument - are oriented towards the winter solstice sunset and sunrise, respectively, exactly the same solar event marked by the sunrise on Peña del Cambrón. On the other hand, the results for the necropolis of Cerro Ortega and motilla of El Azuer, although much less striking, point to the same astronomical event, the winter solstice. It seems reasonable to propose that the winter solstice played an important role in the funerary– perhaps also religious - symbolism of the Bronze peoples of La Mancha. It is well known that winter solstice had enormous significance for other cultures in different epochs. This moment of the year marks the time of year when the length of the day begins to increase with respect to the night. It is when the Sun symbolically defeats the darkness of winter, a time of rebirth of nature.

Alignments and markers toward the sunrise or sunset at the winter solstice are rather common in European megalithic structures such as the megalithic complex of Stonehenge (Wiltshire, UK), whose oldest structures date back to around 2500 BCE and has its symmetry axis oriented to the winter solstice sunset, a moment of the year when large multitudinous celebrations were made according to the latest archaeological data collected in the area (Parker Pearson 2012). Another emblematic site is the impressive megalithic tomb of Newgrange (Donore, Ireland, O'Kelly 1982) built between 3300 and 2900 BCE. Its long corridor shows a perfect alignment towards the sunrise at the winter solstice at the time of its construction.

The Iberian Peninsula is very rich in megalithic monuments. There are many different typologies and their chronology range from sixth to third millennia BCE. The orientation of over one thousand megalithic burial monument have been measured (see Hoskin 2001; González-García 2009) and most of them – specially in the west and south of the Peninsula – show orientations in the range of declinations encompassing the rising points of the sun or the moon. However, the presence of astronomical markers in the horizon surrounding Iberian megalithic monuments has been hardly reported in the literature. In the case of dolmens, the only examples of orientations of corridors toward prominent mountains of the landscape are in Los Millares (Almería) and in the huge Dolmen of Menga in Antequera (Málaga, Belmonte and Hoskin 2002: 72, 78). Of these, only in the case of Menga the orientation may have some astronomical relation, in particular with the moonrise at the northern major lunastice (J.A. Belmonte, private communication). As far as we know, the only clear astronomical marker reported in an Iberian megalithic monument is in the cromlech of Oyanleku (Guipuzcoa), which major axis is facing the remarkable profile of Peña de Aya, where the moonrise at northern major lunastice took place at 3000 BCE (Belmonte and Hoskin 2002: 45-46). As we can see, Castillejo del Bonete stands especially relevant because of the presence of an astronomical marker that is undoubtedly related to the sun.

The apparent absence of astronomical markers in Spanish megalithic monuments contrast with that is found in similar sites of the British Isles, where many stone circles and standing stone complexes show markers of singular positions of the Sun and Moon (e.g. Thom 1967, 1971; Ruggles 1999). In our opinion, this has two possible explanations: a) the megalithic people of the Iberian Peninsula where not concerned with astronomical markers, b) the archaeoastronomical studies of Iberian megaliths have been biased on measuring orientations of dolmen corridors and not in the analysis of their horizons. Although this last possibility may be true in the case of most Hoskin's studies - that were dedicated to collect the maximum number of precise bearings of dolmens and megalithic sepulchres across the Mediterranean - this is not generally the case of later studies made, for example by J. A. Belmonte and A. C. González-García, much concerned on the landscape of monuments. The relevant aspect behind the presence of markers or not in a given prehistoric monument is that it gives clues about the importance of the precise place where the monuments were built. The presence of a marker indicates that the site was selected prior to the construction of the monument because its astronomical significance. In the absence of a marker we can argue that the monument was constructed in a precise place because any non-astronomical reason and its astronomical relation was established *a posteriori*. Therefore, it seems that Castillejo del Bonete was built in a place selected by the presence of the striking marker of the winter solstice sunrise and its astronomical significance was further reinforced through the alignment of some of its structures. Most probably, all these astronomical relations where understood as elements that increased the sacred character of the site, linking chthonic and cosmic elements: funereal monuments, caves and celestial bodies. An intricate complex of symbolic elements whose precise meaning still eludes us.

Figure 4. Left: Google Earth satellite image of the necropolis of Cerro Ortega area. The tombs are carved on a rock wall perpendicular to the point where the winter solstice sunrise occurs. The inset image shows the point where the winter solstice occurred in 2000 BCE. Right: Motilla of El Azuer seeing from the east. The arrow indicates the position where the measurements were made. The inset shows the area of the horizon where the rising of the winter solstice occurred in 2000 BCE, on the highest point of the mountain range of Sierra de Alhambra.

ACKNOWLEDGEMENTS

This work has been funded by the Consejería de Educación, Cultura y Deportes de la Junta de Comunidades de Castilla-La Mancha and the Instituto Geológico y Minero de España. The authors would especially like to thank Miguel Mejías Moreno and the personnel of the Instituto Geológico y Minero de España for their support and technical cooperation. CE thanks Juan Antonio Belmonte for fruitful discussions.

REFERENCES

Aranda, G., Fernández, S., Haro, M., Molina, F., Nájera, T. and Sánchez-Romero, M. (2008) Water control and cereal management on the Bronze Age Iberian Peninsula: la motilla del Azuer. *Oxford Journal of Archaeology*, 27 (3), 241-259.

Barrio Aldea, C. and Maquedano Carrasco, R. (2000) La necrópolis calcolítica de Cerro Ortega (Villanueva de la Fuente). In *El Patrimonio arqueológico de Ciudad Real. Métodos de trabajo y actuaciones recientes*, L. Benítez de Lugo (Ed.), Valdepeñas, UNED, 67-86.

Belmonte, J.A. and Hoskin, M. (2002) *Reflejo del Cosmos. Atlas de arqueoastronomía del Mediterráneo antiguo.* Madrid, Equipo Sirius S.L.

Benítez de Lugo Enrich, L. 2010. *Las motillas y el Bronce de La Mancha*, Valdepeñas, Anthropos.
Benítez de Lugo Enrich, L., *et al.* (2014a) Aportaciones hidrogeológicas al estudio arqueológico de los orígenes de la Edad del Bronce de La Mancha: la cueva monumentalizada de Castillejo del Bonete (Terrinches, Ciudad Real, España). *Trabajos de Prehistoria*, 71 (1), 76-94.
Benítez de Lugo Enrich, L., *et al.* (2014b) Castillejo del Bonete (Terrinches, Ciudad Real): complejo tumular prehistórico de la Cultura de las Motillas en el Alto Guadalquivir. *Menga, Revista de Prehistoria de Andalucía*, 5, 151-174.
Esteban, C. and Moret, S. (2006) Ciclos de tiempo en la cultura ibérica: La orientación astronómica en el templo del Tossal de Sant Miquel de Llíria. *Trabajos de Prehistoria*, Vol. 63, no. 1, 167–178.
González-García, A.C. (2009) Statistical Analysis of Iberian Peninsula Megaliths Orientations. In *Cosmology Across Cultures*, J.A. Rubiño-Martín, J.A. Belmonte, F. Prada and A. Alberdi (eds.), ASP Conference Series, 409, 354-358.
Hoskin, M. (2001) *Tombs, temples and their orientations*. Bognor Regis, Ocarina Books.
Magny, M., Vannière, B., Zanchetta, G., Fouache, E., Touchais, G., Petrika, L., Coussot, C., Walter-Simonnet, A.V. and Arnaud, F. (2009) Possible complexity of the climatic event around 4300-3800 cal. BP in the central and western Mediterranean. *The Holocene*, 19, 823-833.
Mejías Moreno, M., Benítez de Lugo, L., López-Sáez, J.A. and Esteban, C. (2015) *Arqueología, Hidrogeología y Medio Ambiente en la Edad del Bronce de La Mancha: la Cultura de las Motillas*. Madrid, Instituto Geológico y Minero de España.
Molina, F., Nájera, T., Aranda, G., Sánchez, M. and Haro, M. (2005) Recent fieldwork at the Bronze Age fortified site of Motilla del Azuer (Daimiel, Spain). *Antiquity*, 79 (306), 223-235.
O'Kelly, M. J. (1982) *Newgrange: Archaeology, Art and Legend*, London, Thames and Hudson.
Parker Pearson, M. (2012) *Stonehenge: Exploring the Greatest Stone Age Mistery*. London & New York, Simon & Schuster.
Ruggles, C. (1999) *Astronomy in Prehistoric Britain and Ireland*. New Haven and London, Yale University Press.
Thom, A. (1967) *Megalithic Sites in Britain*. Oxford, Oxford University Press.
Thom, A. (1971) *Megalithic Lunar Observatories*. Oxford, Oxford University Press.

8

ASTRONOMICAL ALGORITHMS OF EGYPTIAN PYRAMIDS SLOPES AND THEIR MODULES DIVIDER

Hossam M. K. Aboulfotouh[1,2]

[1]*Faculty of Environmental Design, King Abdulaziz University, Jeddah, KSA*
[2]*Faculty of Fine Arts, Minia University, Minia, Egypt*

Corresponding author: (fotouh@mail.com & haboulfotouh@kau.edu.sa)

ABSTRACT

This paper is an attempt to show the astronomical design principles that are encoded in the geometrical forms of the largest five pyramids of the fourth Egyptian dynasty, in Giza and Dahshur plateaus, based on using the pyramids' design-modules that are mentioned in the so-called Rhind Mathematical Papyrus. It shows the astronomical algorithms for quantifying the slopes of pyramids, with reference to specific range of earth's axial tilt, within spherical co-ordinates system. Besides, it decodes the design of the ancient Egyptian cubit rods with reference to the systems of measurement used by Herodotus and Diodorus, and proves that Herodotus never talked about Giza pyramids and he meant only the pyramids of Memphis in Dahshur. In addition, it proves that the Egyptian arrow of 22.5cm, or half the mean cubit of 45cm, is the largest common divider of the design-modules of these pyramids, and accordingly it retrieves their intended astronomical design dimensions.

KEYWORDS: Giza pyramids, Dahshur pyramids, obliquity, axial tilt, Egyptian cubit, bent pyramid, red pyramid.

1. INTRODUCTION

Methods of design and construction of the Egyptian pyramids were, and still are, in the forefront of scientific debates. In Egyptian mathematics, the so-called Rhind Mathematical Papyrus-RMP that was first translated by Thomas Eric Peet (1923), and then by Arnold Buffum Chace *et al* (1927-1929), is the oldest mathematical source book, where four of its eighty-four problems (RMP#56-59) show the design-sketches and slope-ratios of two pyramidal design models, i.e., the pyramid's height to half its side-length. In the architectonic and archaeological realms, there are five large pyramids found in Egypt from the days of the fourth dynasty, which still retain their original form, and are grouped in two sites: the three Giza pyramids, and the red and bent pyramids in Dahshur. According to Petrie (1883, 1887), the slopes of these pyramids are between 43° and 55°. Retrieving the design method that was used to set the pyramids' slopes with reference to spherical co-ordinates became the interest of modern scholars[1]. In archaeoastronomy, Belmonte and Magli (2015) proposed an idea similar to the shed design method, where they suggest that the slopes of these pyramids match the elevation angles of the ascending sun when it crosses the first vertical east-west plane and thus illuminate their four faces for few minutes, in specific days. Contrary to this opinion, previous publications by the author showed that the prime aim of the pyramids designers was to encode and preserve their scientific knowledge for the coming generations. For example, Aboulfotouh (2007) discussed the concept of shrinking contour circles that models the relative change of earth's axial tilt (obliquity), from an assumed lower limit (~21.672°), as a frame of reference, in order to retrieve the ancient Egyptian relativistic algorithms that most likely were used to reckon the tilts of the entrance passages of these five pyramids. Related to this, Aboulfotouh (2005) showed that the cotangent of edge angle of the great pyramid in Giza equals the result of subdividing the obliquity angle of time ~24.10° by the lower limit. However, the astronomical parameter of earth's obliquity-range, proposed by the ancient Egyptian designers, was never discussed in relation to the design modules of the large pyramids of the fourth dynasty, or in relation to the design modules of the four pyramids problems in RMP. Besides, in history, Herodotus (440BCE, pp.426-437) mentioned the dimensions of the four pyramids of Memphis, where one of them he called the great pyramid. Nowadays scholars of Egyptology believe that Herodotus mentioned the dimensions of the largest four pyramids in Giza Plateau; despite he did not talk about the Sphinx. In geography, Strabo (1932-1967, p.89) said that those pyramids are about 40 stadia (~ 6.3km) south of Memphis[2], which matches the location of the four pyramids in Dahshur. On the other hand, Diodorus (60BC-30AD, p.32, 63-3) mentioned the correct location of the largest three pyramids, in the Giza plateau, as 120 stadia from Memphis and 45 stadia from the Nile, but the dimensions he mentioned are less than those of the Giza pyramids, and are similar to those of the pyramids of Memphis. Perhaps he cited the dimensions given by Herodotus[3]. Today, scholars find difficulties in verifying the dimensions given by those historians with the survey data of Petrie (1883, p111), and many anti-Herodotus criticisms were published accordingly. For example, How *et al* (1913-1927) pointed out that Herodotus's figures of Giza pyramids' dimensions are inaccurate; despite, many parts of Herodotus's text indicate that he studied civil engineering. Besides, in stating the heights of the Egyptian pyramids, ancient and mediaeval historians alike were mentioning instead either the height along the inclined edge of the pyramid, or the height along the line from the crest of the pyramid to the midpoint of its side[4], which did mislead some scholars. The accurate lengths of measurement units used by Herodotus and Diodorus and their correlation to the Egyptian cubits are still unknown. Hence, the paper discusses four key points. Firstly, it shows the most likely astronomical design algorithms for quantifying the slopes of pyramids, and in relation to the pyramids-modules in RMP. Secondly, it shows that Herodotus never talked about, and defiantly never saw the Giza pyramids; and he meant only the pyramids of Memphis, in Dahshur plateau. Thirdly, it decodes the design of the ancient Egyptian cubits, and identifies the metric value of measurement units used by Herodotus and Diodorus. Fourthly, it identifies the largest common divider of the design-modules of pyramids of the fourth Egyptian dynasty, and retrieves their intended astronomical design dimensions.

2. PYRAMIDS DESIGN MODULES

Fig.1 shows the hieratic text of two of the four pyramids problems in RMP, namely: RMP#56-59. Based on reviewing its hieratic text (see: Chace *et al*, 1927-1929, plates 78-81) and the translation by Clagett (1999, pp.166-168), the data of pyramids' height h and their side-length l in cubits are: in RMP#56, h=250 and l=360, in RMP#57-58, h= (93+1/3) and l=140. However, in RMP#59, the hieratic text states that h=12, and l= 8, and not the contrary. Accordingly, the first observation is that these problems show two design models, and not only one as some scholars believed. Besides, in the hieratic text, the pyramids' slopes are not represented as a number of palms[5], but as abstract modules, written under the modular arc notation (see: Aboulfotouh, 2012, p.135). In RMP#56-58, the number of modules of pyramid's height h_m is constant and equal 7, and the number of modules in each side of the pyramid's square base l_m is variable and is less than 7 modules in these problems. Hence, as known, $l_m/2$ indicates the slope β of the pyramid's four faces; i.e., $\tan \beta = 2h_m/l_m$. In the hieratic text, $l_m/2$ is denoted by the Egyptian Arabic word *Saqet*[6] that we use in descriptive geometry and means the horizontal projection (run) of the inclined line. In RMP#56, $l_m/2 = (5+1/25)$, which imply $\beta = 54° 14' 46.01''$; and in both RMP#57 and RMP#58, $l_m/2 = (5+ 1/4)$, which imply $\beta = 53° 7' 48.37''$. Contrary to opinions of scholars, e.g., Gillings (1982, p.186) and Clagett (1999, p.168), RMP#59 shows different model, where the number of modules of the pyramid's side l_m is constant and equal 7; and h_m is variable, and in this case it equals (10+1/2). Hence, $\tan \beta = 10.5/3.5$, which imply $\beta = 71° 33' 54.18''$. It is similar to the pyramidal part at the top of an Egyptian obelisk, or like the pyramid of Khonsu in Deir el-Medina, with its steep slope.

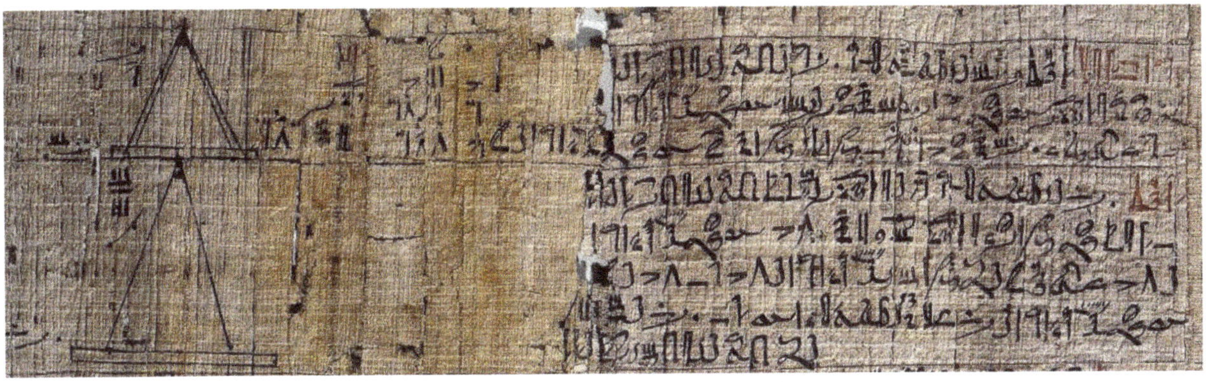

Figure 1: The hieratic text of RMP#56 upper, and RMP#57 lower; the pyramid diagram with horizontal rectangular base is explained in fig.2; and the double inclined lines in the upper diagram imply a pyramid with concave faces, where the outer line is its main slope; it is used with the permission of the British Museum.

3. PYRAMIDS DESIGN AND ASTRONOMICAL PARAMETERS

Aboulfotouh (2007) retrieved the relativistic equations for reckoning the tilt a of the entrance passages of the largest five pyramids of the fourth Egyptian dynasty. The equation of the great pyramid in Giza plateau is: $\sin a = (\sin \lambda * \sin O_m)/(1-(O_i^2/O_t^2))^{1/2}$; where its $a = \sim 26.56°$. Here, λ is the latitude ~30° at which the pyramid stands. Additionally, Aboulfotouh (2002, 2005, 2014) showed that the encoded axial tilt[7] O_t (or ε) in the design of the great pyramid and the site plan of the horizon of the Giza pyramids is ~24.10°; where its designer used an obliquity range from $O_i = \sim 21.672°$ as lower limit to $O_x = \sim 24.30°$ as upper limit, and $O_m = \sim 22.986°$ as a mean. Besides, using the arcs' lengths of the obliquity-angles as radii of contour circles, Aboulfotouh (2007) showed that $O_i/O_t = \cos \varphi$, where φ is the angle of the horizontal deviation of time that equals ~25.93° for $O_t = 24.10°$, which is used as tilt of the upper entrance passage of the 2nd pyramid in Giza plateau. The tilt of its lower entrance passage was found equal to $O_i = \sim 21.672°$.

Furthermore, using the data of Petrie (1883, p.43), Aboulfotouh (2005) showed that $O_i/O_t = h/r$, where h of the great pyramid is ~146.51m; and r is radius of the circle that encloses the pyramid's square-base, i.e., $r = (l/2^{1/2})$. For l = 230.35m., r = 162.88m. Besides, it was shown that r is also the radius of the spherical co-ordinates system of the pyramid (see fig.2) upon which its internal-design was created. That imply, $\cos \varphi = h/r = \tan \psi$, where ψ is the pyramid's edge angle. Using the equation: $h = r\,(O_i/O_t)$, if we intend to construct another great pyramid to encode $O_t = 23.452145°$, i.e., ε of the epoch 1900AD (Wittmann, 1979), for r =162.88m, its h would be ~ 150.51m.

Fig.2 shows north-south cross section in a spherical co-ordinates system, where point C represents the earth, line Ys-Yn represents the earth's axis of rotation, line P_1-P_2 represents the plane of the ecliptic, and line Xe-Xw represents the plane of the equator. Ahoulfotouh (2005, 2014) showed that this cross section was used for creating the internal design of the great pyramid as well as the site plan of the horizon of Giza pyramids.

In fig.2, the generated main-rectangle P_1-D_1-P_2-D_2 includes the triangle cross-section of the pyramid M_1-Z-M_2 and the residual-rectangle P_1-G_1-G_2-D_2, similar to the hieroglyphic sign that denotes a pyramid △. The horizon line H-H divides the main-rectangle into two equal parts; the upper half is V_1-D_1-P_2-V_2. If we assign names for the vertical distances P_2-V_2, V_2-G_2 and G_2-D_2 as: j, k, and f, respectively; and taking into consideration that $J= f+k$, the pyramid's height $h= k+J$. Moreover, the altitude k of center of the pyramid's spherical system C (and the horizon line H-H) from its zero ground-level in the construction site; and the tilt of its entrance passage a will vary according to the latitude λ where the pyramid stands, and also according to the used O_t. The altitude k could be reckoned using the equation: $k= h- (sin\ (90- \lambda- O_t)* r)$. Therefore, this pyramid design-model is latitude and axial-tilt identifier, i.e., a space-time pyramid. In the great pyramid, f= ~44.5m., k= ~51m., and J= ~95.5m, where its k is close to the level of the highest point of its grand gallery (Aboulfotouh, 2005). If we imagine that this pyramid will not collapse and keep standing during one complete obliquity cycle, over time, it is expected that k will fluctuate between two limits. The upper limit is k_x for O_x and the lower limit is k_i for O_i; e.g., its k_i=45.45m for $O_t = O_i$ =21.672°, and its k_x =51.42m for $O_t = O_x$ =24.30°.

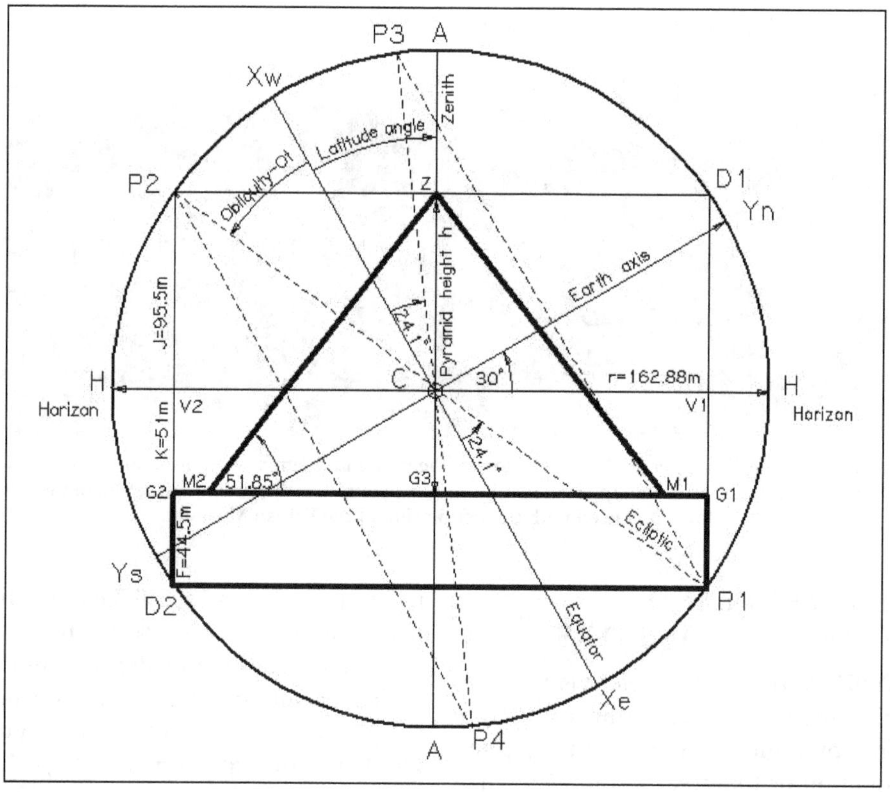

Figure 2: North-south cross section in a spherical co-ordinates system that shows the astronomical design parameters of the Great Pyramid in Giza Plateau, while looking due west.

Based on reviewing the diverse slopes of the Egyptian pyramids, the author identified four design models. Their astronomical slope-ranges are symbolized here below as: β_1, β_2, β_3, and β_4. These ranges of pyramids slopes are reckoned, using the assumption of the ancient Egyptian designers concerning the descending values of earth's axial tilt O_t (or ε), from ~24.30° as upper limit to ~21.672° as lower limit.

In the first model, since $h/r=O_i/O_t$; $l/2= r/(2^{1/2})$; and $tan\ \beta_1= 2h/l$; then, $\beta_1= Atan\ (O_i*(2^{1/2}))/O_t$. Accordingly, the lower limit of β_1 is ~51° 35' 26.81", and its upper limit is 54° 44' 8.2". Using the data of Petrie (1883 & 1887), the slopes of the three Giza pyramids and the lower part of the bent pyramid are within the range of β_1. The slopes of pyramids models in RMP#56-58 are also within this range.

In the second model, the pyramid's slope β_2 could be set equal to the edge angle ψ of the first pyramidal model. That is, by using the diagonal cross-section of the first model as the north-south cross-section of the second model, taking into consideration that, the real radius r of the circle-enclosing the pyramid's square base, and its spherical co-ordinates system, is variable. In this case, $\beta_2= Atan\ (2h/l)= Atan\ (O_i/O_t)$. Accordingly, the lower limit of β_2 is ~41° 43' 41.63" and its upper limit is 45°. Using the data of Petrie (1887, pp.27-32), the slope of the red pyramid and the upper part of the bent pyramid are within the range of β_2.

In the third model, the pyramid's slope β_3 could be set equal to the angle that complements the deviation angle of time φ, i.e., β_3 = $Asin$ (90°- φ)= $Asin$

(O_i/O_t). Accordingly, the lower limit of β_3 is 63° 6' 24.77" and its upper limit is 90°. For example, β_3 = 70° 33' 41.16" for O_t= 22.986°. The slope of the pyramid-model in RMP#59 is within the range of β_3.

In the fourth model, the pyramid's slope could be set as β_4= *Asin* (O_i/φ), where φ= *Acos* (O_i/O_t). This case implies that the deviation angle of time φ is used as radius of shrinking circle, and is being referenced to the circle that its radius is the lower limit of earth's obliquity O_i (see Aboulfotouh 2007). For O_i =~21.672°, the possible range of O_t is from ~24.30° to ~ 23.3205°. Accordingly, the lower limit of β_4 is 53° 41' 34.0" and its upper limit is 90°. The recorded slope of pyramid of Unas[8] is within the range of β_4; where it is close to 56° 39' 52.77" that corresponds to O_t=24.10°.

Petrie measured the side length l of the three pyramids in Giza plateau and reckoned the height of each using what he recorded as mean slope for each pyramid. He recorded 230.35m and 215.26m as l of the great and 2nd pyramids in Giza plateau, respectively; and recorded 51° 52' ±2' and 53° 10' ±4' as their β_1, respectively (Petrie, 1883, p.43 & p.98). His data confirm the values of O_t for these two pyramids, as 24.10° and 22.986°, respectively. Regarding the 3rd pyramid, he recorded 105.5m as l of its base. However, either due to the demolition of the granite casing of the 3rd pyramid[9] during the 13th century AD, or due to its concave faces, as in RMP#56 in fig.1, Petrie (1883, p.111) recorded four different slopes: 50° 57', 50° 42', 51° 0', and 51° 58'. Accordingly, he assumed that 51° 0' ±10' is its most probable slope. None of his four records encodes the obliquity angles that are used by designers of the fourth dynasty. Most probably, the intended β_1 of the 3rd pyramid is 51° 33' 12.41" that corresponds to O_t=24.30°, as what is encoded in the tilt of its entrance passage (Aboulfotouh, 2007).

Furthermore, according to Petrie (1887, p.27&p.32) the slope of the core masonry of red pyramid is 44° 36', and he said Vyse stated this as 43° 36' 11". Perhaps, Petrie did not notice the concaveness in its four faces[10]. Despite 44° 36' is within the astronomical range of β_2, it does not imply encoding any of the obliquity values that are used by designers of the fourth dynasty. In addition, 45° is not valid because it will increase its h for l=220m. Most properly, its intended β_2 is 43° 18' 52.7" that corresponds to O_t=~22.986°, as what is encoded in the tilt of its entrance passage (Aboulfotouh, 2007). Besides, Petrie (1887, pp.28-30) measured the slope of the lower part of the bent pyramid and recorded diverse values from 54° 36' to 55° 1'. Most probably, the intended design slope of the lower part is the upper limit of β_1, i.e., 54° 44' 8.2". He recorded 189.45m as l of its base, and concerning the horizontal-projection of the slope of its lower part, he recorded two distances: 32.7m and 33.0m. Concerning the slope of the upper part of the bent pyramid, he recorded 43° 0' and in the west side, he said, "it is perhaps 42° 39' higher up". Most probably the intended slope of the upper part is 42° 54' that encodes O_t=~23.3205°, and identifies the lower limit of the deviation of angle of time φ, i.e., at this limit φ=O_i=~21.672°.

By reviewing the design of the bent pyramid, it seems that its designer combined the first and second models in one pyramidal form. The final product shows only the lower part of the model β_1, and the upper part of the model β_2. Despite the rest of the two models are not observed, their r, l, and h could be identified, using the survey data of Petrie (1887, pp.28-30). In this special case, apparently, its designer used other altitude k_φ related to the deviation angle of time φ, where k_φ =r– (*sin* (90– λ– φ)* r). In fig.2, the distance, AZ will equal k_φ, when O_t is put equal φ; and k_φ will be measured from the ground point G_3. The altitude of the bent line[11] of the bent pyramid would be ~46.5m, for the lower horizontal-projection, between 32.7m and 33.0m, and the intended β_1= 54° 44' 8.2". The altitude of the bent line matches k_φ=46.51m of its model β_1, for r=h =133.875m, λ = 29.788°, and φ = 0.0°. Besides, in its model β_2, most probably its intended radius r=158.19m, and its intended h= 103.95m, where it is the observed height of this combined pyramid; and the observed l of its base= 133.875*$2^{1/2}$= 189.33m. Table.3 shows the intended design data for this combined pyramid.

4. PYRAMIDS DESIGN AND MEASUREMENT UNITS

In the realm of architecture, designers state the dimensions of buildings in three ways: (i) in terms of design modules, (ii) in terms of specific measurement-unit to be used for implementation, and (iii) in terms of other measurement-unit that might was used for survey, after implementation. In the realm of megalithic construction, modern surveys, using any measurement unit, belong to the third category; and it might not imply knowing the original design module or the measurement unit used for design and implementation. In all cases, particularly after thousands of years, the results will include some deviations from the original design dimensions, in drawings. The first is at least the allowable implementation tolerance, if not exceeded it. The second is due to other vertical or horizontal force of seismic waves that might have caused, e.g., displacement, cambering, deflection, twisting, and/or tilting of all or part of the megalithic structure. The third is partial demolition due to human vandalism or by natural effects that create the second cause. In pyramids design realm, it is necessary to understand the design principles of the ancient Egyptian cubit rods. Sarton (1936, pp.399-402) noted out that the cubit rod of 28 digits is 52.5cm, and referring to plate-I by Lepsius (1866, p.18), from right to left, he said the first 16 digits are 18.75mm long, the eight following are 17.19mm long, and the last four are 21.87mm long. Correlating these numbers, the last two figures should be 17.1875mm and 21.875mm, to become 7/6 and 11/12 of the mean-digit δ (18.75mm) respectively. Hence, its design length[12] is exactly 52.5cm. Besides, starting from the right side, the first fifteen mean-digits are subdivided to create mini-intervals, starting from 1/2, 1/3, 1/4,..., down to 1/16 of δ=18.75mm, respectively. By reviewing the design of four types of ancient Egyptian cubit-rods of the same size, it seems that all cubit-rods contain three types of digits; one of them is the mean-digit δ (18.75mm). The diversity remains in changing the sizes of the other two types of digits: above and below the mean δ, as well as their number and location in the cubit-rods[13]. Fig.3 shows the design components of the ancient Egyptian measurement rod of 52.5cm, the so-called the royal cubit of Torino.

Besides, in some cubit rods, we see special hieroglyphic marks that identify the limit, or the domain, of lengths, e.g.: the bird's claw implies a foot; claw with sparrow imply a short foot; an arrow looks like reed pen imply half a cubit; and hand with sparrow imply a short cubit. In general, a palm is four digits; an arrow is 3 palms, a foot is 4 palms; 1.5 foot or 2 arrows make a cubit. Thus, the number of digits in any cubit is only 24. The 28 mean digits and their 15 types of mini-intervals are the design grids for creating diverse sets of 24 fingers that form the lengths of diverse types of cubits, including their related feet, arrows, and palms, in the same measurement rod. For Example, (i) cubits of 39.375cm, 41.25cm, 45cm, and 52.5cm, (ii) feet of 26.25cm, 27.5cm, 30cm, and 35cm, (iii) arrows of 19.6875cm, 20.625cm, 22.5cm, and 26.25cm, (iv) palms of 6.5625cm, 6.875cm, 7.5cm, and 8.75cm. In this context, the cubit of 45cm long is the mean cubit[14], which is composed of two arrows of 22.5cm. In some rods, the arrow symbol takes the digit number 12, to mark its mean length of 22.5cm; and in other types, it takes the digit number 13, to become 24.375cm long, or ½ a cubit of 48.75cm.

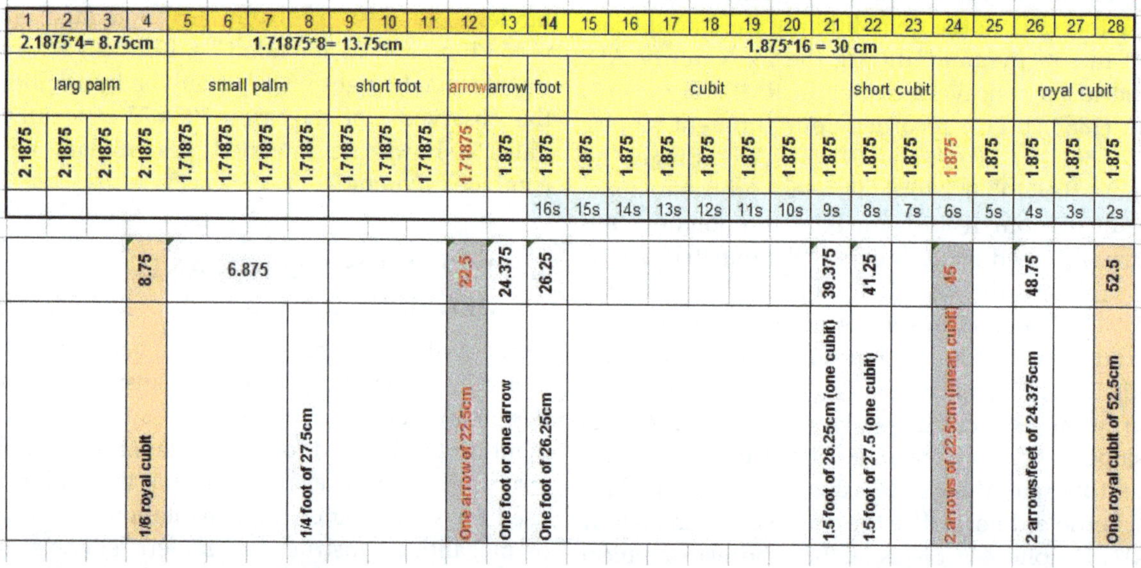

Figure 3: The architectonic design of the Egyptian measurement rod that contains diverse types of cubits, foots, arrows, palms, and fingers; its cubits starts from the cubit of 39.375cm up to the royal cubit of 52.5cm, where each cubit contains only 6 palms. Its design direction is from left to right, and for geometrical use, it will be turned upside-down; in order to make the first subdivided digit of 2s in the left side of the architect/engineer, and below the intended drawing line. This integrated design of the Egyptian measurement rod of diverse cubits implies that it might was used in many parts of the known world in their days, similar to the rod of diverse-scale ratios that we use today.

Correlating the design principles for producing the Egyptian cubit rod that mentioned above with reliable historic reference, might reveal some ancient measuring systems. Herodotus (430BCE, p.559, 2-149) correlated the measurement units that he used as follows: (i) one hundred fathoms is exactly a stadium (stade) of six hundred feet; (ii) the fathom is six feet in length, or four cubits; (iii) a cubit measures six palms, and a foot is four palms; (iv) parasang is thirty stadia; and (v) schoene is sixty stadia. Using these correlations between measurement-units, table.1 shows the possible metric values of each, based on using four types of feet that are included in the design of the Egyptian measurement rod. Besides, Herodotus (430BCE, p.285, 2-9) said, "the distance between Heliopolis and the sea is fifteen hundred stadia, and sailing up the river from Heliopolis to Thebes, the distance is eighty-one schoenes". Using Google Earth Data, of 2015, these two distances, along the course of the Nile River are about 250km and 670km, respectively. This implies that Herodotus used the measurement system based on the foot of 0.275m, i.e., he used the stade of 165m, and the schoene of 8,250.0m.

Moreover, Diodorus (60BC-30AD, p.24, 47-1) said, based on citing Hecataeus, "ten stadia from the first tomb of concubines of Zeus (i.e. entrance of the valley of the Queens), stands a monument of the king known as Osymandyas (i.e. Ramesseum temple), and its first pylon is two plethra in breadth". Using Google Earth data (2015), this walking distance, along the road, is about ~1.89Km, and the breadth of the first pylon of the Ramesseum temple is ~63m. Besides, Diodorus (60BC-30AD, p.26, 51-5) said, "ten schoene above Memphis king Moeris (Seti-I) excavated a lake". Using Google Earth data, the distance from Memphis to the old entrance of Lake Moeris (beside Lahun Pyramid) is ~78km. It implies that Diodorus used the measurement system based on the foot of 0.2625m, i.e., he used the stade of 189m, the plethra of 31.5m, and the schoene of 7,875.0m. In this regard, Nallino (1910, pp.268-275) said[15], "Eratosthenes used the stade of Alexandria of 157.5m long", which equals 600 feet, each is 0.2625m long.

Table 1: The possible metric values of diverse measurement units mentioned by Herodotus.

Name	(mean) Length in meter	(large-royal) Length in meter	(small) Length in meter	(short) Length in meter	Design intervals
Finger (δάκτυλος)	0.01875 (mean δ)	0.021875 (7/6 of δ)	0.0171875 (11/12 of δ)	0.01640625 (7/8 of δ)	1/16 foot (1/24 cubit)
Palm (παλάμη)	0.075	0.0875	0.06875	0.065625	1/4 foot (1/6 cubit)
Arrow	0.225	0.2625	0.20625	0.196875	3 palms (1/2 cubit)
Foot (ποδες)	0.30	0.35	0.275	0.2625	4 palms (2/3 cubit)
Cubit (πήχηεἰς)	0.45	0.525	0.4125	0.39375	6 palms
Fathoma (οργιά)	1.8	2.1	1.65	1.575	6 feet
Plethra (πλέθρον)	30	35	27.5	26.25	100 feet
	36	42	33	31.5	120 feet
	39	45.5	35.75	34.125	130 feet
Stade (στάδιον)	150	175	137.5	131.25	500 feet
	180	210	165	157.5	600 feet
	216	252	198	189	720 feet
Parasang (παρασάγγες)	4,500.0	5,250.0	4,125.0	3,937.5	30 Stadia of 500 feet
Schoene (σχοίνον)	9,000.0	10,500.0	8,250.0	7,875.0	60 Stadia of 500 feet

Furthermore, Herodotus (430BCE, pp.426-437) mentioned the dimensions of the pyramids of Memphis[16], that Strabo (1932-1967, p.89) said, "they lie about forty stadia (6.30 Km.) south of Memphis". This location is the plateau that includes the pyramids of Dahshur, where its elevation is about a hundred feet high[17], or 27.5m. Despite Herodotus did not mention the Sphinx and the remaining small pyramids in the Giza plateau; scholars of Egyptology have thought that he describes the Giza Pyramids. Herodotus called the bent pyramid, as the pyramid of Cheops[18]. Concerning the dimensions of its existing causeway (of 690m in Google Earth data 2015), he said, "it is five stadia in length (137.5m*5), ten fathoms wide (16.5m), at the highest part, eight fathoms; and the pyramid is a square, eight plethra[19] each way, and the height (along its inclined edge) the same"[20]. The pyramid of Cheops's daughter (satellite pyramid), which lies directly south of, and close to, the bent pyramid, and each side of it measures one and half plethra[21]. The red pyramid, he called the pyramid of Chephren, and said, "it was built of the same bigness as that of Cheops, save that it falls forty feet (40*0.275=11m) short of it in height (along its inclined edge)"; and he added, "I have myself measured it". It implies that the length that Herodotus assigned to the side of the pyramid of Cheops (bent pyramid) is the result of measuring the base of the red pyramid, assuming that their square-bases are equal in size. The black pyramid (now called pyramid of Amenmhat-III) he called the pyramid of Mycerinus or Rhodopis[22] and said, "the length of each side of its square base falls short of three plethra by twenty feet"[23]. Diodorus (60BC-30AD, pp.32-33, 63-3 & 64-2,7) said, "the lengths on each side of the bases of pyramids of Chemmis (Cheops), Cephres (Chephren), and Cherinus (Mycerinus) are: seven plethra, a stade, and three plethra, respectively". Concerning the names of builders of pyramids of Memphis, Diodorus (60BC-30AD, 64-13) said, "with regard to the pyramids there is no complete agreement among either the inhabitants of the country or the historians; for according to some the kings mentioned above (in his text) were the builders, according to others they were different kings". Herodotus (440BCE, p.431, 2-128) also said, "people call these pyramids after the shepherd Philitis". Table.2 shows the dimensions of the pyramids of Mimphes (in Dahshur), by Herodotus and Diodorus, using different types of plethra, in comparison with the available data by Petrie and other scholars.

Table 2: Bases dimensions of Dahshur Pyramids by ancient and modern scholars.

Pyramids of Memphis, in Dahshur	Herodotus		Diodorus		Petrie[24] & Rossi[25]
	Plethra (πλέθρων)	meter	Plethra (πλέθρων)	meter	meter
Red Pyramid	**8** * (27.5m)	220	**7** * (31.5m)	220.5	220.5
Bent Pyramid	Assumed = to Red pyramid	--	**6** * (31.5m) or (stade)	189	189.56
Satellite Pyramid	**1.5** * (35.75m)	53.625	--	--	52.42
Black Pyramid	(3*35.75m) – (20 *0.275m)	101.75	**3** * (34.125m)	102.375	105

5. THE LARGEST DIVIDER OF PYRAMIDS MODULES

Based on the above discussion, the common divider of pyramids of the fourth dynasty could be reckoned by dividing *h* of each pyramid by 7 and then finding out what is the size of the palm that divides the intended design modules of the five pyramids. Trying out diverse types of palms, the result was the mean-palm of 7.5 cm. This imply that each module in the five pyramids is composed of a number of mean-palms; if it is not a prime number, there might be a divider larger than the size of the mean-palm. For example, each module in the great pyramid is composed of 279 mean-palms, where 3 is the only integer from 2 to 7 that divides it. That is, the Egyptian arrow of 0.225m is the largest module divider *d* for designing the great pyramid, where it is 2/3 of the mean-foot of 0.30m and ½ of the mean-cubit of 0.45m. The case is also similar in the other four pyramids. Table.3 shows that the Egyptian arrow of 22.5cm is the largest common divider *d* of the design modules of pyramids of the fourth dynasty. It shows also, the design modules *m*, the number of module divisions n_d, the related intended design lengths (*h*, *l*, *k* and κ_φ), and the astronomical design parameters (O_i, O_t and φ) of each pyramid. Similarly, one can identify *d* in each of the four pyramids problems in RMP, using the royal cubit of 0.525m, in order to identify the metric lengths of pyramids in the-

Table 3: The intended design dimensions and astronomical parameters of pyramids of the fourth dynasty.

Pyramid		Modules of *h* and *l* (*m*)	Divisions in one module (n_d)	Intended design length ($m*n_d*d$) d=0.225m in meter	Astronomical parameters for O_i=21.672°; where O_t, and φ are in degrees; and κ and κ_φ are in meter			
					O_t	φ	κ	κ_φ
Great Pyramid	*h*	7	93	146.475	24.10°	25.93°	50.9	71.6
	l	11		230.175				
2nd Pyramid	*h*	7	91	143.325	22.986°	19.46°	51.8	53.2
	l	10+1/2		214.9875				
3rd Pyramid	*h*	7	42	66.15	24.30°	26.89°	22.9	33.8
	l	11+1/10		104.895				
Red Pyramid	*h*	7	66	103.95	22.986°	19.46°	9.66	54.2
	l	14+1/2+1/4+1/10		220.5225				
Combined or Bent Pyramid	$h_{\beta 1}$	7	85	133.875	21.672°	0.0°	50.4	**46.5**
	$l_{\beta 1}$	9+1/2+1/3+1/15		**189.3375**				
	$h_{\beta 2}$	7	66	**103.95**	23.3205°	21.672°	36.8	42.1
	$l_{\beta 2}$	15+1/20+1/63		223.7282143				

se problems that are mentioned in section-1 above, taking into consideration that in RMP#56-58, h_m=7, and in RMP#59, l_m=7. In RMP#56, d=0.15cm (1/2 mean-foot), n_d=125, and h= 0.15*125*7= 131.25m. In RMP#57-58, d=0.35m (royal foot), n_d=20, and h=0.35*20*7=49m. In RMP#59, d=0.30m (mean foot), n_d=2, and l= 0.30*2*7= 4.2m.

6. CONCLUSION

This paper showed four astronomical design algorithms for quantifying the slopes of pyramids, in relation to the pyramids design modules of RMP; and that could be used on earth or elsewhere, using diverse obliquity limits. It proved that Herodotus never talked about, and defiantly never saw, the Giza pyramids; and he meant only the pyramids of Memphis, i.e., Dahshur pyramids. It decoded the design of the ancient Egyptian cubit rod of 52.5cm, and showed that it includes other types of cubits, feet and arrows. It showed that Herodotus used the measuring system based on the foot of 27.5cm, and Diodorus used other system based on the foot of 26.25cm. It also, showed that the arrow of 22.5cm (half cubit of 45cm) is the largest common divider of the design modules of pyramids of the fourth Egyptian dynasty. In general, this paper showed an additional proof to support that defiantly there was consistent astro-mathematical theory in pyramids design (the Horizon theory), which was formulated and architectonically encoded by designers of pyramids of the fourth dynasty in ancient Egypt.

REFERENCES

Aboulfotouh, Hossam (2014), The Architectonic Encoding of the Minor Lunar Standstills in the Horizon of the Giza Pyramids, *Mediterranean Archaeology and Archaeometery*, Vol. 14, No. 1, pp.343-352.

Aboulfotouh, Hossam (2012), The Geometric Grids of the Hieratic Numeral Signs, *Mediterranean Archaeology and Archaeometery*, Vol. 12, No. 1, pp.129-138.

Aboulfotouh, Hossam (2007), the Relativistic Tilts of Giza Pyramids' Entrance Passages, *Mediterranean Archaeology and Archaeometery*, Vol. 7, No. 1, pp.23-37.

Aboulfotouh, Hossam (2005), Horizon Theory-Part-II: Internal Design Concept of the Great Pyramid, 1st international Conference of the Department of Restortion, Faculty of Fine Arts at Minia University, March 2005, http://www.geocities.com/fotouh28/H-Aboulfotouh-Horizon-II.pdf.

Aboulfotouh, Hossam (2002), The Horizon Theory, Part-I: Original Concept Plan of the Pyramids Plateau, 1st International Conference of the UIA Work Programme "Architecture and Heritage," Region-V, on "Architecture and heritage as a paradigm for knowledge and development: lessons of the past, new inventions and future challenges", Bibliotheca Alexandrina, March 2002. http://www.geocities.com/fotouh28/H-Aboulfotouh-Horizon-I.pdf

Al-Maqrizi (1364–1442AD) *Al Mawaes Wal A'atebar Bezeker Al-khetat Wal Asar* (Lessons and Consideration by Mentioning Areas and Old Buildings) Dar Al-Tahrier, Bolaque Edition, Cairo, 1849.

Belmonte; Juan Antonio & Magli; Giulio (2015) Astronomy, Architecture, and Symbolism: The Global Project of Sneferu at Dahshur, *Journal for the History of Astronomy*, Vol. 46(2) pp. 173–205.

Nallino, Carlo (1911) *Arabian Astronomy: its History During the Medieval Times*, Oriental Papers, Arabian House Bookshop, First Ed. Rome, Second Ed. Beirut, 1993 (in Arabic).

Diodorus the Sicilian (60BC-30AD), *Library of History*, Book-I, made English by G. Booth, Printed by Edw. Jones, London, MDCC (1700 AD), p.1-52, Ch-63.

Eyre, C. J. (1991) Book Review "Herodotus Book II Commentary 99-182 by Alan B. Lloyd", *The Journal of Egyptian Archaeology*, Vol. 77, pp. 223-225.

Gillings, R. J. (1982) *Mathematics in the Time of the Pharaohs*, Dover Publications Inc, New York.

Herodotus (484 - 425 BC) *The Histories:* Books I-II, with English Translation by A. D. Godley, Harvard University Press, London, William Heinemann LTD., 1920-1975.

How, W. W.; Wells, J., (1913–1927) *A Commentary on Herodotus*, Oxford University Press, ii. 124-36, p.249, (Electronic Edition).

Laskar, J., (1986) Secular Terms of Classical Planetary Theories Using the Results of General Theory, *Astronomy and Astrophysics*, Vol. 157, p.59-70.

Lepsius K. R. (1866) *Die altägyptische Elle und ihre Eintheilung; Philol. Abhandle. der Akademie der Wissenschaften zu Berlin*, 67 p., 5 pl. Berlin.

Manetho, (1940-1964), Aegyptiaca Αἰγυπτιακα (Epitome), with an English Translation by W. G. Waddell, Cambridge, Massachusetts, Harvard University Press, London, William Heinemann Ltd, MCMLXIV.

Meeus, Jean, (1991), *Astronomical Algorithms*, Willmann-Bell Publications Inc., USA.

Petrie, W.M.F., (1883), *Pyramids and Temples of Gizeh*, London.

Petrie, W.M.F., (1887) *A Season in Egypt*, the Leadenhall Press, London.

Rossi, Corinna (2003), *Architecture and Mathematics in Ancient Egypt*, Cambridge University Press, New York.

Sarton, George (1936), On the Curious Subdivision of the Egyptian Cubit, *ISIS*, Vol. 25, No. 2, pp.399-402.

Scott, Nora E. (1942), Egyptian Cubit Rods, *The Metropolitan Museum of Art, New Series*, Vol. 1, No.1, pp.70-75.

Scott, R. B. Y. (1958), The Hebrew Cubit, *Journal of the Biblical Literature*, Vol. 77, No. 3, pp.205-214.

Shaltout, M. & Belmonte, J. A., (2005) On the Orientation of Ancient Egyptian Temples I: Upper Egypt and Lower Nubia" *JHA*, Pre-print series: PP 03/2005.

Spence, Kate (2002) Book Review "Roger Herz Fischler: The Shape of the Great Pyramid", *ISIS*, Vol. 93, No.1, pp.83-84.

Strabo (36BC-24AD), *Geography*, Book XVII, with an English translation by Horace Leonard Jones, Harvard University Press, London, William Heinemann Ltd, 1932-1967 (Pyramids 17-1-33).

The Penny Cyclopedia of the Society for the Diffusion of Useful Knowledge (1840), Charles Knight and Co., London, MDCCCXL, Vol. XVIII and Vol. XIX.

Wittmann, A. (1979) the Obliquity of the Ecliptic, *Astronomy and Astrophysics*, 73, pp.129-131.

NOTES

---

[1] Without some basic knowledge in spherical astronomy, it is hard to retrieve any mathematical method for designing the slopes (angles of inclination) of the Egyptian pyramids. RMP includes mere numerical ratios without the design method. In her review, Spence (2002) disputed the strange conclusion by the mathematician Herz-Fischler who said in his book, "there may have been no true theoretical basis for the angles of inclination of pyramids".

[2] Strabo said, "on proceeding forty stadia from the city (from Memphis heading south as he walks), one comes to a kind of mountain-brow; on it are numerous pyramids the tombs of kings, of which three are noteworthy; and two of these are even numbered among the Seven Wonders of the World, for they are a stadium in height."

[3] Eyre, C. J. (1991, pp.223-225) said Herodotus's text mercilessly plagiarized by Diodorus Siculus.

[4] How *et al* (1913-1927, p. 250, ii 124) called it «height along sloping side»; in the pyramids chapter, Al-Maqrizi (1849) said, based on citing Abdulrhim Alquise, "the perimeter of the Great Pyramid's base is 2000 black cubits and its height is 500 black cubits", implying the height along its inclined edge.

[5] In the opinion of scholars, e.g., Gillings (1982, p.186) and Clagett (1999, pp.166-168), a cubit measures 7 palms, and thus they thought it represents the rise of the pyramids slopes in RMP. That is despite, Sarton (1936, pp.399-402) had noted out that the 7 palms shown in the cubit rods are not equal, and Herodotus (430BCE, p.559, 2-149) said, "a cubit measures only 6 palms".

[6] Gillings (1972, p.185) pronounced it as *Seked*, and said, "it means the slope of the sides of a pyramid".

[7] Using flat mutation of 0.47" per year from the epoch 2000 AD (23.45°), the obliquity 24.10° meets ~ 3055-3065 BCE; and using, M. Bessel's Eq.: $\varepsilon = 23° \ 28' \ 18'' + 0.48368'' \ T - 0.00000272295'' \ T^2$ (Nallino, 1911, p.270) & (The Penny Cyclopedia, 1840, Vol. XVIII, p.495), 3065BCE meets ε= 24.109°; where ε is O_t, and T is the years elapsed from the epoch 1750AD. Bessel's equation is the only polynomial fit that gives value close to the record of Chou Li in 1100BC, ε=23.881° (data in: Wittmann 1978). The polynomial fits that are set to finite limits, e.g., between 22.5° and 24.5° (Meeus, 1991, p.135) or between 22.61° and 24.23° (Laskar, 1986, p.86), give ε=~24.027° for 3070BCE; and fail to give ε = ~23.922° for the azimuth 296.75° of winter solstice of the axis of the main Karnak temple, at λ=25.72°, in ~ 1620BCE (data in: Shaltout *et al*, 2005, p.15).

[8] Its recorded slope is 56° 18' (Rossi, 2003, p.249; based on citing, Lauer, Observations, p.95).

[9] Al-Maqrizi (1849), based on citing Abdelatif Al-Bugdady, said, "the demolition of the granite casing of the small pyramid was by the order of king Al-Aziz Osman in 593H, whose illiteracy made him imagine that there are treasures beneath it, but the lack of finance obliged the workers to discontinue after 8 months of demolition work".

[10] Since, its h= ~104.61m that is composed of 154 steps (The Penny Cyclopedia, 1840, Vol. XIX, p.152), and its l=220.5m (Rossi, 2003, p253), any record above that of Vyse 43° 36' 11" is invalid.

[11] Petrie (1887, p.32) reckoned it as follows: *tan* 55° 1' * 33m= 47.15m.

[12] Over time, shrinking might cause slight reduction in the length of wooden rods; thus, any royal cubit length other than 52.5cm, e.g., 52.375cm or 52.3cm (Scott, 1942, p.70), is incorrect and misleading.

[13] Further, 13/12 of the mean δ produces the cubit of 48.75cm; and 9/8 of δ produces the cubit of 50.625cm; 10/9 of δ produces the cubit of 50.0cm; and 61/60 of δ produces the cubit of 45.75cm that 2/3 of it is the English foot of today.

---

14 Scott (1942, p.70) called it a short cubit, and Rossi (2003, p60) called its half (arrow) of three mean palms as small-hands. Scott (1958, pp.205-214) also mentioned the lengths of other cubits, e.g., 39.6cm and 50.0cm.

15 Carlo Nallino (p.274) referred to the works of the German scholar Fr. Hultsch.

16 In Greek: πυραμίδας κατατμηείσαι τας ἐν Μέμφι (Herodotus, 430BCE, p.282, 2-8).

17 Herodotus (430 BCE, p.430-431, 2-127) perhaps mentioned the elevation from the plain of Memphis or from the water surface of the Nile River; and considering the metric value of the foot he used, Google Earth data of 2015 show the same elevation.

18 Despite, Africanus had cited Manetho and said, "the fourth dynasty kings are: Soris, Suphis (i), Suphis (ii), Mencheres, etc." (Manetho, 1940-1964, p.47), Egyptologists thought that Cheops (or Chmmes) is the 2nd king of the 4th dynasty. Herodotus (430BCE) put Cheops the 4th king after Sesostris, and Diodorus (60BC-30AD, p.32) put him the 5th king after Sesostris, who most probably is the so-called Ramses-III, where Sesostris is a royal or military title before the king's cartouche. The names: Chemmis, Cephres, and Cherinus, match the cartouches of the so-called Ramses-VIII, Ramses-IX, and Ramses-XI, respectively.

19 In Greek: ὀκτώ πλέθρα (Herodotus, 440 BC, p.426, 2-124).

20 Strabo (36BC-24AD, p.89, 17-I, 33) said, "high up approximately midway between the sides, it has a movable stone, and when this is raised up there is a sloping passage to the vault"; related to this, see Fig-XI: Door of the South Pyramid of Dahshur, by Petrie (1883).

21 In Greek: ὅλον καὶ ἡμίσεος πλέθρον (Herodotus, 440 BC, p428, 2-128).

22 Most probably, Rhodopis is the so-called Nubheteti-Khered that her tomb is found under this pyramid; the correct pronunciation of the last word  in her name is Radophi (or Pharida) without the Greek "ς"; the upper letter is *Ph* not *Kh*, which becomes the last letter if we follow the ancient Egyptian spiral sequence of writing the letters of words.

23 In Greek: εἴκοσι ποδῶν καταδέουσαν κῶλον ἕκαστον τριῶν πλέθρων (Herodotus, 440 BC, pp.436-437, 2-134).

24 Data of the bent and satellite pyramids are by Petrie (1887, p.28).

25 Data of the red and black pyramids are by Rossi (2003, p.253, based on citing, Stadelmann, MDAIK 39, p.253, and Arnold, Amenemhet III., p.9, respectively).

9

EQUINOCTIAL MARKERS IN PROTOHISTORIC IBERIAN SANCTUARIES

César Esteban[*,1,2]

[1] *Instituto de Astrofísica de Canarias, La Laguna, Tenerife, Spain*
[2] *Departamento de Astrofísica, Universidad de La Laguna, La Laguna, Tenerife, Spain*

Corresponding author: César Esteban (cel@iac.es)

ABSTRACT

This paper summarizes the most relevant results of a thorough archaeoastronomical study conducted since 1997 in several tens of Iberian archaeological sites, most of them located in the southeast of the Iberian Peninsula. We found that a substantial fraction of the sanctuaries studied, all of them dedicated to a feminine goddess, shows a marker of the sunrise or sunset on conspicuous elements of the horizon at the equinoxes or a date very close to it, possibly the temporal midpoint between solstices. This result suggests the existence among the Iberians of rituals associated with seasonality, agrarian festivities of fecundity similar to other well-known ones of the ancient Mediterranean. The earliest chronology of the sanctuaries related to the equinoxes indicates that they became operational around mid-IV BCE, coincident with a period of major ideological changes in the Iberian society and increased Punic influence in the area. The analysis of the results suggests a possible influence of Greco-Punic beliefs on the astronomical aspects of Iberian ritual.

KEYWORDS: Archaeoastronomy, Protohistory, Iberian Peninsula, Iberian Culture, Sanctuaries, Equinox.

1. INTRODUCTION

The Iberians were a group of peoples who inhabited the Iberian Peninsula from the sixth century BCE up to practically the change of the era, by which time the Romans were fully established in their territory. They occupied the Mediterranean facade and part of the centre of the Iberian Peninsula, as well as the French Languedoc region (see Harrison 1988 for an introduction to Spanish protohistory in English). The main Iberian deity was apparently a fertility goddess that also had strong funereal associations. Her iconography is often a more or less direct reflection of the aristocratic Iberian female image, but is also influenced by exogenous models and sometimes represented with attributes of Eastern goddesses such as Astarte, Tanit, Artemis or Demeter (Moneo 2003: 427-439). Iberian sanctuaries were usually in locations that favour the manifestation of the sacred, such as on the top of mountains, within caves or in proximity to springs (see Blázquez 1975: 148-166). They mostly consist of open-air deposits or temples housing a statue of the divinity and a large number of offerings, similar to the Greek *thesauroi*.

2. ARCHAEOASTRONOMICAL FIELDWORK IN IBERIAN SITES

Since the end of the 1990s and in collaboration of archaeologists, I have surveyed 33 sanctuaries (27 open-air sites and 6 caves) across the territory once occupied by the Iberian Culture (see Figure 1). The measurements have been made with a theodolite and a precision compass. The typology of the sanctuaries is fairly variable; there are open-air sites without any trace of construction, others consist of urban or isolated rural sacred buildings and there is an important fraction of caves. The fieldwork at each site is based in the measurement of the orientation of the standing constructions, the entrances and main axes of the caves and - in all the cases - the astronomical analysis of the horizon surrounding the sites.

The description of the characteristics of the 33 sanctuaries surveyed would be too wordy to be included in this necessarily brief paper. The information of the different sites can be found in Esteban (2002, 2003, 2013 and 2015), Esteban & Benítez de Lugo Enrich (2016), Esteban & Cortell Pérez (1997), Esteban & Moret (2006), Esteban & Ocharan Ibarra (2016) and Esteban, Rísquez and Rueda (2014). All these papers can be found in the academia.edu webpage of the author.

One of the main results of the survey is that the orientations of the entrances of Iberian temples show a non-random pattern. This pattern has been presented and discussed in Esteban (2003, 2015). In fact, most of the temples are facing the zone of the horizon where the sun (or moon) rises along the year. As it was discussed by Esteban (2003, 2015), the orientation pattern of the Iberian sanctuaries is different from that shown by Roman, Etruscan or mainland or Aegean Greek temples but similar to that of Punic sanctuaries and Greek temples of Magna Graecia and Sicily.

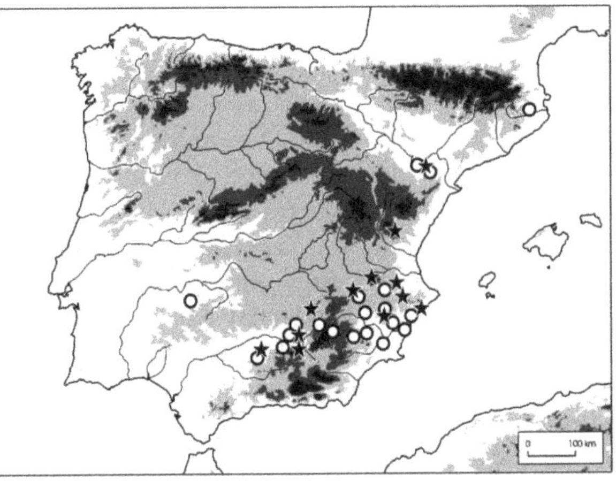

Figure 1. Location of the Iberian sanctuaries studied in our fieldwork campaigns. Black stars represent the sites that show equinoctial markers.

The most remarkable result of the archaeoastronomical fieldworks in Iberian sanctuaries is the discovery of equinoctial markers in 12 of them, a 36% of the surveyed sites (see Figure 1). This result clearly indicates that equinoxes or dates close to them were milestones of the Iberian ritual. The markers have been found in an extended geographical area, in sites located in the present-day provinces of Teruel, Valencia, Ciudad Real, Albacete, Alicante, Jaén and Córdoba.

We find other possible astronomical relations in an important fraction of the sanctuaries that do not show equinoctial markers. Three of them show temples oriented close to the sunrise at equinoxes and four are oriented to the solstices. Five show solstitial markers over topographic features and one (perhaps up to three) may be related to Venus setting at its southernmost position (see Table 1 of Esteban 2013). Finally, eight of the surveyed sanctuaries do not show apparent astronomical relations.

3. THE EQUINOCTIAL MARKERS

The feature that defines a sanctuary having an equinoctial marker is that the sunrise - or sunset in fewer cases - at the equinoxes or a date very close to them occurs exactly on a topographic feature of the horizon surrounding the site. The peak of a distant mountain is the most common topographic feature. In some of these sanctuaries that contain a temple, the building is also oriented to the equinox, as in the cases of Sant Miquel de Llíria, La Malladeta, Cerro de las Cabezas or Cerro de San Cristóbal. Another kind of marker is based on illumination effects at equinox sunset – as in the case of the cave-sanctuary of La Lobera in Castellar (Esteban, Rísquez and Rueda 2014) – or at sunrise – as in the entrance sanctuary of Puente Tablas (Jaén; Pérez Gutiérrez et al., these proceedings).

One of the important aspects to elucidate is the exact astronomical target of the markers. Although we generically call them "equinoctial markers", this may not be strictly true. Our modern concept of equinox is defined as one of the two moments of the tropical year when the centre of the solar disc crosses the celestial equator. This is an abstract concept that only has a practical sense within the framework of a geometric model of the celestial sphere. It is thought that Babilonian or Greek astronomers discovered the concept of the equinox in the IV or II centuries BCE. At the same time or even somewhat later when our Iberian sanctuaries were actually in use.

Another possibility of defining something similar to the equinox is using the concept of the temporal midpoint between solstices (hereinafter TMPS; also known as "megalitic equinox", introduced by Alexander Thom, 1967). This corresponds to the day just in the middle between the exact dates of summer and winter solstices. Because the Earth's orbit is not circular, this day does not coincide with the equinox. It occurs between one or two days after spring equinox or before the autumn equinox. The declination of the sun is between +0.3° and +1° (see Ruggles 1997) in the moment of the sunrise closer to the TMPS. Therefore, the TMPS permits to divide the year in four equal parts that coincide with our seasons at intermediate latitudes, and seems to be a much simpler and intuitive concept with more practical utility than the equinox.

One of the first equinoctial markers we found was at the Iberian settlement of El Amarejo (Bonete, Albacete; Esteban 2002), located on a flat-topped hill. The sanctuary is just on the eastern edge of the settlement, with a clear and impressive view of the eastern horizon. Although the presence of a temple has not been clearly established, the sacred element of the site is a ritual deposit in the form of a well of about four metres depth. The deposit is located exactly in the east edge of the hill. The contents of the deposit indicate that the sanctuary was dedicated to a goddess protector of the traditional feminine activities. The highest and most conspicuous feature of the horizon is mountain Chinar, at a distance of about 3 km and a height of about 200 m above the surrounding plain. Figure 2 shows the sunrise as seen from the ritual deposit at the autumnal equinox of 2004. We can see that the sun appeared at the bottom of the northern slope of the peak and climbed over the slope as it went up in the sky. This striking "sun-climbing" phenomenon cannot be seen in the same way one day before or after the equinoxes. However, at the TMPS, where the centre of the sun is around $\delta \approx +0.5°$, the sun rises about a solar diameter northerly touching slightly the slope when going up in the sky. As we see, in El Amarejo, the preferred astronomical event would seem to be the equinox, as long as the proper foresight of the phenomenon is the exact position of the ritual deposit. Considering the closeness of mountain Chinar, displacements of the foresight of about few metres may change the date of the striking "sun-climbing" event. The archaeologist that excavated the sanctuary of El Amarejo found that the offerings of the ritual deposit were burned and distributed in apparently periodical layers of ashes inside the well. The most common vegetal offerings were acorns and their degree of growth indicated that they were collected still unripe at the beginning of autumn. Therefore, this independent clue suggests that the ritual was probably held at autumnal equinox or a date very close to it. The spectacularity of this marker suggests that the ritual associated with this phenomenon might have a public dimension, perhaps related to pilgrimage activities – attested in other Iberian sanctuaries – to present offerings to the divinity and to witness the sunrise at the precise day of the ritual.

The sanctuary of Tossal de Sant Miquel (Llíria, Valencia; Esteban and Moret 2006) is a temple inside a relatively large Iberian settlement situated on a hill. In contrast to the rest of the buildings of the village, the temple is oriented very close to the east-west direction, and its entrance is facing east (see Figure 3). The slope of the hill falls rather steeply in front of the entrance and the line of sight of the temple was – perhaps deliberately – free of buildings, allowing one to see the distant eastern horizon from the sanctuary. From the temple, the sunrise at the TMPS takes place just on the top of a distant and isolated small hill of the horizon (indicated in Figure 3). The entrance and the long axis of the temple are facing exactly towards that point. It is interesting to note that the southern half of the east horizon is a featureless coastal flat plain and that the aforementioned hill is the first topographic feature that the sun touches when coming from its southernmost position at winter solstice. The sunrise is produced at the south edge of the hill at the exact dates of the equinoxes, but on the top of the hill at TMPS. Therefore, we consider that the marker is actually more precise indicating the TMPS than the equinox. The precision of the marker and its lack of spectacularity – in contrast to what happens in El Amarejo, for example – indicates that it should be better used as a practical tool for pinpointing the calendar and not for a public ritual.

Figure 2. Sunrise at the Iberian sanctuary of El Amarejo (Bonete, Albacete) on the autumnal equinox of 2004, September 21st. The Sun had a declination of = +0°09′. Left photograph shows a general view of mountain Chinar at the moment that the first sunrays emerge behind it. The three photographs on the right show a sequence of the solar path just after sunrise.

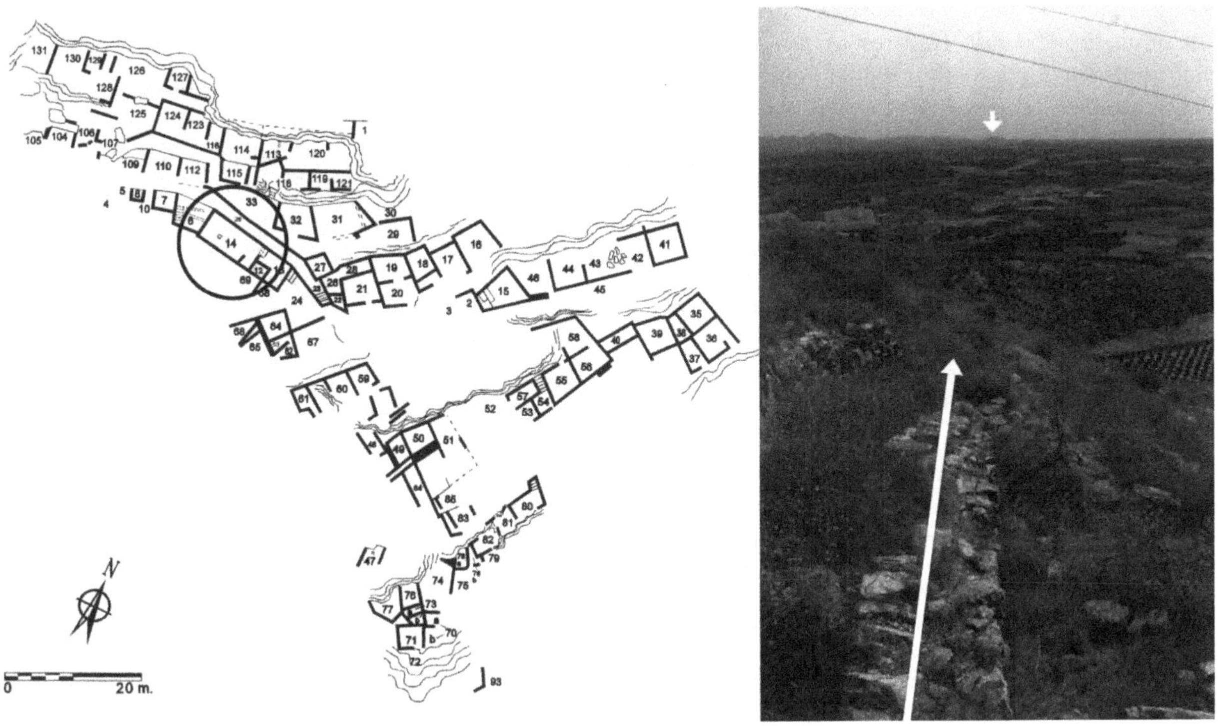

Figure 3:. Left: plan of the Iberian settlement of Tossal de Sant Miquel de Llíria (adapted from Bonet 1992). The circle indicates the position of the urban temple (department no. 14). Right: View of the east horizon from the temple. The long arrow indicate the orientation of the north wall of the temple, which is facing due east, just to a small hill (indicated by the short arrow) where the sunrise at the TMPS takes place.

Similar markers of the sunrise at dates close to the equinoxes on the top of non-conspicuous hills have been found in La Carraposa (Rotglá i Corbera, Valencia; Pérez Ballester and Borredá Mejías, 2004), La Serreta (Alcoi, Alicante; Esteban and Cortell Pérez 1997), Cueva Negra (Fortuna, Murcia; Esteban and Ocharan Ibarra, in preparation), Cerro de las Cabezas (Valdepeñas, Ciudad Real, Esteban and Benítez de Lugo Enrich 2016) and Torreparedones (Morena López and Abril Hernández, 2013). On the other hand, at El Pajarillo (Huelma, Jaén, Esteban *et al.*, in preparation) and Cueva del Moro (Ayora, Valencia; Esteban and Ocharan Ibarra, in preparation) the equinoctial markers are related to the sunset instead of the sunrise. This may be related to the different chararacter of these two sites with respect to the other Iberian sanctuaries with equinoctial markers. El Pajarillo is interpreted as a rural *heroon* or a funerary monument and Cueva del Moro is a cave–sanctuary. It is interesting to note that other cave-sanctuary with clear relation with the equinox, Cueva de la Lobera at Castellar (Esteban, Rísquez and Rueda 2014) is also related to the sunset at equinoxes or TMPS. Perhaps, this fact may be due to the chthonic aspects of the cult at the Iberian cave-sanctuaries and the *heroon* of El Pajarillo.

As in El Amarejo, there are some Iberian sanctuaries that show striking equinoctial markers on conspicuous topographic features of the horizon that may indicate rituals with possible public dimension. These are the urban shrine at Cerro de San Cristóbal (Mazaleón, Teruel; Esteban and Chavarría Fores, in preparation) and La Malladeta (La Vila Joiosa, Alicante; Esteban and Espinosa Ruiz, in preparation).

The sanctuary of La Malladeta shows perhaps the most striking equinoctial marker found so far in an Iberian sanctuary. It is located on the top of a seashore promontory, with a perfect view of the horizon and surrounded by a complex of auxiliary buildings distributed on the slopes of the promontory, which have been interpreted as warehouses and temporal dwellings perhaps related to periodical celebrations. There are few surviving walls of what is supposed to be the temple but, toward the east, these walls are oriented facing the distant Isle of Benidorm (Figure 4). Several campaigns of in situ observations made by the archaeologist Antonio Espinosa and local photographers of Fotocine La Villa, have revealed that the sunrise at the TMPS takes place over the southern slope of the isle (see Figure 5). Moreover, at the equinoxes, the sun does not touch the isle. For us, this fact is a strong argument to propose that the TMPS was the most probable astronomical target of the marker. Very close to the temple building, just at its east side, the excavations have discovered an open-air area plenty of ashes and dedicated to cremation rituals (Moratalla et al. 2014). In addition, the excavations have revealed an abnormal large number of lucernes at the sanctuary, indicating that nocturnal activities were commonly carried out at the site (Espinosa and Marcos 2014). We speculate that the lucernes might be used to illuminate gatherings held before dawn to witness the TMPS sunrise. All these archaeological and astronomical elements suggest the celebration of public rituals at and before the spectacular sunrise at the TMPS.

Figure 4. View of the east horizon from the sanctuary of La Malladeta. The arrows indicate the orientation of one of the remaining walls of the temple, which is facing due east, just to the Isle of Benidorm (at the centre of the circle), where the sunrise at the TMPS takes place (see Figure 4).

As we can see, most of the results of this study suggest that the TMPS was the most probable target of the Iberian equinoctial markers. In this sense, there is revealing epigraphic evidence from the cave-sanctuary of Cueva Negra (Fortuna, Murcia), one of the sites where we have found a possible marker of the TMPS. The cave has several Latin inscriptions dated from I CE painted in their walls. Four of the inscriptions give dates and of those only two are complete. These two inscriptions tell about ritual activities made precisely at the same date: March 27th

(Stylow and Mayer 1996; Velázquez and Espigares 1996) that would corresponds to March 26th in the Gregorian calendar, a date closer to the TMPS - that would correspond to March 24th at I CE - than to the equinox.

It seems clear that the existence of astronomical markers on topographic elements of the horizon should be a decisive criterion for placing some Iberian sanctuaries, especially those located in isolated environments like La Malladeta, La Carraposa or El Amarejo, for example. In the case of the urban temple of Sant Miquel de Llíria, the visibility of the marker would determine the orientation and place of the building and also the fact that the area located opposite the facade was free of buildings (Esteban and Moret, 2006).

Figure 5. Sunrise from the Iberian sanctuary of La Malladeta at March 21st, 2013. The declination of the sun was $\delta = +0°18'$. The sunrise at the equinox does not touch the Isle of Benidorm. In the TMPS, the sun emerges at the south slope of the isle.

Another aspect of special interest is the comparison of the chronology of our sanctuaries. The four sanctuaries that show earlier dating: El Amarejo, Cueva de La Lobera, Coimbra del Barranco Ancho and La Malladeta indicate that they became operational around mid-IV BCE, while others as La Serreta, La Carraposa and Sant Miquel de Llíria, have later chronologies, between III and II BCE (see Esteban 2013 for references). The absence of sanctuaries of this kind with earlier dating could be indicating the date *post quem* such rituals appeared in the Iberian world. This moment could be related to ideological changes in the society and/or the effect of alien influences. According to different authors, the beginning of the IV BCE was a moment of consolidation of the aristocratic system and territorial expansion of the Iberian urban settlements, as well as the emergence of an ideological model based on the figure of a heroized ancestor (e.g. Rueda, 2011).

All excavations carried out in the sanctuaries showing equinoctial markers, show clear evidences that they were dedicated to a fertility goddess (see references in Esteban 2013). In this context, it seems clear that the rites celebrated in the sanctuaries should be related to seasonality and the cycles of nature. Festivities related to agricultural fertility were common in the ancient Mediterranean, such as those dedicated to the "resurrection" or *egersis* of Melkart that were held in Tyre - and perhaps also in Gades, the westernmost Phoenician city, in south Spain - which, according to some authors might have taken place the new moon nearest the spring equinox (Cohen, 1993: 401). Among the most famous celebrations of antiquity were those related to the Greek goddess Demeter, whose Great Mysteries at Eleusis were represented in early autumn (Espejo Muriel, 1995: 95). These mysteries represented the annual growth cycle through the myth of the descent and return of Kore from the underworld. A similar mythic narrative, where the protagonist is a possible hero-god of vegetation may also be represented in the reliefs of the Iberian funerary monument of Pozo Moro (Moneo, 2003: 416). The symbol of the natural cycle of death and resurrection might be inspired in the annual solar motion on the celestial sphere. Throughout the year, the sun changes its position of sunrise and sunset on the horizon as well as the maximum height that can reach at noon. It is around the equinoxes when the equilibrium is reached: the day has the same duration that night, sunrise and sunset occur at almost equidistant points to the solstices. These astronomical moments could symbolically represent the moments of death and descent of the divinity to the underworld - autumnal equinox - and her subsequent rebirth or return to earth - spring equinox. The presence of illumination effects at those dates in the cave-sanctuary of Cueva de La Lobera (Esteban, Rísquez and Rueda 2014) seems also to agree with this relation of the goddess and her ascent/descent to the underworld.

Equinoctial markers similar to those discussed herein - especially those found at Sant Miquel de Llíria and La Malladeta, where we have the combination of orientation of a temple and equinoctial marker - have been found in the temple of Apollo at Maktar (Tunisia). This temple was built over an ancient Libyan-Punic sanctuary dedicated to Baal Hammon and reused in Roman times as the temple of the sun god Apollo (Esteban, 2003). Another very similar example is the temple C of the Greek-Punic city of Selinunte, where Belmonte and Hoskin (2002: 203-206) found that the building points to a striking conical mountain located just to the east. This temple is dated around the early VI BCE - being a possible precedent of the Iberian cases - and was dedicated to Saturn in Roman times.

It is interesting to note that the geographical distribution of the Iberian sanctuaries showing equinoctial marker is very similar to that presented by the

Iberian tower-form monuments studied by López Pardo (2006: 199), which coincides, in turn, with the areas of more intense Punic influence in the Iberian Peninsula. After the defeat of the Punic general Himilco, during the attempted conquest of Syracuse in the early IV BCE, Carthage suffered heavy political upheavals affecting even religion (Prados Martínez, 2008: 39). As discussed by Prados Martínez (2008: 230) at that time, there was an increase in the sensitivity of the Punics to the concept of underworld that some authors associate to the introduction of the Greek cults of Demeter and Kore at Carthage (Ribichini and Xella, 1994: 39) as atonement for the aforementioned Punic defeat in Sicily. The two goddesses worship was widespread in Sicily and Magna Graecia where, according to Diodorus (XIV, 63) two annual festivals in spring and autumn were held in their honour. One of the most common kinds of offerings of this worship were terracotta figurines of female busts emerging from the ground, also present in many Iberian sanctuaries in the area in question (see Ruiz de Arbulo, 2000). All these archaeological and historical evidences seem to favour a possible Greek-Punic inspiration for the use of equinoctial markers in Iberian ritual.

The aforementioned characteristics of the equinoctial markers inclines us to think that the pinpointing of the Iberian calendar was made tracking the position of solar risings and settings over topographic features of the horizon, using something known as an "horizon calendar", a technique developed by different cultures of the world (Ruggles, 1999: 152). This possibility recalls us the controversy about the existence of priests or priestess in Iberian religion. It is clear that the use of astronomical markers and to fix the exact dates of celebrations through horizon calendars need personnel in charge of the follow-up of the sun position along the year. As we said before, the equinoctial markers seem to appear at IV BCE, a moment of socio-ideological changes in the Iberian Culture and the emergence of a more popular forms of worship. The control of the ritual and seasonal calendar by the observation of the celestial bodies should provide the aristocratic elites with a demonstration of having some divine power as guarantors of the welfare of the community. In the cave-sanctuary of Cueva de La Lobera, we noted the resemblance between the shape of the projected patch of light at equinox sunset and the typical representation of the Iberian goddess (Esteban, Rísquez and Rueda 2014). There is evidence that the shape of the opening that projects the patch of light at La Lobera was manipulated. Therefore, the illumination phenomenon could be recreated with the purpose of producing a convenient atmosphere for the ritual. Moreover, it may be interpreted as an intentional dramatization of the appearance of the goddess in a moment of annual renewal of nature. The study of the interconnection between astronomy, ideology, and ritual as an interdisciplinary field of research seems to have an exciting future.

ACKNOWLEDGEMENTS

During all the years of fieldwork, I have enjoyed the collaboration of a large number of archaeologists and interested people that would be very long to enumerate, my deepest gratitude for all of them. Part of this work has been funded by the project *Arqueoastronomía* of the Instituto de Astrofísica de Canarias.

REFERENCES

Belmonte, J.A. and Hoskin, M. (2002) *Reflejo del Cosmos. Atlas de arqueoastronomía del Mediterráneo antiguo*, Madrid, Equipo Sirius S.L.

Blázquez, J. M. (1975) *Diccionario de las religiones prerromanas de Hispania*, Madrid, Ediciones Istmo.

Cohen, M.E. (1993) *The Cultic Calendar of the Ancient Near East*, Bethesda, CDL Press.

Bonet, H. (1992) La ceramica de Sant Miquel de Llíria: su contexto arqueológico. In *La sociedad ibérica a través de la imagen*, Madrid, Ministerio de Cultura, 224-236.

Espejo Muriel, C. (1995) *Grecia: sobre los ritos y las fiestas (2ª edición corregida)*, Granada, Universidad de Granada.

Espinosa, A. and Marcos, A. (2014) Los materiales muebles. In *Villajoyosa Antique (Alicante, Espagne). Territoire et topographie. Le sanctuaire de La Malladeta*, P. Rouillard, A. Espinosa and J. Moratalla (eds.), Madrid, Casa de Velázquez, 108-155.

Esteban, C. (2002) Elementos astronómicos en el mundo religioso y funerario ibérico. *Trabajos de Prehistoria*, Vol. 59, no. 2, 81-100.

Esteban, C. (2003) Equinoctial markers and orientations in pre-Roman religious and funerary monuments of the Western Mediterranean. In *Ad Astra per Aspera et per Ludum. European archaeoastronomy and the orientation of monuments in the Mediterranean Basin*, A.-A. Maravelia (Ed.), Oxford, BAR International Series. Archaeopress, 83–100.

Esteban, C. (2013) Arqueoastronomía y religión ibérica. In *Santuarios iberos: territorio, ritualidad y memoria*, C. Rísquez and C. Rueda (eds.), Jaén, Asociación para el desarrollo rural de la Comarca de El Condado, 465-484.

Esteban, C. (2015) Iberian Sanctuaries. In *Handbook of Archaeoastronomy and Ethnoastronomy*, C. L. N. Ruggles (Ed.), New York, Springer Science+Business Media, 1163-1168.

Esteban, C. and Benítez de Lugo Enrich, L. (2016) Orientaciones astronómicas en el *oppidum* oretano del Cerro de las Cabezas (Valdepeñas, Ciudad Real), *Trabajos de Prehistoria*, in press.

Esteban, C. and Cortell Pérez, E. (1997) Consideraciones arqueoastronómicas sobre el santuario ibérico de La Serreta. *Recerques del Museu d'Alcoi*, 6, 131-140.

Esteban, C. and Moret, S. (2006) Ciclos de tiempo en la cultura ibérica: La orientación astronómica en el templo del Tossal de Sant Miquel de Llíria. *Trabajos de Prehistoria*, Vol. 63, no. 1, 167–178.

Esteban, C. and Ocharan Ibarra, J. A. (2016) Winter Solstice at the Iberian Cave-Sanctuary of La Nariz. In *SEAC 2014 Proceedings: The Materiality of the Sky*, F. Silva, K. Malville, T. Lomsdalen and F. Ventura (eds.), Ceredigion, Sophia Centre Press, in press.

Esteban, C., Rísquez, C. and Rueda, C. (2014) An Evanescent Vision of the Sacred? The Equinoctial Sun at the Iberian Sanctuary of Castellar. *Mediterranean Archaeology and Archaeometry*, Vol. 14, no. 3, 99-106.

Harrison, R. J. (1988) *Spain at the Dawn of History. Iberians, Phoenicians and Greeks*, London, Thames and Hudson.

López Pardo, F. (2006) La torre de las almas. Un recorrido por los mitos y creencias del mundo fenicio y orientalizante a través del monumento de Pozo Moro, *Gerión*, Anejo X, Madrid, Universidad Complutense de Madrid.

Moneo, T. (2003) *Religio Iberica. Santuarios, ritos y divinidades (siglos VII-I a.C.)*, Madrid, Real Academia de la Historia.

Moratalla, J., Marcos, A. and Gailledra, E. (2014) Estratigrafía de cada uno de los sectores. In *Villajoyosa Antique (Alicante, Espagne). Territoire et topographie. Le sanctuaire de La Malladeta*, P. Rouillard, A. Espinosa and J. Moratalla (eds.), Madrid, Casa de Velázquez, 49-93.

Morena López, J.A. and Abril Hernández, J.M. (2013) Estudio arqueoastronómico del santuario ibero-romano de Torreparedones (Baena, Córdoba). *Cuadernos de Prehistoria y Arqueología de la Universidad de Granada*, 23, 293-321.

Pérez Ballester, J. and Borredá Mejías, R. (2004) La Carraposa (Rotglá i Corbera - Llanera de Ranes). Un lugar de culto ibérico en el Valle del Canyoles (La Costera, València). *Madrider Mitteilungen*, 45, 274-320.

Prados Martínez, F. (2008) *Arquitectura púnica. Los monumentos funerarios*, Anejos de Archivo Español de Arqueología, XLIV, Madrid.

Ribichini, S. and Xella, P. (1994) *La religione fenicia e punica in Italia*, Roma, Libreria Dello Stato.

Rueda, C. (2011) *Territorio, culto e iconografía en los santuarios iberos del Alto Guadalquivir (ss. IV a.n.e.-I d.n.e.)*, Textos CAAI nº 3, Jaén, Universidad de Jaén.

Ruggles, C.L.N. (1997) Whose equinox? *Archaeoastronomy (Supplement to the Journal for the History of Astronomy)*, 22, 45-49.

Ruggles, C.L.N. (1999) *Astronomy in Prehistoric Britain and Ireland*, New Haven and London, Yale University Press.

Ruiz de Arbulo, J. (2000) El papel de los santuarios en la colonización fenicia y griega en la Península Ibérica. In *Santuarios fenicio-púnicos en Iberia y su influencia en los cultos indígenas*, B. Costa and J.H. Hernández, XIV Jornadas de Arqueología Fenicio-Púnica, Eivissa, 9-56.

Thom, A. (1967) *Megalithic Sites in Britain*, Oxford, Oxford University Press.

Stylow, A.U. and Mayer M. (1996) Los tituli de la Cueva Negra. Lectura y comentarios literario y paleográfico. *El balneario romano y la Cueva Negra de Fortuna (Murcia). Antigüedad y Cristianismo*, XIII, 367-406.

Velázquez, I. and Espigares A. (1996) La traducción al castellano de los textos de la Cueva Negra. *El balneario romano y la Cueva Negra de Fortuna (Murcia). Antigüedad y Cristianismo*, XIII, 453-475.

10

A DIACHRONIC ANALYSIS OF ORIENTATION OF SACRED PRECINCTS ACROSS JORDAN

A. César González-García*[1], Juan A. Belmonte[2] and Andrea Polcaro[3]

[1]*Instituto de Ciencias del Patrimonio, Incipit-CSIC, Santiago, Spain*
[2]*Instituto de Astrofísica de Canarias, Tenerife, Spain*
[3]*University of Perugia, Italy*

Corresponding author: (cesar.gonzalez-garcia@incipit.csic.es)

ABSTRACT

Located in the southern part of the Levant, the territory of present day Jordan functioned as a cultural crossroad since times immemorial. The state of preservation of the different remains and the quantity of the different stages makes Jordan a perfect area to perform a study on the evolution along time of the orientation customs. During our field campaign in December 2011, we collected data on a number of different cultic structures throughout the country, from Bronze Age temples, to megalithic monuments, Iron Age Nabataean monuments, Roman and Hellenistic cities and temples and Byzantine churches with a small number of Muslim mosques. This sample of over 300 structures of different periods allows a diachronic comparison of the orientation of cultic buildings for the last 5000 years in this area of the Levant. We find a consistency to orientate the cultic structures in accordance to similar areas of the horizon. This similarity is striking when comparing the megalithic monuments found along the whole country with the Nabataean monuments of nearly 2000 years later. This consistency appears despite the chronological gap, the cultural differences and also possibly the different ethnic components of these societies. The consistency seems to be broken after the Roman conquest, especially with the introduction of Christianity and definitely after the expansion of Islam. A comparison with other neighboring areas of the Near East is sketched in order to compare with contemporary monuments for each epoch.

KEYWORDS: Bronze Age Jordan, Iron Age Jordan, Nabataeans, Diachronic analysis, Temporality

1. INTRODUCTION

The study of the orientation of cultic structures provides relevant information on the spatial concepts of the societies that designed and built them. This is obviously part of a wider field of study, that of Landscape Archaeology, where the location and mutual relations of the different archaeological remains and natural features of a given area provide valuable input to testify the territory apprehension and space and time incorporation of a given society within its cultural mind-set (see, e.g. Ingold 1992; Tilley 1994; Criado-Boado 2012).

The orientation of cultic buildings has been demonstrated to be linked in many instances to the manifestation of heavenly bodies (see, for instance, several works at the Handbook for Archaeoastronomy and Ethnoastronomy, Ruggles 2015). Orientation, in this sense is somehow related to the spiritual regime but leaves a define fingerprint in the archaeological record that can be measured and compared for different epochs in a given region, as any other item in that record.

The present paper focuses on how the sky was materialized in the orientation of the cultic buildings of different epochs in Jordan. To do so, several different religious and funeral buildings are analyzed. The aim of the paper therefore is to investigate how the cultural and social changes were reflected in the orientation customs of their buildings.

The basic hypothesis behind that objective is that the architectural remains as part of the material record from those societies and thus socially constructed, can provide information on different aspects of the culture that built them. In particular, the location and orientation of the building is directly related with the spatial conceptions of such societies. In those cases where orientation can be safely related to the rise or set of heavenly bodies the rhythmic cycles of those bodies could be connected to specific moments and concepts of temporality in those societies. Then, orientation and perhaps also location could be connected to temporality. The concept of time, as socially understood is a cultural fingerprint (Gell 1992). In particular, given the material record we manage, such times are most probably related to cultic practices, and therefore to the religious sphere. Thus studying the persistence or change of orientation traditions we may follow the footsteps of possible changes and persistence in the religious sphere of a certain area through time.

The approach developed here has been tested and used before. On a pioneering work, González-García & Costa Ferrer (2011) studied the diachronic evolution of the orientation of a large number of monuments in the Spanish town of Mérida. This city is known for housing some of the most remarkable Roman remains in the Iberian Peninsula. Mérida was founded as a Roman settlement of army troops in the region after the conquest of the Iberian NW. However, the comparison showed that the orientation of the Roman city is not alien to what is was customary in the area centuries before its foundation. In particular this exercise could identify two moments of clear break with previous customs, being the clearest one the introduction of Christianity.

Further on, the method was applied to study the alleged persistence of orientation customs in the megalithic monuments of Sardinia (Hoskin 2001; González-García et al 2014). There, the comparative approach was expanded to include several statistical tools. The results verified the proposed shift from homogeneity of orientations along the island at an early stage in the late Neolithic, represented by the Domus de Janas, to a change in orientation customs after the introduction of classical megalithic monuments (dolmens), possibly following commercial routes from the northern and western shores of the Mediterranean. Thereafter, the orientation custom seems to divide the island into two at latitude of 40°. Such systematic division appears even until the Phoenician colonization of the southern parts of the island.

Finally, González-García & Belmonte (2014) devised a diachronic comparison of the orientation of cultic structures throughout the Mediterranean from the Bronze Age until Roman times. They indicate that in the Mediterranean Iron Age there seems to be basically three orientation customs: one mostly related to equinoxes and solstices, with a predominance of the last. Such is mainly present in the Near East. A second group, with mostly Greek temples, included monuments with a larger appearance of equinoctial orientations (almost no solstice) and a third with orientations mostly in the meridian line, mostly towards south, appeared in the central Mediterranean. Some of the orientation groups seem to have an origin in certain areas in earlier times, like the solstitial group in Egypt and Anatolia, or the equinoctial on the Minoan cultic areas. In this last exercise, the authors included the Nabataean temples, and it appeared that they belonged to the first family that was apparently common in most of the Near East.

One of the motivations for the present study is therefore see if such scheme was present in previous epochs of Jordanian history. The territory occupied by present day Jordan has been a cultural cross-road and the cradle of different societies throughout history. Such fact resulted in the numerous and well preserved archaeological remains of the country (MacDonald, Adams and Bienkowski 2001).

During our field campaign in the area, at the end of December 2011, and in coincidence with Winter Solstice we collected measurements on over 300 cultic structures in the region and from different epochs. Ranging from late Neolithic, to Bronze Age (see Polcaro et al. 2013), megalithic monuments (Belmonte et al. 2013a), Iron Age Nabataean monuments (Belmonte et al. 2013b), Roman and Hellenistic cities (Rodríguez-Antón et al. this volume) and temples and Byzantine churches (González-García et al. in preparation) and a small number of Muslim mosques.

Data were collected using precision compasses and clinometers and corrected for magnetic declinations. Magnetic alterations are not expected in the Nabataean territory, where most of the terrain is limestone or sandstone. The individual measurements have an average error of ¼° in azimuth and ½° in horizon altitude, which translates into an error ~¾° in declination. The declination histograms used hereafter were calculated using a density distribution with an Epanechnikov kernel with a pass band of 1½°

In the following we show the orientation customs of the different structures studied so far trying to unveil the persistence and change along history. Section 2 includes a brief summary of the archaeology of the different sites explored in this campaign and provides the archaeoastronomical data, 3 presents our comparison and section 4 provides the discussion and conclusion of such analysis.

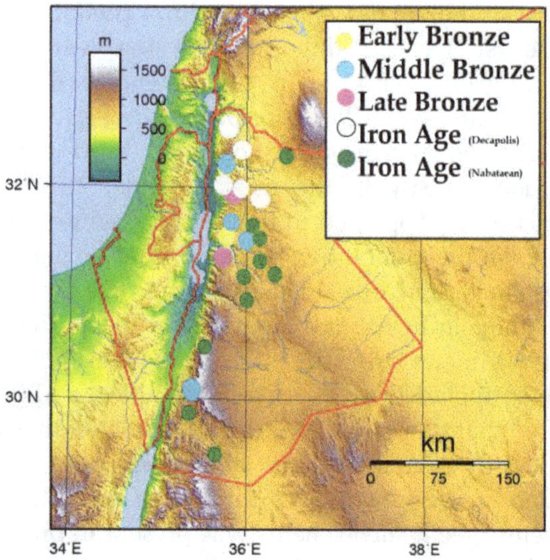

Figure 1. Location of the different sites referred to in the text. The colour coding presents the different periods (see legend). Besides these monuments, byzantine churches were measured in several sites widely similar in distribution to those of the Iron Age sites.

2. ARCHAEOLOGY AND ARCHAEO-ASTRONOMY OF JORDANIAN SITES

2.1. *Neolithic-Chalcolithic-EBA*

Neolithic and Chalcolithic sites are frequent in Jordanian lands, there are however few measurable remains. The earliest site we could measure was the Early Bronze Age (EBA) of Bab edh-Dhra (Southern Gor Region). The archaeological remains attest what has been interpreted as a pastoralist nomadic occupation for EBA IA (attested by a large funerary area) and a first settlement for the EBA IB - EBA II phases (strata III). The next phase, dated to the EBA III phase (strata II), witness the erection of large city walls encircling a large settlement. At the end of this phase there are remains of destruction and a different occupation for the last phase (EBA IV). The main temple, dated from 2900 to 2400 BC, has a typical plan of the Bronze Age, with a single rectangular pillar room and the entrance on the long side (Rast-Schaub 2003: 157-166). As many Chalcolithic and Bronze Age temples of Southern Levant, the temple of Bab edh-Dhra had an external circular altar in front of the door, used for animal sacrifices (the most impressive case is the round altar of Megiddo; Loud 1948). The temple is the last phase of different layers, testifying the first construction of the sanctuary at least in the 3100 BC, when the first permanent settlement was built on the Eastern side of the Dead Sea. We do not know the deity venerated in this temple, but from the decorated vessels and other templar paraphernalia recovered in the excavation area it is possible to presume a cult of a main Levantine deity similar to the one worshipped in other temples of Jordan in the Early Bronze Age (in particular, the cultic vessels decorated with applications of serpents recovered also in the sacred areas of Khirbet Zeraqon or Jebel al-Mutawwaq; Ajlouny et al. 2001; Polcaro et al. 2014).

The temple has an orientation compatible with the equinox (see Table I and Fig. 2). That orientation was also the orientation of the main entrance to the city walls. The altar in front of the temple had a small angle with respect to the orientation of the temple.

2.2. *EBA-Middle Bronze Age*

This period of Jordanian history is characterized by the appearance of large megalithic fields (see e.g., Scheltema 2008). Dolmens appear for the first time in the Southern Levant at the end of the Late Chalcolithic Period (c. 3800/3500 BC based on C14 date from the abandonment phase of the Tuleilat al Ghassul sacred area). During the Early Bronze Age I (EBA IB which ended between c. 3000 and 2900 BC) the pastoral and agricultural communities of the

Transjordan Plateau used these megalithic monuments for funerary purposes (large family tombs for primary or secondary depositions). This period was a critical moment in the urbanization process in the Levant, as mentioned earlier. Some populations settled in more fertile areas practicing horticulture and seasonal agriculture; meanwhile, especially in the semi-desert areas, there prevailed a pastoral style of life.

Table I. Measurements of several individual elements commented upon in the text. The first three lines are Bronze Age data. The next nine lines give the data for the Hellenistic/Roman temples and the last two are the orientation of the mosques. For the collective measurements of dolmens, the Nabataean structures, Roman cities and Byzantine churches see the references in the text. The columns provide the site or structure measured, the azimuth (A) and horizon altitude (h) measured with a tandem (error estimated to be ½°), and the calculated astronomical declination (δ).

Site	A	h	δ	Structure
Bab edh-Dhra	269½	0¾	−1	Temple
	261		−7¼	Altar
Deir Alla	91½	2	−0¼	Sanctuary
Amman	76	−0½	11¼	Hercules temple
Gadara	356	−0½	56	Temple
	195½	5½	−49½	Theatre temple
Madaba	216	blocked	−44	Roman temple
Pella	277½	−1	5¼	Acropolis temple
	340¾	0½	52¾	Small temple
Jerash	54	3½	31½	Zeus temple
	55½	3½	30½	Altar
	115½	1	−20½	Artemis temple
Amman	179¾	2½	−55¾	Umayad Mosque
Pella	163½	3½	−51	Mamluk Mosque

Dolmens are funerary monuments marking the landscape, both in fertile and in semi-desert areas, following the course of the seasonal rivers running across Jordan. The material culture from the dolmen fields and landscape analysis of the phenomenon indicate these monuments represent the creation of a communal ideology, connecting the ancestor cult with the territory, probably very useful as a social and political binding agent between different communities (Polcaro 2013).

In our campaign in 2011 we could measure the orientation of 185 dolmens in five necropolis: Juffain, Damiyeh, Wadi Jedid, Al Mureighat and Tawaniyeh (Belmonte et al. 2013a). The local sacred topography seems to play a relevant role in the location and orientation of several dolmens at some of the sites, with double topographic and astronomical orientations, and a tendency of several dolmens to be facing each other. But the basic result was that the overall distribution of the dolmens seemed astronomical (see Figure 3a). There are basically four main orientations. The largest concentration would be towards north, while the other three tend to be around the extreme positions of the sun and the moon and the equinoxes. Indeed, the astral relationships seem important, and could tell us something on the belief system of these communities.

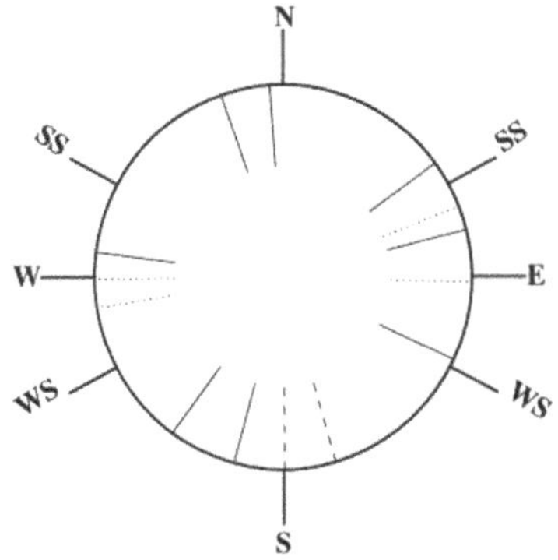

Figure 2. Orientation diagram for the temples in Table I. The circle indicates the horizon, and the short strokes outside label the cardinal directions and the solar extremes, marked as SS for summer solstice and WS for winter solstice. Each temple orientation is indicated by a short stroke inside the circle. Bronze Age sanctuaries are indicated by a dotted line. Hellenistic-Roman temples by a solid line and mosques by dashed strokes.

2.3. *Middle and Late Bronze Age*

We could measure two structures from this epoch: the sanctuary at Tell Deir Alla and the Baal temple at Pella.

Pella is famous for its Roman remains, but in the lower acropolis there are the remains of a large sacred building of the XIX century BC. This Middle Bronze Age temple was continuously used and restored until the Iron Age. The temple has a typical Southern Levant Bronze Age architectonic style with a single elongated room with the entrance on the short side flanked by two antis. The remains found in the temple point towards Baal as the possible deity worshiped there (Polcaro, González-García and Belmonte 2013).

The orientation of the building, consistent with summer solstice sunrise a moment that in Near East cultures was linked with the festivity of dying gods associated with harvesting and a moment of confluence of the underworld and the upper world (Polcaro et al. 2013).

Tell Deir Alla is famous for the collection of clay tables that had been recovered from the different rooms of the so called sanctuary. The sanctuary, dated from the XVI century BC to the XII century BC, consists on several rooms and a cella (Franken 1992). The complex was built on a high platform and the cella was located another meter above the other rooms. West of the rectangular cella, there was the complex of storerooms; in one of them were recovered the 12 clay tablets indicating that a temple archive had been recovered (Strange 2008: 297). During its last phase, the building was destroyed by an earthquake (Franken 1992). The entrance to the Templar complex has an almost perfect equinoctial orientation, in this case towards east (Table I).

2.4. *Iron Age: Hellenistic Period*

The Iron Age, from the turn of the first millennium is characterized by the establishment of several kingdoms in the area, more specifically those of Ammon, Moab and Edom, and the expansion of other regional powers such as the Assyrian, Neo-Babylonian and Persian empires. We could not measure any structures from these times, despite the fact that at the archaeological excavations of some of the temples of later times archaeologists testify earlier Iron Age occupations (see for instance, Najjar 1993, for the Hercules temple in Amman).

From the second half of the millennium, the whole Levant suffered a process of cultural influence from Greece especially after the expansion of the Macedonian troops of Alexander and the stablishing of the several Hellenistic kingdoms (Schmid 2008: 353). The northern part of Jordan, along with neighbouring areas in Syria and Israel were part of the Decapolis, a region of difficult definition, perhaps some sort of federation of ten cities with strong Greek influence (Schmid 2008; 353-360). We could measure eight temples from this period, possibly all of them re-erected at later Roman times, at the sites of Philadelphia, Jerash, Gadara, Abila and Pella. The number of temples is admittedly low but there seems to be no clear pattern (if any at all). It could be interesting, though, to comment some specific cases. Jerash is one of the best-preserved Hellenistic and Roman cities in Jordan, housing two magnificent temples that dominated the ancient city. The Zeus sanctuary (Raja 2013), on top of a hill, had several different phases and it is interesting that there seem to be a change in orientation. Its oldest phase presents an orientation whose perpendicular could be consistent with winter solstice sunrise. However, both the main axis of the temple and the altar located in front of it have an orientation that could be related either with some bright stars (such as Arcturus, Pollux or even Algol), or perhaps with the northern major lunistice. It is interesting that this temple could have later integrated the Roman imperial worship. The other temple of the city, the sanctuary of Artemis could be a different case. It appears that it was originally built as an extra mural complex that later became part of the Roman town (Raja 2015). It has an orientation that would not match any bright star at the time, but could be consistent with the rise of the lunar disk at the minor southern lunistice. It would therefore be interesting that the two main temples of the city, supposedly built at different times, could have been related with the moon.

2.5. *Iron Age: the Nabataeans*

The Nabataeans enter history in this region after the victory of Obodas I over the Seleucid Empire in the first century BC. Possibly originally a nomadic Arab people, they settled in the western part of present day Jordan and over the next couple of centuries controlled the caravan routes towards Arabia. They built several cities and temples along the country being those in Petra the most famous and best preserved (see, e.g. Belmonte et al 2013b). Their religion possibly with an animistic character had a strong astral flavour. The images of the gods were kept in baetylus stones. The principal male divinity was the god Dushara or Dushares, very probably an astral god with a hypothetical lunar or solar character. The main female divinity is much discussed as she is named Allat in the north while in the south appears as Al-Uzza. They could be identified both with Venus, perhaps in her aspect as evening star.

We could measure 50 sacred precincts throughout the country. The results are shown in Figure 3b. The main peaks in the histogram could be identified with those yielded by astral bodies, most notably the sun and perhaps the moon. We also witnessed and put forwards several illumination events at the interior of some of the most remarkable carved structures in Petra (Belmonte et al 2013b).

2.6. *Roman and Byzantine remains*

After the annexation of the Nabataean kingdom by Emperor Trajan on 104 AD, the country became part of the Roman province of Arabia. The *via nova Traiana* articulated the country from south to north, bringing the products of the Red Sea towards the Mediterranean. Beside, a number of military camps and garrisons were built at the *limes arabicus* to protect those cities from the nomadic tribes. We could measure the orientation of several cities and camps. They are not treated here (see however, Rodríguez Anton et al. this volume). The Roman temples of this time are treated under the Hellenistic part.

After the IV century AD, a large number of churches were built along those same cities. According to the church fathers, the priest and the community must face east when praying, and this premise has been mostly investigated in Mediterranean countries but little research has been devoted to such matters in the Holy Land (González-García 2015). We could measure 66 churches, concentrated mainly in several towns, as well as some eremitic monasteries (González-García et al. in prep.). The results are shown in Figure 3c. All churches but two are inside the solar range. In particular it is interesting to note that most churches follow the urban grid of the cities they were built upon. This is at odds to what we have observed customary in the Christian west (for instance, in Mérida we see that the urban grid is broken by the eastern orientation of the churches very early on, González-García and Costa Ferrer 2011), but apparently it could be common in the East (see Dallas 2015 for the orientation of churches in Thessaloniki).

2.7. *Muslim mosques*

After the Yarmuk battle in 636 Jordan became part of the Muslim world. Islam expanded with its own particular orientation customs so that mosques must be orientated to the Sacred Mosque in Mecca (i.e. the Ka'aba) according to Quran (II, 19) prescriptions. However, this was not so easy to achieve and different approaches to the problem were accepted (Rius, 2000), including the possibility of orientating oratories like Ka'aba (Hawkins and King, 1982) instead than to Ka'aba.

We could measure two mosques from ancient times. They are the one at the acropolis of Amman, from Umayyad times, and the second from Pella, from the Mamluk period (Table I, Fig. 2). It is interesting that the first has an orientation almost perfectly towards due south, while the second is deflected several degrees from such orientation.

Perhaps an explanation for such difference may come from the periods of their construction. The first is from Umayyad period, an epoch when Muslim power resided in Damascus, in modern Syria. The people from Syria were known to orientate their mosques almost due south as this was an approximate orientation for the direction towards Mecca in Damascus and also of the rising position of Canopus ("like Ka'aba", King, 1995). However, the Mamluk domain was mostly connected with Egypt and Cairo, where the astronomical determination of the Qibla was almost a rule in that period. If we follow such prescription, the mosque at Pella should be facing an azimuth of 160°, in good agreement with its actual orientation.

3. A DYACHRONIC COMPARISON

3.1. *Within Jordan*

One of the most striking results is the consistency in orientations. If we directly compare the orientations in Figure 3, Bronze Age temples and dolmens share broad orientations towards the extreme positions of sun and moon and towards cardinal directions. This is strongly maintained in the Iron Age, especially during Nabataean times. In particular it is highly striking the similarity between the two histograms of the megalithic monuments and the Nabataean temples and shrines. Even more so if we consider that there is a span of more than 15 centuries in between!

The Hellenistic and Roman temples however do seem to break such consistency. This should not be surprising, as a large fraction of the population of those cities of the Decapolis were under a strong Hellenic cultural influence or were directly of Greek origin.

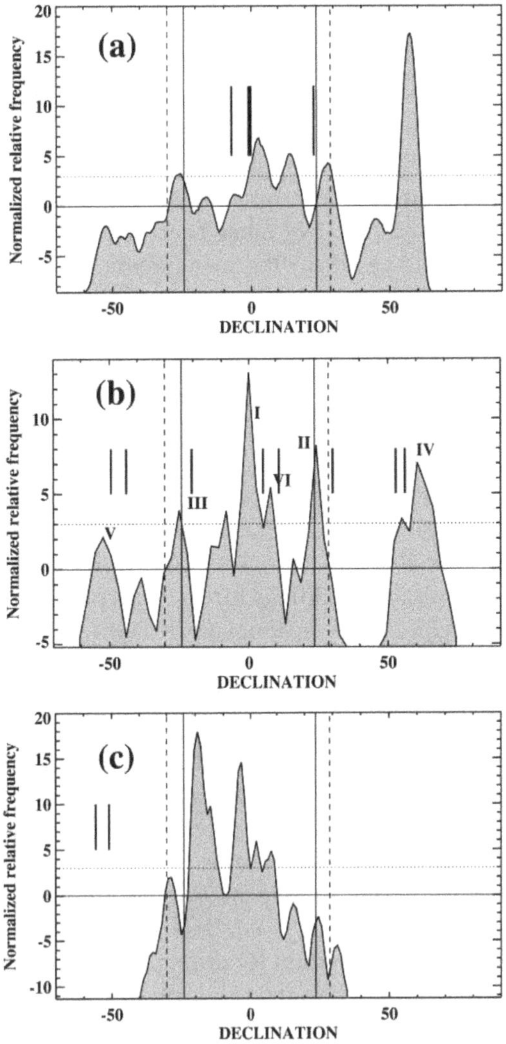

Figure 3. Declination histograms for three periods of Jordanian history. Vertical solid lines indicate the extremes of the solar range -the solstices- and dashed lines the extremes of the lunar movements. The horizontal dotted lines give the 3sigma value for each histogram. (a) Histogram for the Jordanian Bronze Age. The shaded curvigram indicates the dolmen KDE. The short vertical strokes are the BA temples in Table I plus the one at Pella. (b) Histogram for Nabataean monuments. The Roman numerals indicate the concentrations with significant astronomical targets. The vertical strokes are the 8 Hellenistic/Roman temples presented in Table I. (c) Histogram for Byzantine churches in Jordan. The two short vertical strokes are the orientation of the mosques indicated in Table I.

The Byzantine churches do not follow any of the above patterns, with a larger tendency to be oriented according to the Roman city grids, and mostly inside the limits of the solar movements.

Finally the Muslim mosques tend to be oriented towards south, something practically alien to Jordan until this period, with just some exceptions in the Nabataean period (Belmonte et al. 2013b).

We thus find that persistence in orientation is a key feature of Jordan orientation, especially during Bronze and Iron ages. This is true that there could be a process of 'definition' of the targets: while in Bronze age megalithic monuments the orientation peaks seem rather broad and thus the orientations appear quite diffuse, these are better defined in the case of the Nabataean monuments. Such process has been suggested before for the Mediterranean (González-García 2013). We must bear in mind that this might be the result of the measurement process itself: the dolmens are rather small and thus the orientation might be less well defined than in the larger Nabataean temples. However, one could speculate if such could also have been the intention of the builders: to have a less defined orientation in the megaliths, where multiple relations appear, both topographic and astronomical with a much better one in the temples if only - or mainly – the astronomical target was the key. A way to achieve such is by erecting suitable buildings for such purpose.

The break of this custom comes first with an alien culture: the Hellenistic world and later with the Roman expansion. The pattern observed in the Hellenistic temples could be consistent with that found in other areas of the Greek world at the time (see, e.g. Boutsikas 2009; González-García and Belmonte 2014). Such break is further confirmed by the orientation of churches and mosques.

Does this exercise help us in determining the intent of the Bronze Age builders of megaliths? We could try an explanation for dolmens and the Bronze Age temples based on the similarities found with the Nabataean temples. For the Pella temple we observed that the orientation towards summer solstice could be connected with the death of a fertility god such as Baal (Polcaro et al. 2013). For Petra we saw that such kind of god was reborn at winter solstice (Belmonte et al. 2013b). In any of the two cases we see that the solstices are clearly a time related to death and the afterlife, resurrection and the end-beginning of cycles. This could also have been the case in Bronze Age. Dolmens are often located in hills and in many occasions overlooking a central structure that might have acted as a sanctuary or sacred area as in Al-Mureyghat. The central place is surrounded by liminal structures, boundaries with the after-world and the afterlife, places of ending and beginning, in close resemblance of the solstices. The centrality of some of those areas, though, could be indicated by the equinoctial/cardinal orientations of the temples perhaps as a way to connect with the cardinal directions and the ordered structuring of the world.

3.2. *Outside Jordan*

Figure 4 presents the orientation of several cultic structures in different areas of the Near East that have been collected from our previous works or from the literature. First we compare for the Early and Middle Bronze Age. We have data from Egyptian temples (Belmonte et al. et al. 2009), megalithic dolmens from the Golan Heights (González-García & Belmonte 2010) and the megalithic funerary structures from the land of Magan (present day UAE and Oman; Belmonte & González-García 2014). For all the four sites we observe that the equinoctial orientations are important. This is specially so for Egypt, where it is by far the most frequent orientation. We must remember all the mythology written in the Pyramid Texts and other sources regarding the journey of the sun and its resurrection in the East. This could, perhaps also explain the equinoctial orientation for instance in Tell Deir Alla, and the group found in the Jordanian and Golan megaliths. Magan presents also a peak that might be called equinoctial, but there cardinal orientations seem to be preferred, especially towards south.

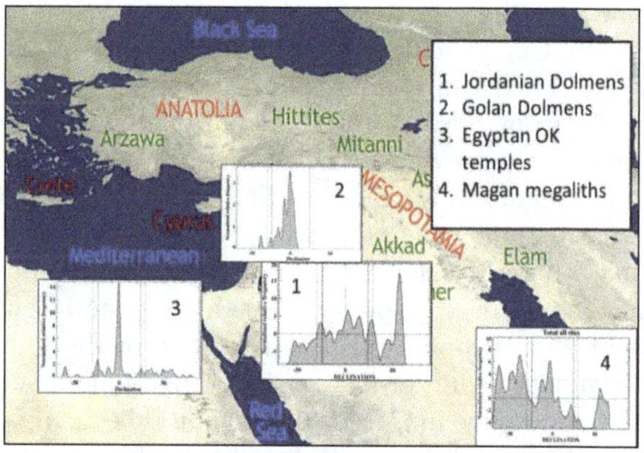

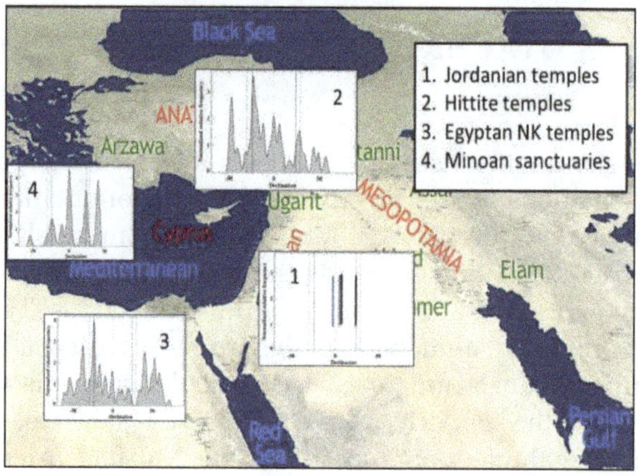

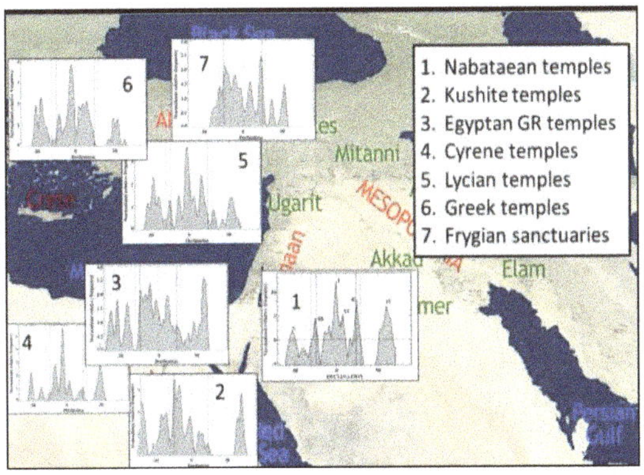

Figure 4. Comparison of the orientation of the different cultic structures in the Near East. Top panel includes data for the early Bronze Age. The middle panel includes the data for the middle and late Bronze Age. Lower panel includes the data for the Iron Age. For details see text.

Late Bronze Age presents results from Egypt (Belmonte et al. 2009), the Minoans (Henriksson & Blomberg 2008) and the Hittites (González-García & Belmonte 2011). We have included here the Jordanian temples from these periods for comparison. Egypt and the Hittites present a striking similarity that has been indicated before: both had religions whose deities presented a strong solar character, and both had the largest concentration of cultic orientations towards winter solstice.

There are other directions of interest such as the equinoxes and perhaps some stellar orientations as well. This is in contrast with Minoan sites, where the equinox is by far the most important orientation, perhaps - as it has been indicated - in relation to a luni-solar calendar. There is also a relation to summer solstice. In this sense the temples in Jordan do seem closer to the Minoan tradition, although the number is so low that no conclusive trend could be indicated here. In fact, given the tendencies observed for instance in the dolmens, the results for the temples are fully compatible with those results.

We have much more results to compare with for the Iron Age. There we have measurements from Egypt and Kush (Belmonte et al. 2010), Cyrene, Lycia (González-García & Belmonte 2014) and Phrygia (González-García & Belmonte 2011) and the Greek temples (Boutsikas 2009). Greek, Lycian and Cyrenean temples do present a similarity already put forward before (González-García & Belmonte 2014).

This is based on a concentration of orientation in the cardinal directions and the absence of significant numbers of temples towards the solstices. Phrygian temples present concentrations towards the solstices but there seems to be an absence towards the equinoxes, something that is observed also in the groups of temples from Kush and Egypt.

Interestingly, Nabataean temples do present characteristics from both groups. We cannot ascertain if such a result is the outcome of an aboriginal evolution or the syncretism of both Near Eastern and Hellenistic traditions. The fact that we had the clear similarity with previous orientations in Jordan may prompt us to favour the first choice but the second should not be overlooked. A third factor is that the influences from both sources could have been well received by a population already inclined to orient their cultic structures similarly to those from abroad.

ACKNOWLEDGEMENTS

This work was partially financed under the framework of the project P31079 "Arqueoastronomía" of the IAC and AYA2011-26759 "Orientatio ad Sidera III" of the Spanish MINECO. ACGG is a Ramón y Cajal fellow of the Spanish MINECO.

REFERENCES

Ajlouny, F., Douglas, K., Khrisat, B. (2011) Spatial Distribution of the Early Bronze Age Figurative Pieces from Khirbet ez-Zeraqon and its Religious Aspects, *Ancient Near Eastern Studies*, 48, 88-125.

Belmonte, J.A., González-García, A.C., Polcaro, A., 2013a, Light and Shadows over Petra: Astronomy and landscape in Nabataean lands, *Nexus Network Journal*, 15, 3, 487-501.

Belmonte, J.A., González-García, A.C., Polcaro, A., 2013b, On the Orientation of Megalithic Monuments of the Transjordan Plateau: New Clues for an astronomical interpretation, *Journal for the History of Astronomy*, 44, 429-455.

Belmonte, J.A., Shaltout M. and Fekri M. (2009) *Astronomy, landscape and symbolism: a study on the orientations of ancient Egyptian temples*. In J.A. Belmonte and M. Shaltout (eds.), In search of cosmic order, selected essays on Egyptian archaeoastrono-my. Supreme Council of Antiquities Press, Cairo, Egypt. 213-84.

Belmonte, J.A., Fekri, M., Abdel-Hadi, Y.A., Shaltout, M. & González-García, A.C. (2010) On the orientation of ancient Egyptian temples: (5) testing the theory in Middle Egypt and Sudan. *Journal for the History of Astronomy*, 41, 65-93.

Boutsikas, E. (2008) Placing Greek Temples: An Archaeoastronomical Study of the Orien-tation of Ancient Greek Religious Structures. *Archaeoastronomy, Journal of Astronomy in Culture* 21, 4-19.

Criado-Boado, F. (2012) *Arqueológicas: La Razón Perdida*. Bellaterra, Barcelona.

Dallas, T. (2015) On the Orientation of Byzantine Churches in Thessalonike. *MAA*, 15 (3), 213-224.

Franken, H.J. (1992) *Excavations at Tell Deir Alla: The Late Bronze Age Sanctuary*. Peeters, Leuven, BE.

Gell, A. (1992) *The anthropology of time*. Berg, London, UK.

González-García, A.C. and Belmonte, J.A. (2010) Statistical analysis of megalithic toms orientations in the Iberian Peninsula and neighbouring regions. *Journal for the History of Astronomy* 41, 225-38.

González-García, A.C. and Belmonte, J.A. (2011) Thinking Hattusha: Astronomy and Landscape in the Hittite Lands. *Journal for the History of Astronomy* 42, 461-494.

González-García, A.C. and Costa Ferrer, L. (2011) *The diachronic study of orientations: Mérida, a case study*. In C.L.N. Ruggles (ed.), Archaeoastronomy and Ethnoastronomy: Building bridges Between Cultures, Cambridge University Press, Cambridge, U.K., 374-381.

Hawkins, G. S. and King, D. A.(1982) On the Orientation of the Ka'aba, Journal for the History of Astronomy, Vol. 13, 102-117.

Henriksson, G. and Blomberg, M. (2008) Orientations at the Minoan peak sanctuary on Philioremos near Gonies (Maleviziou), Crete. *Archaeologia Baltica* 10, 125-30.

Hoskin, M. (2001) *Tombs, temples and their orientations*. Ocarina Books, Bognor Regis, UK.

Ingold, T. (1993) The temporality of the Landscape. *World Archaeology*, 25, 2, 152-174.

King, D. (1995) Discussion Article: The Orientation of Medieval Islamic Religious Architecture and Cities, Journal for the History of Astronomy, Vol. 26, 253-274.

Loud, G. (1948) *Megiddo 2. Seasons of 1925-1939: Text and Plates*. Oriental Institute Publication 62, The University of Chicago Press, Chicago, US.

MacDonald, B., Adams, R. and Bienkowski, P. (2001) *The archaeology of Jordan*, Sheffiled, Sheffield Academic Press.

Najjar, M. (1993) Amman Citadel Temple of Hercules Excavations Preliminary Report. *Syria*, 70, 220-225.

Polcaro, A., Muniz, J., Alvarez, V., Mogliazza, S. (2014) Dolmen 317 and Its Hidden Burial: An Early Bronze Age I Megalithic Tomb from Jebel al-Mutawwaq (Jordan), *Bullettin of American Schools of Oriental Research*, 372, 1-17.

Polcaro, A. (2013) The Stone and the Landscape: the phenomenon of megalithic constructions in Jordan in the main historical context of Southern Levant at the beginning of the 3rd Millennium BC. In L. Bombardieri et al. (eds.), *Identity and Connectivity. Proceeding of the 16th Symposium On Mediterranean Archaeology, 1st-3rd March 2012 Florence, Italy*. BAR series 2581, Oxford, UK. 127-136.

Polcaro, A., González-García A.C., Belmonte, J.A., 2013, The orientation of the Bronze Age Temple of Pella, Jordan: the dying god Baal and the Rituals of the Summer Solstice, *Anthropological Notebooks*, XIX, Supplement, 481-492.

Raja, R. (2013) Changing Spaces and Shifting Attitudes: Revisiting the Sanctuary of Zeus in Gerasa. In T. Kaizer et al. (eds.), *Cities and Gods: Religious Space in Transition*. Leuven. 31–46.

Raja, R. (2015) Complex Sanctuaries in the Roman Period. In R. Raja and J. Rüpke (eds.), *A Companion to the Archaeology of Religion in the Ancient World*. Wiley Blackwell, Oxford. 307-319.

Rast, W.E. and Schaub, R.T. (2003) *Bab Edh-Dhra. Excavations at the Town Site (1975-1981)*. Eisenbrauns, Winona Lake, US.

Rius M. (2000) *La alquibla en al-Andalus y al-Magrib al-Aqsa*. Anuari de Filologia xxi, B-3. Universitat de Barcelona.

Ruggles, C. (2015) (Ed.) *Handbook of Archaeoastronomy and Ethnoastronomy*. Springer-Verlag, New York, US.

Scheltema, G. (2008) *Megalithic Jordan: An introduction and field guide*. Amman.

Schmid, S.G. (2008) The Hellenistic Period and the Nabateans. In R.B. Adams (ed.), *Jordan. An Archaeological Reading*. Equinox, London, UK. 353-411.

Strange, J. (2008) The Late Bronze Age. In R.B. Adams (ed.), *Jordan. An Archaeological Reading*. Equinox, London, UK. 281-310.

Tilley, C., 1994, *A Phenomenology of Landscape. Places, Paths and Monuments*. Oxford: Berg.

11

REFRACTION AND PRECISE LUNAR ALIGNMENTS

Thomas T. Gough

Independent Scholar, Parkhead Farmhouse, Ballindalloch, Scotland AB37 9BJ

**Corresponding author: (tt_gough@btinternet.com)*

ABSTRACT

During a major standstill the maximum declination of the moon at successive lunations varies by only a small amount. Observing the extreme north/south position of the moon would require an alignment of high precision. Alexander Thom believed that this had been done in the Early Bronze Age. Most of the claimed sites are in Scotland.

Atmospheric conditions, especially variation in temperature and pressure can cause variation in atmospheric refraction, particularly at very low altitude. Large variations have been found to be common in Chile and North America as result of which Bradley Schaefer has stated that the results claimed by Thom would not be possible. However there is no reason *per se* why atmospheric conditions in other parts of the world should be relevant to Scotland. Investigations in Scotland have found that refraction variation, even down to one degree altitude, is typically less than one minute of arc. It is concluded, therefore, that it is unreasonable to claim that refraction variation elsewhere would prevent the possible existence of precise lunar alignments in Scotland. Recent investigations by the present author, provides evidence for the possible existence of such alignments.

KEYWORDS: Low level refraction, Chile, Scotland, temperature gradient, temperature inversions.

1. INTRODUCTION

Schaefer and Liller (1990) analysed values for refraction variation at low altitude from a number of different locations in the Americas finding much larger values than expected. As a result they concluded that determination of accurate alignments was not possible; in particular the claim by Alexander Thom for the existence of precise lunar alignments. Their results are considered later following a discussion of the apparent movements of the moon.

The moon comes to its major standstill every 18.6 years. During the standstill the maximum monthly declinations, both north and south, lie in the region of about ± 29°, which is about five degrees greater than that of the sun at the solstices (Figure 1).

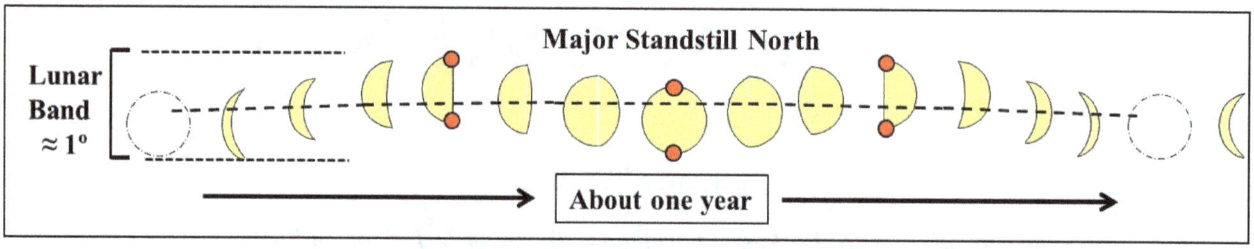

Figure 1. The extreme positions of the moon at a major standstill

The images in Figure 1 are one month apart. Between each, the moon moves to its southern maximum declination, where the lunar phase is opposite, before returning north. The reverse scrolling phase pattern continues in a snake-like pattern after the standstill. At the next major standstill the phase pattern is displaced about three months to the right.

During the standstill period the maximum declination of the moon, both north and south, lies within a range of about 20′ for about a year. Only the lunar positions marked by a red dot, when the moon is at the top or bottom of the 'wobble', are unique. These positions are the 'key' declinations. The moon at these key declinations is always at the quarters or full moon; or new moon which cannot be observed. The moon's movements, the Lunar Band and related matters are described by Gough (2016).

By using a suitable horizon feature as a foresight the 'key' lunar positions, when viewed from a marked backsight, could, in principle, be found. However given the small variations in declination, measurement of the order of ± 2′ of arc precision would be necessary. Alexander Thom believed that he had found a number of precise lunar alignments and deduced that they had been set up in the Early Bronze Age (Thom, 1971).

Light rays from celestial objects are refracted by the atmosphere, increasingly so as the horizon is approached. Tables of standard corrections to be applied are published. However these cannot allow for random deviations in refraction which can occur at very low altitude, and which are usually caused by variations of temperature and pressure. Such effects can lead to significant refraction deviations at altitudes below one degree and particularly so for negative altitudes.

Using the results of measurements made from sites in Chile together with a number from elsewhere, Bradley Schaefer claimed that the large deviations found in refraction near the horizon showed that Thom's results were not possible. He wrote "...the historically important claim by A. Thom that British megalithic sites were used as accurate lunar observatories is shown to be wrong because the needed accuracy is much greater than can be obtained for long averaging intervals." (Schaefer and Liller, 1990). The Chile results were based on timed observations of the setting sun from Cerro Tololo, 2215m, and Vina del Mar, 120m, both near the west coast of Chile. These were combined with observations from North America by other observers. The overall conclusion was that refraction deviation was much greater than previously thought, with a rms value of about 4′ of arc but with much larger values occurring. Deviations in refraction will result in errors in calculated declinations with corresponding errors in azimuth on the horizon (Sampson, 2000). Schaefer and Liller (1990, 804–5) argue that such large refraction deviations would prevent the claimed azimuth accuracy required for precise lunar alignments as proposed by Thom.

Exton (1992) expressed reservations concerning the validity of Schaefer and Lillers' assertions. He discussed how they had allowed the, not unexpected, very large refraction deviations observed from Cerro Tololo to dominate their findings. However, this has not prevented use of their results by others to doubt the possibility of significant precision in archaeoastronomy: "...recent work indicates that variations in atmospheric conditions can alter the apparent declination of an observed low-altitude object in Britain or Ireland by several arc minutes, and possibly by as much as half a degree. (Ref. Schaefer and Liller, 1990)." (Ruggles, 1999, 25).

2. ATMOSPHERIC REFRACTION

2.1. *REFRACTION MEASUREMENTS IN SCOTLAND AND THEIR IMPLICATION*

A fundamental question concerns the applicability of the above results to Scotland where the atmospheric conditions are likely to be very different from the regions considered by Schaefer and Liller. Five sets of observations of the setting sun from three different sites were made by the present author in 2008 to 2015. Timed theodolite measurements were made of the upper and lower limbs of the setting sun. Most were in the altitude range from 4° down to 1° altitude, but with one site near the sea to 0° altitude. The difference between the observed and the calculated refraction gives the deviation (Figure 2).

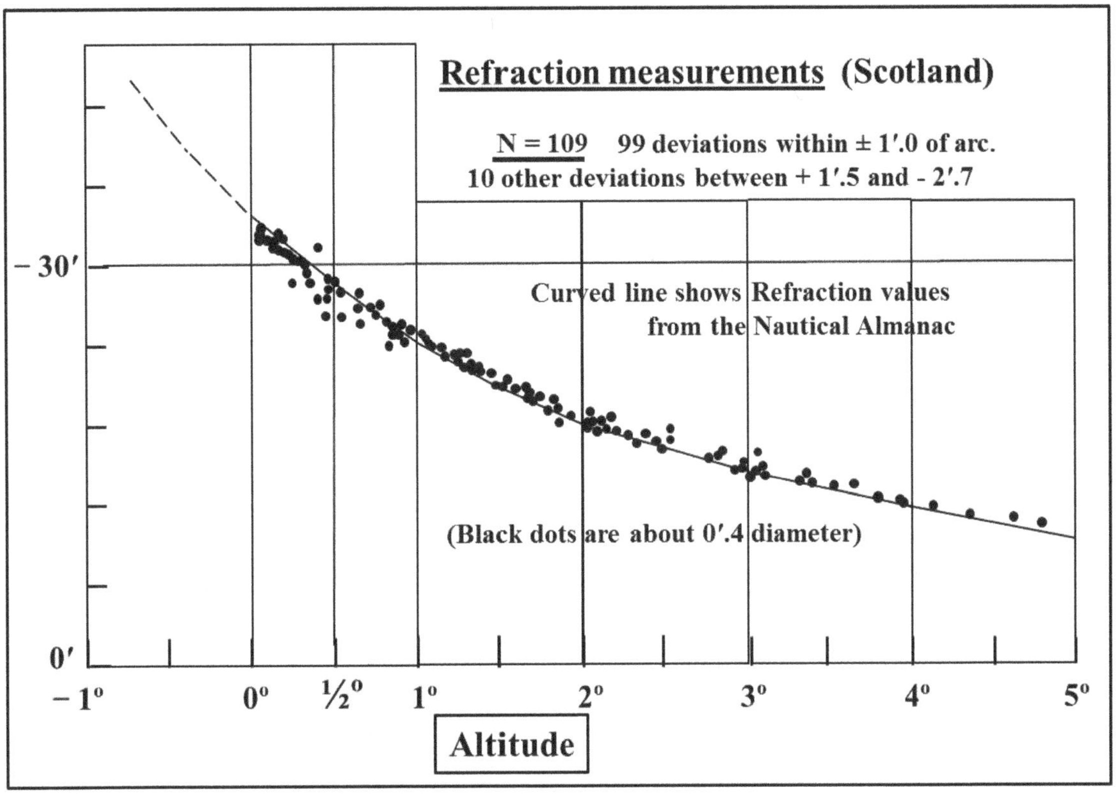

Figure 2. Refraction measurements in Scotland

The large deviations given by Schaefer and Liller, which are not supported by observations in Scotland, can be explained as follows:-

- They consider observations of the sun at the horizon or adjusted to the horizon.
- The Chile observations have negative altitude.
- The sightlines are generally long – especially so from Cerro Tololo.
- The west coast of Chile at the latitude of about 30° south commonly has a temperature inversion layer.
- Day/night temperature variations in the plains of Nebraska are likely to be significant (Schaefer and Liller, 1990, 801).

Schaefer acknowledged that the Chile observations were likely to be affected by temperature inversions. However he claimed that temperature inversions were ubiquitous "...Schaefer and Liller (1990) have shown that significant thermal inversions are ubiquitous and will drastically change the refraction near the horizon." (Schaefer, 1993, 157).

Such a claim, as a general statement, is questionable. In Scotland temperature inversions are not common. When they do occur it is usually on clear, calm nights and in hilly regions. The cold air falls down the hillsides to pool in the valley bottom leaving warmer air above. The inversions are almost always short lived and are easily dispersed by wind or sun.

2.2. *LOW ALTITUDE ATMOSPHERIC REFRACTION*

In order to avoid as far as possible atmospheric effects, astronomers usually observe celestial objects near culmination. Only with the relatively recent advent of archaeoastronomy, with observations commonly below 5° altitude and even down to 0° altitude, has an interest been taken in low level refraction. Alexander Thom investigated refraction at low altitude. From his home at an elevation of about

140m in south west Scotland there are good views of distant hills, mountains and two lighthouses. In the 1950s he made some 600 theodolite measurements of altitudes in a wide variety of conditions. He was mainly concerned with the effects on refraction of changing temperature gradients. He concluded that the effects could be large but that they were mostly the result of larger than usual temperature gradients near the observer and mainly in the first 100m height above the observer (Thom, 1958, 1971). The variable refraction values for the two lighthouses discouraged him from placing much reliance on potential sites with negative altitudes.

Overall, Thom concluded that while refraction deviation could, on rare occasions, give anomalous results, under normal conditions observations down to an altitude of about 1° were reliable to ± 0′.5 of arc with almost no problems above about 4° altitude. (Thom, 1971, 61; 96—7). Later investigations led him by the mid-1970s to the view that lines below 1° altitude could be reliable, writing: "An examination of the very full table of refraction published by the Nautical Almanac shows that, providing the observed altitude is not negative, there is not much trouble about refraction. We draw attention to work done on refraction measurement (Thom, 1958). At very low and negative altitudes the refraction becomes significant but only a few lines like this exist." (Thom and Thom, 1982, 79—80).

Thom used his many refraction measurements of observed distant features in the altitude range of about 0°.4 to 1°.0 to determine values for the refraction coefficient. This he defined as $K = rT^2/LP$ where r = observed (terrestrial) refraction in seconds of arc, T = absolute temperature in °F, L = length of ray in feet and P = barometric pressure (inches of mercury) (Thom, 1958; 1971). , Thom found an average value for K of 7.5 The values varied with the time of day and the seasons; the extreme range of individual results was 5 to 13. The values averaged for time of day were in the range 6.5 to 9.0.

Schaefer (1990, p. 802) analysed Thom's results. He writes:- "If the sight lines that Thom measured are typical for the atmosphere, then K will be proportional to the astronomical refraction." He then discusses how K can be used to estimate the variance in R_o. (R_o is the standard refraction value at 0° altitude and is about 0°57 or 35′ of arc.)

He then uses a programme he had written (Schaefer, 1989) and continues:- "Hence, the observed variation in K must be at least doubled so as to get a lower limit on variations in R_o." He goes on to deduce that:- "...the standard deviation of the variation ranged from 0°.05 to 0°.09...." And:- "As discussed earlier, these variations are likely to be at least doubled when the line of sight extends to outside the atmosphere. Hence Thom's data give a variation in R_o of 0°.18 or larger." Finally:- "...for the conditions in Scotland, the astronomical refraction on the horizon has a standard deviation of >0°.18, while the range is >0°.62."

The assumptions that Schaefer uses to obtain this result may or may not be valid, but it does not agree with what is found in Scotland.

It is recognised that the temperature gradient can cause changes in refraction values at low altitude (Schaefer and Liller, 1990, 801; McCluskey, 2017). A study of timed observations of the setting and rising sun was carried out from two very different sites; Edmonton and Barbados (Sampson et al., 2005). Edmonton has a continental climate and experiences large daily temperature variations, whereas Barbados, being surrounded by the sea, has a much smaller temperature variation. Both had nearby research stations making regular weather balloon temperature measurements. The refraction standard deviations at sunset found for Edmonton and Barbados were 0°.118 and 0°.012 respectively. These values correlated closely with the standard deviation surface temperature gradients which were 0.0405 $C^{o}m^{-1}$ for Edmonton and 0.0065 $C^{o}m^{-1}$ for Barbados.

2.3. *PRECISE LUNAR ALIGNMENTS*

The region of Argyll and the islands of Mull and Islay in western Scotland was investigated. There are 92 standing stone sites in this region. All except four remote sites were visited. Forty two sites could not be assessed due to trees, stones fallen or stones with no indicated direction. The remaining 46 sites were assessed. Standing stones with a flat face, stone pairs and short rows can give an indicated direction.

Fourteen of these sites had no alignment. Eleven of these sites can be explained as probable 'waymarkers'; for example five stones on Mull near the medieval pilgrimage island of Iona (Ruggles, 1984). The other three sites that gave no alignment consisted of a pair, a short row; and a stone that is probably medieval.

The declinations found for the remaining 32 assessable sites, each with an indicated direction, were compared with the key declinations to be expected for the sun and moon in the late Neolithic/Early Bronze Age. Sixteen of the sites gave one or more calendrical alignments (Thom, 1967, 107—117). The remaining sixteen sites all gave lunar alignments. Five of these lunar alignments were very short, each being less than 700m in length, and so may not be reliable. The eleven remaining sites all gave a precise lunar alignment (Gough, 2016). The site of Ford in Argyll is used as an example. Lat. 56^0 10′.5 (NM 8668 0333) Canmore ID 22802 (Figure 3).

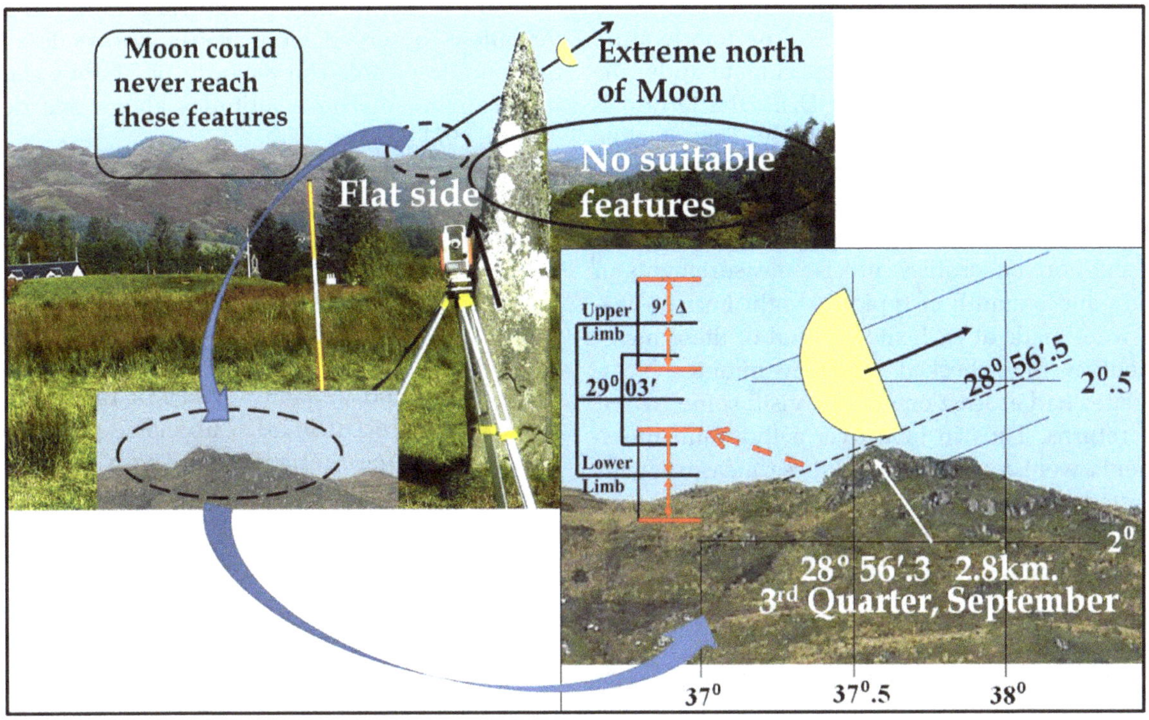

Figure 3. Precise lunar alignment at Ford, Argyll

The flat left flank of the menir indicates the region of a small prominent rocky feature. There are no other suitable horizon features in the vicinity. The moon, *c*. 1700 B.C.E., is at its extreme north with positive 'wobble'. Use of the lower limb would ensure that the moon would be clearly visible before the alignment was reached.

All eleven of the sites had at least one return visit and the details re-measured. The azimuth range determined for each site was typically less that ± 0′.3 of arc. This value includes sun/theodolite measurements used to convert the plate bearing to the true azimuth. In a separate investigation, a total of 1500 degrees of hilly horizon was assessed in order to test for the likely occurrence of chance alignments. These regions contained 30 lunar bands. Three chance alignments were found which is 1 in 10. Thus the possibility that the eleven alignments found could be due to chance is vanishingly small ; $p \approx 10^{-11}$ (Gough 2016).

2.4. *THE IMPLICATIONS OF PRECISE LUNAR ALIGNMENTS*

As briefly discussed earlier, the claim that Thom's precise lunar alignments were not possible depends upon the large refraction deviations found in Chile and North America which would result in significant azimuth variation and hence declination changes (Schaefer and Liller, 1990, 804 – 5; Sampson, 2000).

In a later paper Schaefer (2000, p. 126) repeats his claim regarding the effect of azimuth uncertainty with specific mention of Scotland. "For the latitude of Scotland, low horizons will yield a full range of azimuth variations (for a given declination) comparable to a full degree. In essence, for many sites, refraction variation rules out the possibility of high-accuracy alignments."

It should be noted that the eleven precise lunar alignments in the region were all within *c*. ± 2′ of arc of one of the four 'key' declinations. There were no 'near misses'. The measurements made to determine these precise alignments depend upon there being small azimuth variations. It follows that refraction deviations at the time of the measurements must also have been small. Furthermore, if we accept that these alignments were intentionally set up in prehistoric times, with presumably night-time observations, then we can infer that refraction deviations were also small in prehistory; otherwise the alignments could not have been set up.

3. SUMMARY

Refraction deviations sufficiently large to prevent the measurement of precise alignments do occur at altitudes of about 0°, and particularly at negative altitudes. The large refraction deviations found by Schaefer and Liller depends almost entirely upon their observations from high elevation sites and hence with markedly negative altitudes. Thus Cerro Tololo (2215m) and Mauna Kea (4205m) together gave a standard deviation of 0°.27 (16′.2 of arc) while the lower site at Vina del Mar (120m) in the same geographical region as Cerro Tololo, gave 0°.048 (3′.0 of arc) (McCluskey, 2017, 331).

The small refraction deviations found in Scotland by the author are consistent with low temperature gradients. Scotland has a maritime climate under the influence of the North Atlantic Drift, the terrain is generally hilly and it is usually windy. These factors would be expected to have the effect of smoothing temperature anomalies.

During the site assessments, in order to determine the horizon profiles, precise measurements of altitude and azimuth of typically eight horizon features were made at each site. Several of these measurements were re-checked after a few minutes. Nearly all sites had at least one return visit; some two or three returns. Thus in total over a thousand measurements were made. None of these measurements suggested anomalous refraction values.

All but one of the eleven alignments had an altitude above one degree; an altitude region where, in Scotland in the present investigation, the refraction deviations observed were nearly always less than one minute of arc. The majority of Thom's claimed lunar alignments have altitudes above one degree (Thom, 1971, 76).

4. CONCLUSIONS

Schaefer and Liller based their claim that precise alignments in Scotland were not possible on a number of assumptions. Namely: that the large refraction deviations found in some regions of the world can be expected to occur in other unrelated geographical regions including Scotland; that Thom's alignments were of sufficiently low altitude; and that temperature inversions are ubiquitous. The evidence found in Scotland does not support their claim.

ACKNOWLEGEMENTS

The author wishes to thank the two anonymous reviewers for their useful comments.

REFERENCES

Exton, H. (1992) A fresh analysis of some recent data on atmospheric refraction near the horizon with implications for archaeoastronomy. *Archaeoastronomy*, No. 17, Supplement to *Journal for the History of Astronomy*, Vol 23, pp. S57–8.

Gough, T. T. (2016) Evidence for the existence of Solar and Lunar Alignments in western Scotland: the contrasting nature of backsights, foresights and alignments, F. Silva, K. Malville, T. Lomsdalen and F. Ventura (eds.), *The Materiality of the Sky* (University of Wales, Ceredigion: Sophia Centre Press, 2016), pp. 145–153. (Available on Academia.edu May 2018).

McCluskey, S. C. (2017) Archaeoastronomy and Refraction near the Earth's Surface. *Journal for the History of Astronomy*, Vol. 48, No. 3, pp. 329–345.

Ruggles, C. L. N. (1984) *Megalithic Astronomy: a new archaeological and statistical study of 300 western Scottish sites*, BAR British Series 123, London, p. 139.

Ruggles, C. L. N. (1999) *Astronomy in Prehistoric Britain and Ireland.* New Haven and London: Yale University Press.

Sampson, R. D. (2000) Astronomical refraction and the equinox sunrise, *Journal of the Royal Astronomical Society of Canada*, Vol 94, pp. 26–29.

Sampson, R. D., Lozowski, E. P. and Machel, H. G. (2005) Variability of low-altitude astronomical refraction (LAAR) from different geographical locations: progress towards a global map of LAAR variability. *Applied Optics*, Vol 44, No 27, pp. 5652–5657.

Schaefer, B. E. 1989, Sky and Telescope, No. 77, p. 311.

Schaefer, B. E. and Liller W. (1990) Refraction near the horizon *Publications of the Astronomical Society of the Pacific*, Vol. 102, pp. 796–805.

Schaefer, B. E. (1993) Basic research in astronomy and its applications to archaeoastronomy, C.L.N.Ruggles (ed.) *Archaeoastronomy in the 1990s*, Group D Publications, Loughborough, pp. 155–177.

Schaefer, B. E. (2000) New Methods and Techniques for Historical Astronomy and Archaeoastronomy. *Archaeoastronomy*, Vol. 15, pp. 121–136. University of Texas.

Thom, A. (1958) An empirical investigation of atmospheric refraction, *Empire Survey Review*, Vol 14, Issue 108, pp. 248–262.

Thom A. (1971) *Megalithic Lunar Observatories* Oxford: Oxford University Press.

Thom, A. and Thom, A. S. (1982) Statistical and Philosophical arguments for the astronomical significance of standing stones with a section on the Solar Calendar, D. C. Heggie (ed.) *Archaeoastronomy in the Old World*, Cambridge University Press, pp. 53–82.

12

THE BEAR AND THE YEAR: ON THE ORIGIN OF THE FINNISH LATE IRON AGE FOLK CALENDAR AND ITS CONNECTION TO THE BEAR CULT

Marianna P. Ridderstad

University of Helsinki, Finland

Corresponding author: Marianna P. Ridderstad (marianna.ridderstad@pp.inet.fi)

ABSTRACT

The role of the bear as a calendric deity among the Finnic and Finno-Ugric peoples echoes it general importance in all aspects of the northern cultures. The Finnic peoples divided their year into four parts by two Bear Days in January and July, and the Summer and Winter Nights between those. In the earliest wooden folk calendars in Finland that can be traced to the 13th century, those calendric marker days had already been fixed into the Julian year, but in earlier times, they would have been determined by a lunisolar or lunar calendar. The calendric system of the late Iron Age Finns and other Finnic peoples was lunisolar with intercalation, and its structure and main marker days indicate that it had several layers of different age: the 'Bear Year' seasonal division, the idea of the eight-divided solar year -probably of Indo-European origin-, and a basic lunisolar intercalation calendar with twelve or thirteen months. The Finnish celestial lore related to the bear can also be connected to the seasonal positions of the Big Dipper asterism. While the seasonal calendric significance of the bear in Eurasian cultures is probably very ancient, possibly even Paleolithic, the connection of the Finnic Bear Year seasonal division and lunisolar calendars can be traced at least to the late Neolithic period.

KEYWORDS: Calendars, Bear Year, Ursa Major, bear cult, Finnic religion.

1. INTRODUCTION

The role of the bear as an important deity in the ancient Finnic and Finno-Ugric religion is well known, but the calendric connection of the bear has not been that widely researched a subject. In this paper, the origin of the Finnish 'Bear Year', i.e. the ancient Finnic seasonal calendric division governed by the bear and its connection to what is known of the Finnish late Iron Age bear cult are discussed.

2. THE BEAR CULT IN ANCIENT FINLAND

The ancient Finnish bear cult can be best traced through the folklore and poetry recorded in Finland from the late 17th century to the early 20th century. In analysing the information, one must note that the details of the cult are those it had among the Finns of historical times. The cult must have had somewhat different forms in the different regions where people speaking various Finnic languages and dialects lived in the late Iron Age and before that (e.g., Tavastia, Karelia, Estonia). Remarkably, however, the known details of the bear cult show great similarities: not only in different regions in Finland and Karelia, but also the bear cult of the Saami and other Finno-Ugric peoples have much in common with the Finnic variations of the bear cult. In fact, most of the peoples living in the northern latitudes have similarities in their bear cult. While the similarities in ecological conditions often lead to similar cultural practices, there are indications that some widely spread religious beliefs concerning the bear may have their origin as far back in time as in the Palaeolithic bear worship (e.g., Frank, 2008).

The Finns believed that the bear originated in the sky, in the stellar realm, as it is revealed by the following Kalevala-metric poem on the birth of the bear, recorded in Eastern Finland in the early 19th century (SKVR VI2: 5405; transl. by the present author; see also SKVR XII2: 6858):

Miss' on otso synnytetty,
mesikämmen käännytetty?
Luona kuun, tykönä päivän,
Otavaisen olkapäillä.
Sielt' on maahan laskettuna
kultasessa kätkyessä,
vitjoilla hopeisilla.

Where's the bear been given birth to,
the honey-paw turned around?
With the moon, with the sun,
on the shoulders of the Big Dipper.
From there he's been lowered onto the earth
in a golden cradle
with silver chains.

It can be seen in the poem above that the Finns believed that the bear was from Ursa Major, Otava in Finnish. *Otava* means a sort of fish net, but the resemblance of the word to the probable original name of the bear in Finnish, *otso* or *ohto*, is remarkable. The original name of the bear is not known for certain, since the name of the bear was a sacred and partly taboo-like and thus also the modern word for 'bear' in Finnish, *karhu* originally was a term of endearment. Other common names of the bear were, e.g. *kouko*, an old man or an ancestor; *mesikämmen*, honey-paw; *lintunen*, birdy; *metsän ukko*, the old man of the forest; *metsän omena*, the apple of the forest; *metsän (kultainen) kuningas*, the (golden) king of the forest; or *metsä*, the forest itself (see, e.g. Pentikäinen, 2005). The name for the celestial home of the bear may have been subject to similar circumventions, and the name of the fish net may thus have become selected for its similarity to the name *ohto*, *ohtava* (the shape of the fish net in question was rectangular, so the 'tail' of Otava also seems to have no function in that explanation). Indeed, in some occasions the bear is called not only *Otavanpoika*, the son of Otava, but even Otava itself (Pentikäinen, 2005).

In other variants of the verses, it is told that the bear was born "on the back of the seven stars" and also "on the top of *honka*", a pine tree, namely *Pinus sylvestris* (SKVR VI2: 5411; I4: 1419). The latter verse echoes the belief that the *emuu*, the ultimate female ancestor and creator deity of the bear was Hongotar (cf. *honka*). For example, in the well-known ritual text from Viitasaari, it is mentioned that "Hongikosta sinun sukusi, Hongotar sinun sukusi", "From the pine forest [is] your family, Hongotar [is] your family" (SKVR IX4: 1096).

A poem from Piippola tells us that the bear was born and raised "päällä pienen pilven naulan", "on the top of a small cloud's nail" (SKVR XII2: 6464). This is interesting as the pole star was seen by the Finns as the nail of the sky, located on the top of the sky-supporting world pillar *sammas* (Harva, 1943; Siikala, 2012: 188). The ancient Finns may also have obtained from the Indo-Europeans the belief that the sun was located at the top of the sky pillar (Parpola 2005: 50-51; Ridderstad 2014). Thus, the Bear, like the Sun, was a deity who could be seen as sitting on the topmost position of the sky-vault.

As can be seen in the poem, the bear was lowered onto the earth and the human realm in a golden cradle, using silver chains. Gold and silver are epithets commonly related to the Sun, the Moon and the stars in the Finnish poetry (Ridderstad 2014); the special materials emphasize the fact that the bear was divine. The cradle of the bear may have been Otava itself, which has some definite resemblance to the ancient cradles which were little more than small boxes supported by one or two cords by which they were tied to, e.g. a tree branch (see also Figure 1 and Section 3 below).

The already above-mentioned poem from Viitasaari (SKVR IX4: 1096) is the most complete ritual corpus consisting of the songs and rites of a bear hunting ceremony in the 17th century Finland. Only traces of other similar ritual poems have been preserved (e.g., SKVR I4: 1095, 1220). The Viitasaari text describes how a bear is hunted singing certain songs that legitimate its killing, how it is taken to the village and how a feast of *peijaiset* is celebrated in its honor, all parts of the ritual accompanied with the ritual Kalevala-metric songs. Peijaiset was both the wedding and the funeral of the bear, who was addressed as *kouko*, the grandfather, and revered like a human ancestor or a bridegroom. The end part of the ritual consisted of the returning of the soul of the bear to the sky, which was accomplished by taking the skull of the bear to a bear skull pine. The skull of the bear was hung on the branches of the tree and its other bones were buried at the foot of the tree. Some of its teeth and claws may also have been distributed among the hunters. As late as the 17th century there were bear furs hung in the churches of Northern Finland - a practice which can be seen as a continuation of the bear skull tree rite. The bear graves of the Saami also echo the same basic belief of the returning of the soul of the bear back to the otherworld.

Many northern peoples share the belief that the bear was their ancestor, and traces of the myth about the marriage of a human female and the celestial bear are wide-spread in Eurasia (see, e.g. Frank, this Volume). Among the Finno-Ugric peoples for example the Khanty believed that the bear was the son of Numi-Torum, the sky god, and that he was lowered to the earth in a cradle and married a woman, thus becoming the ancestor of humans (see, e.g. Sarmela, 2006). The *koltta* Saami people shared a similar origin myth and it seems that a similar belief once exists among the Finnish tribes, too, based on the way the bear is addressed in the ritual hunting texts and also the belief that women have a special relationship with the bear: a bear would not attack a woman.

In the Viitasaari ritual, the bear is called *Pyhäjoen putaan miesi*, the man of the *pudas*, an anabranch of the Pyhäjoki river, which probably refers to the bear as a master fisherman. Among the rapids of the fast-flowing Pyhäjoki river, the anabranches often formed quieter stream pools. Those were good places for fishing for both bears and humans. Perhaps Pyhäjoki, the Sacred River, was sacred precisely because it attracted so many bears? In this context, it is interesting to note that Väinämöinen, the main male shamanistic deity of the Iron Age Finns also received his name from a quieter part or a stream pool of a river, *väinä*. Perhaps Väinämöinen at some point came to receive some of the properties of an earlier bear deity? Even Orion, the asterism related to Väinämöinen in the Finnish folklore, could have originally been a bear as well an anthropomorphic deity. This would explain why Ursa Major may have been seen as part of an elk or, as mentioned above, a cradle or a fish net, but not a bear. Unfortunately, connections like these may be impossible to prove.

In Finland, bear tooth pendants have been found in many inhumation graves of the late Iron Age (the burials of the earlier periods were usually cremations, which does not preserve small organic objects). Interestingly, most of the pendants seem to have belonged to women (see, e.g. Lehtosalo-Hilander, 1969-1972). The bear tooth and claw pendants can be seen as powerful protective elements. They may have also been connected to fertility, since the bear was also one of the governors of the seasonal cycle (see the next Section).

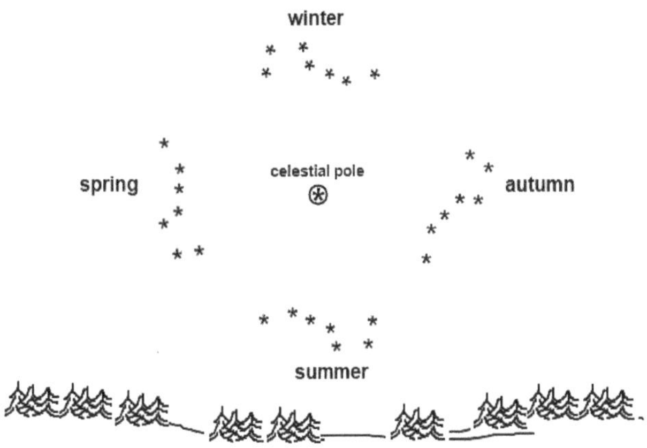

Figure 1. In Southern Finland in the Viking Age, the position of the Big Dipper in the early hours of the morning could be connected to the lifecycle of the celestial bear as presented in the Bear Year seasonal division: in the midsummer, the 'cradle' seemed to come closer to the ground, while in the midwinter it was at its highest position, farthest from the human realm. In the autumn, the cradle carried the bear upwards, and in the spring downwards. Note also that the asterism has been circumpolar for most of Europe for a very long time.

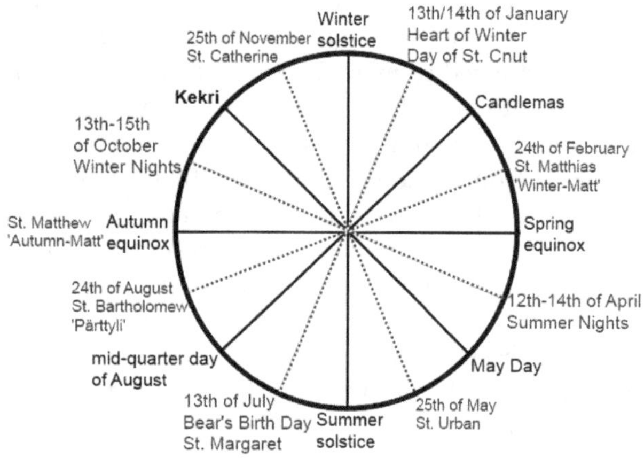

Figure 2. The eight-divided 'Bear Year' and its relation to the eight-divided solar year.

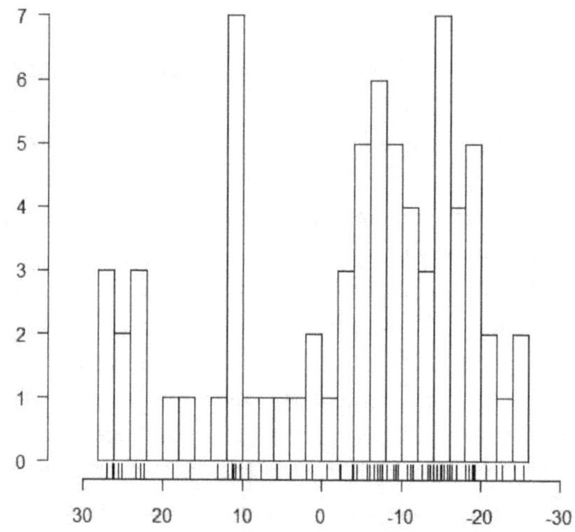

Figure 4. The Neolithic terraced houses or 'longhouses' of Ostrobothnia, Finland, most of which were built 3200-2500 BCE, also had many axis orientations towards the sunrises of mid-April. The doorways were placed at the ends of the house and thus faced the sunrises indicated by the axial orientations. According to Ridderstad (2016a).

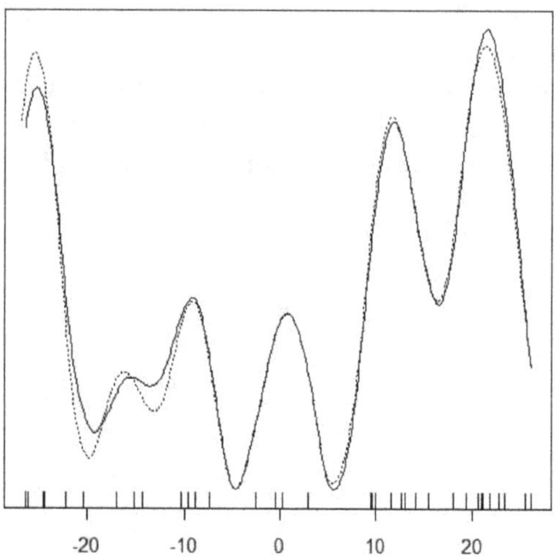

Figure 3. Many of the orientations of the axes and the doorway openings of Giants' Churches of Ostrobothnia, built 3000-1800 BCE, were towards the sunrises about one month after the equinoxes (the declinations of ca. +/-10 degrees) or the solstices (the declinations of ca. +/-20 degrees) . In the figure, the orientations of the western doorways of the enclosures are shown. According to Ridderstad (2015).

3. THE BEAR AS THE GOVERNOR OF SEASONS

It is usually assumed that the earliest calendars of the Finno-Ugric hunter-gatherers were based on the lunar cycle. However, in addition to the solar Julian-Gregorian calendar, there is only evidence of an indigenous Finnic lunisolar calendric system(s) that can be traced back to the Iron Age.

From folklore it is know that the Finns had twelve or thirteen months in their year, which indicates that the folk calendar was a lunisolar calendar with an intercalation system: occasionally, an extra month called *vaahtokuu* was inserted after the month *helmikuu* in winter (Vilkuna, 1950: 55). In the present Gregorian system, helmikuu is the Finnish name for February.

On the other hand, the ancient Finnish year started on the festival *kekri* in late autumn, close to the present All Saints' Day. Kekri was preceded (or in some regions apparently followed) by a period of eleven or twelve days called *jakoaika*, a 'division time' (Vilkuna, 1950: 293). Obviously, the twelve days correspond to the difference between the lengths of one solar year and twelve synodic lunar months. It thus seems that kekri was originally a festival determined by a lunisolar calendar and that it was meant to be celebrated always on the same day of the solar year. Thus, kekri may have originally been a solar festival, and may even have been attached to the mid-quarter day of November, the

Celtic Samhain. However, kekri probably precedes any Celtic influence by thousands of years: the word *kekri* comes from *kekräj*, which is an Indo-European loan word from around 2000 BCE meaning a wheel or a cycle (Koivulehto, 2000; Parpola, 2005; note that around the same period, also the Indo-European word *sammas* meaning the world pillar was loaned). The Finnish word *kehrä*, the disk of the sun, is of the same root as *kekri* (Koivulehto, 2000). Kekri was thus the festival where the wheel of the solar year had turned and the cycle was to start anew.

Both folklore and the oldest preserved Finnish wooden calendar staffs reveal that the Finnic peoples had yet another ancient division of the year: the year was divided into two halves by two 'Bear Days'. The Midwinter or the Turning Day of the Bear on the 13th of January, and the Midsummer or the Birthday of the Bear on the 13th of July. It was believed that the bear, who went to its winter sleep on the St. Matthew's Day on the autumn equinox and woke up on the feast day of St. Matthias, 'Winter-Matt' on the 24th of February, turned from one side to another on the Midwinter (or in Christian times, on the feast day of St Henry) and said "Yö puolessa", "Half of the night has passed" (Vilkuna, 1950: 33, 266). It is interesting to note that during the long northern midwinter nights, the Big Dipper could be seen to move almost a full circle around the celestial pole, turning from one side to another.

The two Bear Days, which in the Gregorian system happen only some weeks after the shortest and longest nights of the year, coincide with the times of the annual temperature minimum and maximum in Finland (Vilkuna, 1950: 183-184). In the Julian folk calendar staffs, the Bear Days were fixed, but originally they would have been lunarly or lunisolarly determined.

The two halves of the Bear Year can be seen to have been further divided into two by the Summer Nights (Finn. *suviyöt*) on the 12th-14th of April and the Winter Nights (Finn. *talviyöt*) on the 13th-15th of October. The fact that both of these calendric marker days lasted three days probably points to their lunar calendric origin: the fullest phase of the moon can be seen to last about three nights.

In Figure 2, the relation of the Bear Year to the eight-divided solar year can be seen. This ancient Indo-European concept of an eight-divided solar year was in use for example among the Celts and the Balts. Interestingly, the many folk beliefs related to the bear that were connected to the days that coincided with the mid-points of the two Bear Days and the Summer and Winter Nights indicate that the idea of the eight-divided year had been applied also to the Bear Year in Finland (Ridderstad 2013). It is difficult to say, however, how old this system was compared to the Indo-European year that started on kekri. It could be as recent as the fixing of the Bear Days into the Julian solar year, or as old as the idea of the wheel of the year among the Finnic peoples.

A calendric system corresponding to the Finnic Bear Year was also in use in Scandinavia, but without any known connection to the bear - however, that is not to say that a connection like that could never have existed. On the contrary: for example the Danish royal house seems to have embraced a legend that traced the origin of the Danish kings to the bloodline of bears (e.g., Shepard & Sanders, 1985: 129). Similar stories, the so-called Bear's Son Tales, as well as the bear as the central character of an important annual festival can be found all around Europe (Frank, 2008; Frank, this Volume).

An annual seasonal division, where a bear was connected to both or one of the main seasons, usually to the summer half of the year (cf. the Finnic Birthday of the Bear) existed among many other Northern Eurasian peoples, too (Konakov, 1994). For example, for the Komi the period from the 22nd of March to the 27th of April, i.e. the period around the Finnish Summer Nights, was the Bear month. For some of the Finno-Ugric peoples, the elk seems to have governed the winter half of the year and the Big Dipper was seen as an elk (Konakov, 1994). There is some indication that the Big Dipper may have been seen as part of a celestial elk among the Finns, too, in addition to the other possible interpretations discussed above. Therefore, it is curious that for the Finnic peoples, the Midwinter was also a Bear Day and not, for example, an Elk Day. Perhaps the relative importance of the bear was the result of interaction between the ancestors of the Finnic peoples and some earlier inhabitants of the area of Finland and the nearby regions.

4. THE ORIGINS OF THE BEAR YEAR TYPE OF SEASONAL CALENDAR IN FINLAND

The origin and age of the Bear Year seasonal calendric system is difficult to trace, but some observations can be made. Before the fixing of the locations of the Summer and Winter Nights in the Julian calendar, they would have corresponded to the full moons of the first summer and first winter months around the present April and May, and October and November in a lunar or lunisolar calendar. In a lunisolar calendar, the Winter Nights could have been for example the time of the first full moon or the full moon of the first full month after the autumn equinox. Similarly, the Summer Nights could have been determined in relation to the spring equinox. The Bear Days were also likely originally determined by the occurrence of a certain phase of the moon in the lunisolar calendar that was used by the Finnic peoples before the introduction of the Julian system. For example the Midwinter or the Turning Day of the Bear could have been determined as the time of the full moon of the first full month after the winter solstice. It is difficult to say which one of the main solar days could have been used as the annual starter day in the above type of system. For the Scandinavians, the winter solstice seems to have been the starter; perhaps the Finnic peoples used kekri, if it indeed was a fixed solar day (as it seems based on the fixed annual preceding period of twelve days). In even earlier times, the Bear Days and the Summer and Winter Nights could have been determined using both the seasonal changes in the appearance of the asterism of the Big Dipper and the lunar phases.

As seen above, the idea of the bear as the governor of the seasons is wide-spread among the Finno-Ugric and other Northern Eurasian peoples, which may indicate that it is very old. However, considering the great age of the loan word kekri, also the interest in the solar year and the use of a lunisolar calendar must have a long history among the Finnic and Finno-Ugric peoples. Thus, if we wish to look for the origin of the Bear Year system, we should probably start in around 2000 BCE, which in Finland corresponds to the end of the Neolithic period.

Due to the ecological similarities experienced by the Northern cultures and the great antiquity of bear worship in Eurasian cultures (see Frank in this volume), the interest in the Big Dipper, and its connection to the Bear deity and the change of seasons in the said cultures can probably be traced to the Mesolithic or even to the Palaeolithic period (see Antonello in this Volume). Therefore, we may assume that when the word kekri and the concept of the wheel of the year were transmitted from the Indo-European cultures to the ancestors of the Finnic peoples in ca. 2000 BCE in the region between the Ural mountains and Finland, the bear was probably already worshiped and revered as the governor of the seasons in the cultures in this area and around it. Then, the concept of the Bear Days as they emerge in the context of the solar year and in a lunisolar calendric system of the kind described above could have been developed among the ancestors of the Finnic peoples. Or, when the Finno-Ugric peoples moved towards the area of Finland between ca. 2000 and 500 BCE, they could have adopted a Bear Year lunisolar calendric system from an earlier Paleoeuropean culture living in Finland. In the latter case, the connection of the Bear and the lunisolar year would likely be even older than ca. 4000 years.

It is very difficult to investigate the possible developments of the Bear Year suggested above. No written sources exist that would tell about the calendric history of the early Finno-Ugric peoples or the Paleoeuropean inhabitants of Finland. The latter, however, have left monuments that can be measured and their orientations investigated: the orientations of the houses, the long cairns and the huge stone enclosures known as Giants' Churches of the late Neolithic culture of Ostrobothnia can be examined to find traces of the calendric practices of their builders. Indeed, it turns out that both the houses and the Giants' Churches had many orientations towards the sunrises about one month after the solstices and/or the equinoxes (Ridderstad, 2015; 2016a; 2016b). In Figures 3 and 4, examples of these types of orientations are shown. The Bronze Age may have seen a massive depopulation of Finland, but interestingly, after that in the Early Roman Iron Age in Southern Finland, the orientations of the early Finno-Ugric graves show orientations towards the sunrises of mid-October (Ridderstad, 2016c). Unfortunately, most of the Iron Age burials in Finland were cremations, which have not left individual graves or monuments to be measured. It can be noted that some of the earliest Christian inhumation graves in Southern Finland show orientations towards the sunrises of mid-April, but these are problematic as that time of year also falls inside the Easter period (Ridderstad, 2016c).

It can be observed that especially orientations to the sunrises about one month after the equinoxes can also be observed in many other European Neolithic and Bronze Age monuments, for example megalithic graves (see, e.g. Hoskin, 2001; 2007; González-García and Costa-Ferrer, 2007; González-García, 2009) and Minoan palaces (Henriksson and Blomberg, 2013). The possible connections of these to the orientations of the Finnish monuments and to the seasonal significance of the calendric marker days related to bear cults remain an open question.

5. CONCLUSIONS

It can be concluded that the role of the bear as a calendric deity echoes its general importance among the Finnic and Finno-Ugric peoples. The Finnic calendric system of the late Iron Age had several layers of different ages; the Bear Year seasonal system was one of these. The connection of the Bear Days and lunisolar calendars can be traced at least to the Neolithic period, while the seasonal calendric significance of the bear is probably much older and may originally have been based on the seasonal positions of the Big Dipper.

ACKNOWLEDGEMENTS

I wish to thank Prof Roslyn M. Frank, Dr Nicoletta Lanciano and Dr Michael Rappenglück for enlightening discussions.

REFERENCES

Antonello, E., this Volume. Sky simulations for the Paleolithic epoch. *Proceedings of SEAC 2015, Rome.* Mediaterranean Archaeology and Archaeometry.

Frank, R. M., this Volume. Sky Bear research: Implications for "cultural astronomy". *Proceedings of SEAC 2015, Rome.* Mediaterranean Archaeology and Archaeometry.

Frank, R. M. (2008) Recovering European ritual bear hunts: A comparative study of Basque and Sardinian ursine carnival performances. *Insula-3* (June 2008), 41-97.

González-García, A.C. (2009) Statistical analysis of Iberian Peninsula megaliths orientations. *Cosmology Across Cultures, Proceedings of SEAC 2008*, ASP Conference Series, Vol. 409, 354-358.

González-García, A.C. and Costa-Ferrer, L. (2007) The orientation of the 'Hünenbetten' of Lower Saxony. *Archaeoastronomy in Archaeology and Ethnography, Proceedings of SEAC 2004*, pp. 29-33.

Harva, U. (1943) *Sammon ryöstö.* Suomen tiedettä 3, Porvoo, WSOY.

Henriksson, G. and Blomberg, M. (2013) The archaeoastronomical results of three Bronze Age buildings at Agia Triada, Crete. *Ancient Cosmologies and Modern Prophets, Proceedings of SEAC 2012*, Anthropological Notebooks, Vol. 19 Supplement, 177-185.

Hoskin, M. (2001) *Tombs, Temples and Their Orientations: A New Perspective on Mediterranean Prehistory* Bognor Regis, Ocarina Books.

Hoskin, M. (2007) Orientations of Neolithic monuments of Brittany: (2) The early dolmens. *Journal for the History of* Astronomy, Vol. 38, 487-492.

Koivulehto, J. (2000) Fest und Zyklus des Jahres: *Jul* und *Kekri. Neuphilologische Mitteilungen*, Vol. 101, 235–252.

Konakov, N. D. (1994) Calendar symbolism of Uralic peoples in the pre-Christian era. *Arctic Anthropology*, Vol. 31, No. 1, 47-61.

Lehtosalo-Hilander, P.L. (1969-1972) *Luistari, Kauttua, Eura, 1969-1972.* Unpublished excavation report, Helsinki, National Board of Antiquities.

Parpola, A. (2005) The Nâsatyas, the chariot, and Proto-Aryan religion. *Journal of Indological Studies*, Vol. 16-17, 1-63.

Pentikäinen, J. (2005) *Golden King of the Forest: The Lore of the Northern Bear* Helsinki, Etnika Oy.

Ridderstad, M. (2013) Orientations of the medieval stone churches in Finland. *Suomen Museo*, Vol. 2012, 5-44.

Ridderstad, M. (2014) The Great Oak: an annual calendric and agricultural fertility myth of the Baltic Finns. *Astronomy: Mother of Civilization and Guide to the Future, Proceedings of SEAC 2013*, Mediterranean Archaeology and Archaeometry, Vol. 14, No 3, 319-330.

Ridderstad, M. (2015) Orientations and other features of the Neolithic 'giants' churches' of Finland from on-site and lidar observations. *Journal of Astronomical History and Heritage*, Vol. 18, No. 2, 135-148.

Ridderstad, M. (2016a) Orientations and placement of the Middle and Late Neolithic housepits in Ostrobothnia: a first investigation based on on-site and lidar observations. *Suomen Museo*, in press.

Ridderstad, M. (2016b) Orientations of the late Neolithic long cairns of Ostrobothnia, in preparation.

Ridderstad, M. (2016c) Orientations of the early Christian graves in Finland. I The Eura-Köyliö region. *Mirator*, submitted.

Sarmela, M. (2006) *The Bear in the Finnish Environment. Discontinuity of cultural existence*, http://www.kotikone.fi/matti.sarmela/bear.html, accessed 1.3.2016.

Shepard, P. and Sanders, B. (1985) *The Sacred Paw: The Bear in Nature, Myth, and Literature* New York, Penguin.

Siikala, A.-L. (2012) *Itämerensuomalaisten mytologia* Helsinki, SKS.

SKVR (2016) *Suomen kansan vanhat runot I–XV*, The Finnish Literature Society, http://dbgw.finlit.fi/skvr/, accessed 1.3.2016.

Vilkuna, K. (1950) *Vuotuinen ajantieto.* 3rd edition, Helsinki, Otava.

13

EARLY MESOPOTAMIAN INTERCALATION SCHEMES AND THE SIDEREAL MONTH

Eshbal Ratzon

Tel Aviv University

Corresponding author: (eshbal@gmail.com)

ABSTRACT

This paper examines the two intercalation schemes found in the Mesopotamian astronomical compendium MUL.APIN from the beginning of the first millenium BCE, and the lunar theory that they imply. It demonstrates that the two schemes do not agree with each other. Two intercalation rules in the second scheme use the conjunction of the moon and the Pleiades. This paper concludes that the intercalation rules are based on the assumption of a 28-day ideal sidereal month. These rules work with a triennial cycle of intercalating one additional month every third lunar year. Two similar intercalation schemes from other compositions, likewise dating from the beginning of the first millenium BCE, are known: a seventh-century intercalation scheme from Babylonia that also assumes a 28-day ideal sidereal month and an intercalation scheme from an unpublished astronomical commentary that, like the scheme in MUL.APIN, uses a triennial cycle. Previous scholars believed that discrepancies exist between the dates of the conjunctions of the moon and the Pleiades across all three schemes. However, this paper proposes that the astronomical assumptions of the three schemes are identical.

KEYWORDS: Sidereal Month, MUL.APIN, Intercalation Scheme.

1. INTRODUCTION

Lunar observations and calculations were the highlight of Mesopotamian astronomy. Their importance was both theoretical and practical, especially for the establishment of the calendar. Nearly all of the civil and cultic calendars in Mesopotamia were luni-solar calendars whose months were determined by the lunar cycle, and were therefore 29 or 30 days long. However, since 12 lunar months have approximately 11 fewer days than solar and sidereal years, an additional intercalary 13th month was added every two or three years. During most of the periods these intercalations were performed ad-hoc.[1]

Because the flexibility of the luni-solar calendar made it highly unpredictable, another schematic calendar was developed for long-term calculations. This calendar contained 12 months of 30 days each, for a total of 360 days a year. It was not used as a civil calendar to determine dates, and was considered an ideal calendar. Evidence of the ideal calendar exists already in third-millennium BCE tablets dealing with administrative calculations (Englund, 1988). Later, beginning in the Old Babylonian period, the ideal calendar was used in astronomical texts such as Enuma Anu Enlil and MUL.APIN (for a discussion of the ideal calendar in all periods see Brack-Bernsen, 2007). This calendar was not only practical for calculations, but also became the standard against which astronomical phenomena were examined in order to predict good or bad omina. In addition, it was used to established intercalation rules. These rules use predictions of cyclic behaviour of certain luminaries throughout the 360-day year. When the relevant celestial bodies appear in the expected position at the expected time during the luni-solar year, it is a normal year. If these phenomena are late according to the luni-solar calendar, the year should be intercalated.

Whether or not these rules were followed in practice, they can teach us about the advancements in astronomical understanding of lunar, solar, and sidereal cycles. One such cycle is the sidereal month, the time period necessary for the moon to return to the same position in relation to the stars. Every night the moon is seen against a background of different stars. After approximately $27\frac{1}{3}$ days, the moon completes a 360° rotation around the earth after which, from the vantage point of an observer standing on Earth, it is seen against a background of the same group of stars as when the cycle began. This paper will demonstrate that, as early as the beginning of the first millennium BCE, Babylonian astronomers not only noticed this phenomenon, but also used an approximation of its length to determine intercalation cycles. However, the texts describing this phenomenon did not attempt to suggest improvements to the computation of the intercalation cycles using observations, as some scholars claim, but were still using schematic cycles.

2. TEXTUAL EVIDENCE: THE INTERCALTION SCHEMES OF MUL.APIN

Two early intercalation schemes were preserved in the astronomical compendium MUL.APIN from the beginning of the first millennium BCE (Hunger and Pingree, 1989).

2.1 *First Intercalation Scheme*

The first intercalation scheme (MUL.APIN II i 9-24) is very simple. It presents predictions of heliacal rising of certain constellations, the positions of the sun and moon in relation to the stars, and the length of day and night for the four cardinal days: the equinoxes and summer and winter solstices (occurring on the 15th day of the first, fourth, seventh, and tenth months), as demonstrated in Table I, below. The intercalation rule states only that a deviation of these celestial bodies from their predicted positions on the named dates or the occurrence of these phenomena on other dates necessitate an intercalation.[2]

[1] This was true until the discovery of the so called "Metonic Cycle," probably during the fifth-century BCE (Hunger and Pingree, 1999; Britton, 2007). The 19-year cycle was put into practice gradually, thus turning the Mesopotamian calendar from a flexible calendar to a fixed one.

[2] Here I follow the common interpretation of MUL.APIN II I 22-24: "On the 15th of month I, on the 15th of month IV, on the 15th of month VII, on the 15th on month X, you observe the risings of the Sun, the visibility time of the Moon, the appearances of the Arrow, and you will find how many days are in excess" (trans. Hunger and Pingree). This interpretation is shared i.e. by (Hunger and Pingree, 1989; Chadwick, 1992; Horowitz, 1998; Watson and Horowitz, 2011), and understands the words "days in excess" to refer to the days that the year is missing. The astronomer is attempting to determine how many days are in excess in order to decide whether or not to intercalate the year. However, (Brown, 2000) contests this assumption, and explains that the purpose of the assessment of the days in excess is divinatory. (Brack-Bernsen, 2005) proposes that an additional purpose of the days in excess is to correct the values given in the scheme. I agree with Lis Brack-Bernsen that we should not reduce the motivations of the Babylonian astronomers to a single one. Therefore, even if Brown's and Brack-Bernsen's interpretations are correct, this section of MUL.APIN may still be an intercalation scheme, as is commonly believed, in addition to serving one or more other purposes.

Table I. MUL.APIN II i 9-24: First Intercalation Scheme

Date	Constellation	Sun	Moon	Day-Night Ratio
IV, 15	Arrow	Lion	-	4:2
VII, 15	-	Middle of Scales	In front of Stars and behind Aries	3:3
X, 15	Arrow	Head of Lion[3]	-	2:4
I, 15	-	In front of Stars and behind Aries, in the west	Middle of Scales, in the east	3:3

The equinoxes and solstices occur in the middle of the month, according to this calendar. On these dates, the sun and the moon are in opposition, i.e. directly opposite each other with reference to the earth. Therefore, it is not surprising that on the two equinoxes, six months apart from each other, these bodies should be in opposite positions in relation to the stars: While on the autumnal equinox the sun is in the middle of Scales and the moon is between the Pleiades (Stars) and Aries, on the vernal equinox the sun is between the Pleiades and Aries and the moon is in the middle of Scales, as illustrated in Figure 1 and Figure 2.

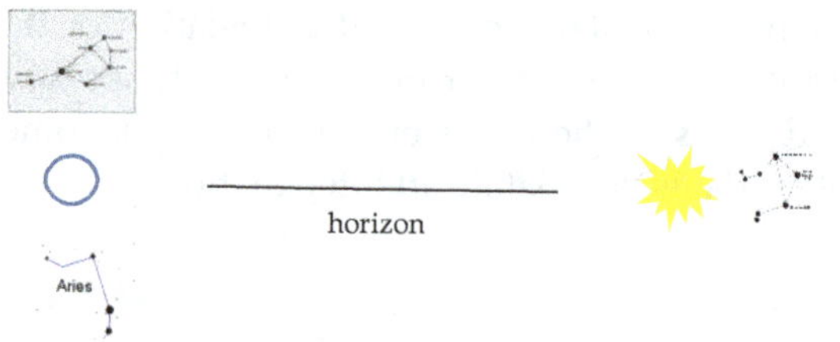

Figure 1. The sun and moon on Tešritu (VII) 15th, during sunrise and moonset. The images are not to scale, and are intended only for demonstration.

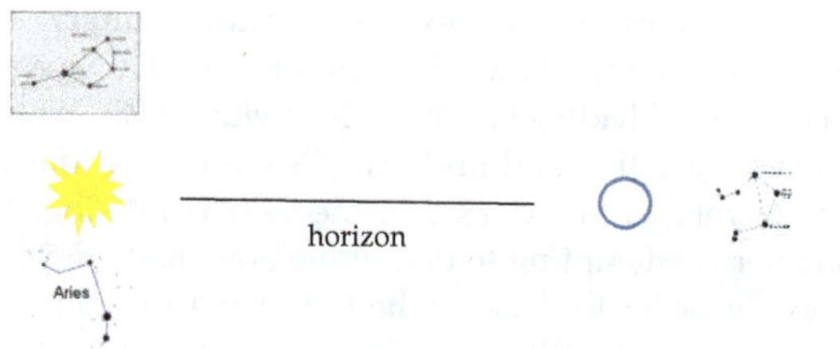

Figure 2. The sun and moon on Nisannu (I) 15th, during sunset and moonrise. The images are out not to scale, and are intended only for demonstration.

While there is much to say about the path of the sun according to this text (see for example Hunger and Pingree, 1989), this paper focuses on the moon. According to this text, the moon's longitude changes by 180° over the course of half a year. All the details here are schematic, and thus the dating of these phenomena is not based on observations. The position of the sun in relation to the stars in the first month is opposite its position in the seventh month, and the moon in the middle of the equinoctial months is opposite that of the sun. However, assuming the ideal calendar used in MUL.APIN, an additional computation of the lunar daily displacement (the rate of change of the position of the moon relative to the stars) is simple. Since the basic calendar in MUL.APIN is a schematic calendar of 12 months consisting of 30 days each for a total of 360 days a year, a period of six months contains 180 days. A sidereal month, during which the moon encircles the earth and returns to the same position in relation to the stars, is shorter than a synodic month, according to which the calendric months are established. Therefore, in six months the moon encircles the Earth and completes a 360° rotation $k \geq 6$ times. The moon's position after six months can be computed using its daily displacement in units of degrees per day (v):

$$360k+180 = 180v;\ k \geq 6$$

$$2k+1=v$$

If k=6, v=13 degrees per day, which is in close approximation to the mean daily angular movement of the moon on the ecliptic. Despite the fact that there is no textual evidence that this further computation was actually performed by the authors of MUL.APIN, these outcomes will be useful below.

2.2 Second Intercalation Scheme

The second intercalation scheme of MUL.APIN (II A 1- II ii 20) is comprised of two sections. The second section (II ii 9-20) is simpler. It presents a triennial cycle, meaning that an intercalary month is added once every three years. An ongoing debate exists among scholars regarding the question of whether the additional month should be added to a lunar year of 354 days or to the ideal year of 360 days. (Albani, 1992; 1994; Horowitz, 1994; 1996; 1998) claim the former, and suggest that the intercalation scheme offers an approximation of a 364-day solar year. In support, Horowitz introduces the seventh-century *ziqpu* star text, "where an annual circuit of the *ziqpu*-stars is measured as 364°." On the other hand, (Koch, 1996; 1998), claims that MUL.APIN always uses a schematic year of 360 days, and the intercalation scheme should not be explained as an exception. He refers to evidence from the third millennium BCE

[3] Although the Head of the Lion (UR.GU.LA) is attested in both manuscripts, it is probably a mistake from mulGU.LA, because the sun cannot be at the same position as it had been half a year earlier (Hunger and Pingree, 1989).

that the triennial cycle had existed alongside the 360-day Sumerian administrative year (Englund, 1988).

In the first section of the second intercalation scheme (II A 1- II ii 8), seven intercalation rules appear, referring to stellar visibility and the conjunction of the moon with "the Stars" (Pleiades). The text of this section offers intercalation rules for each month. Five of these rules use the first visibility of different stars. Thus, if a star is first visible on its ideal date (after a period of absence from the night sky), the year is normal, but if that star is not visible until a month later, that year is a leap year. For example:

"[If] the Fish and the Old Man become visible on the 15th of Addaru, this year is normal.

"[If] the Fish and the Old Man become visible on the 15th of Nisannu, this year is a leap year."

(MUL.APIN II ii 5-6; trans. Hunger and Pingree)

However, most of the dates for the intercalation rules are not fully preserved, and Hunger and Pingree reconstruct them based on other lists from the compendium. The data for the intercalation rules, according to Hunger and Pingree's reconstruction, is provided in **Table II.** All reconstructed data is written between brackets.

Table II. MUL.APIN II ii 9-20: Intercalation Rules

Day	Month	Event	Type of Year
[1]	[I]	Conjunction of the moon and Stars	Normal year
3	[I]	Conjunction of the moon and Stars	Leap year
[1]	[II]	Stars become visible	Normal year
1	[III]	Stars become visible	Leap year
[15]	[IV]	[Arrow becomes visible]	[Normal year]
[15]	[V]	Arrow becomes visible	[Leap year]
[15]	[VI]	ŠU.PA becomes visible	[Normal year]
[15]	[VII]	ŠU.PA becomes visible	[Leap year]
[15]	[VIII]	Conjunction of the moon and Stars	Normal year
15	[IX]	Conjunction of the moon and Stars	Leap year
15	X	Arrow becomes visible in the east in the evening	Normal year
15	XI	Arrow becomes visible in the east in the evening	Leap year
15	XII	Fish and Old Man become visible	Normal year
15	I	Fish and Old Man become visible	Leap year

The reconstruction of the last two columns is obvious. According to (Hunger and Pingree, 1989), the dates of the five pairs of heliacal risings are taken from a list of heliacal risings also preserved in the same compendium (MUL.APIN I ii 36 – iii 12). This seems probable, since two of them are preserved in both lists (X 15 and XII 15) and the rest follows the structure of the table, listing one phenomenon for each month. This naturally leads to the reconstruction of the conjunction of the moon and the Pleiades on Araḫsamnu (VIII) 15 in a normal year and on Kislimu (IX) 15 in a leap year, thus completing the sequence of the months. The same applies to the conjunction of the moon and the Pleiades on Nisannu (I) 3rd in a leap year.[4] The reconstruction of the first date simply assumes that the scheme should begin on the first day of the year.

Unlike all of the rules in the second intercalation scheme, the conjunction of the moon and the Pleiades is only late by two days in the leap year and not by a full month, according to the first rule. Several scholars try to explain the astronomical rationale for this rule. (Schaumberger, 1935) computes the longitudes of the moon and the Pleiades in both cases of conjunction, on Nisannu 1st and 3rd. However, according to his computations, the days in excess during the leap year amount to only about half a month.

[4] (Schaumberger, *1935*) considered the unpublished intercalation scheme found in Virolleaud ACh II Suppl. 19,22 as evidence for reconstructing the month Nisannu. However, the text states an allegedly contradictory statement that if the moon and the Pleiades are in conjunction on Nisannu 3rd, the year is normal. Schaumberger interprets the Akkadian there differently, but see (Hunger and Reiner, 1975) about the improbability of Schaumberger's interpretation. However, see below for a possible solution to the contradiction.

In addition, the heliacal rising of the Pleiades in a normal year is supposed to occur in the middle of Ajjaru (II) rather than on the first day of that month. Schaumberger himself admits that the intercalation rules depending on the moon's conjunction with the Pleiades are less accurate than the intercalation rules depending on the stars.

(Hunger and Pingree, 1989) suggest a different explanation. Following the assumption accepted throughout MUL.APIN (for example, II i 19-21) that the spring equinox of the ideal year occurs on Nisannu (I) 15th, Hunger and Pingree compute that on Nisannu 1st the moon is after the Hired Man and before the Pleiades, and not in conjunction with the Pleiades, as the verb *šitqulū* that describes the relation between the moon and the Pleiades usually means. Therefore, they conclude that the verb *šitqulū* means "closest to" in this context. However, as (Brown, 2000) rightfully points out, the verb *šitqulū* in an astronomical context generally means that two celestial objects are either in opposition or in conjunction. Brown concludes that the verb *šitqulū* should be interpreted as the exact conjunction in this context, and that the difference between the computation and the intercalation scheme indicates that the scheme was not derived from observations, but was instead a simple "ideal intercalation scheme," working with the triennial cycle. Brown accepts Koch's interpretation that the year discussed in the second intercalation scheme is an ideal year of 360 days, in which every month is exactly 30 days. He explains the intercalation rule of the moon and the Pleiades as assuming exactly such a month. Brown uses a daily lunar displacement of 13° per day. He derives this number from "1/30th of [360° (one month's revolution) + 30° (the additional movement of the earth in the same month)]." According to Brown, during the two days between Nisannu 1st and Nisannu 3rd, the moon passes 26°; after an additional day, the moon will have passed more than 30° total. Therefore, after two days the moon has moved approximately 30°. Brown claims that "[t]his is equivalent to saying that a month should be added when the lunar calendar has fallen behind the sidereal year by 30°."

I agree with Brown regarding the interpretation of the word *šitqulū* as the exact conjunction, and with his conclusion that the scheme was merely an ideal intercalation scheme and was not based on observations. However, I have a few reservations about Brown's explanation of the specifics of the intercalation rule.

1. Thirteen degrees per day is a good approximation of reality and does not contradict the first intercalation scheme, as demonstrated above. However, even without using the modern explanation that assumes the earth's movement, Brown never demonstrates that the author of MUL.APIN had any such knowledge of lunar velocity.
2. The time interval between the conjunction of the moon and the Pleaides on Nisannu (I) 1st in a normal year and their conjunction on Nisannu 3rd in a leap year is not two days, as these two occasions occur in two different years. If Brown is correct and the rule is as precise as the rule about intercalation happening every three years, then the time interval is three years and two days, not two days.
3. Brown never offers an interpretation of how this intercalation rule relates to the triennial cycle according to his explanation.
4. Brown's explanation does not clarify the second occurrence of conjunction of the moon and the Pleiades in the same intercalation scheme, in which the year should be intercalated if the conjunction occurs 30 days after its ideal date.

(Britton, 2007) has a different view of the set of intercalation rules. He agrees with Albani and Horowitz that the second intercalation scheme should be applied to the 354-day lunar year, and claims that since the triennial cycle was known to be inaccurate and required more frequent intercalations, the additional intercalation rules "amount to guides for determining when a 2-year intercalation was required." Britton's explanation is possible for the five rules that use the first visibility of the stars and constellations. If a stellar phenomenon occurs 30 days too late, it means that the date, determined by lunar observations, is 30 days too early. Therefore, intercalation of a 30-day month would resynchronize the moon with the stars. However, no such simple explanation can be attributed to the intercalation rules using the conjunction of the moon and the Stars. A simple way to determine whether or not these rules improve the precision of the triennial cycle is to actually do the computation, using the dates given in the text.

According to the scheme for the ideal calendar, the moon should be in conjunction with the Pleiades on Nisannu (I) 1st and on Araḫasamnu (VIII) 15th in normal years. If the moon returns to its original position on the 15th day of the eighth month, it has completed a whole number of cycles (*k* cycles of 360°) in 224 days, using the dates of the ideal 360-day year. Using the simple motion equation of distance equals the lunar daily displacement (*v*) times the time:

$$360k = 224v; k \geq 7$$

In this case the lunar daily displacement is not an integer. Therefore, it is easier to compute the sidereal month, the time it takes the moon to complete a cycle of 360°, which is an integer. In a schematic text like MUL.APIN, it is better to assume the usage of the rounder number of the sidereal month. We can use an approximation of the sidereal month by computing the period of time (T) it takes the moon to complete 360° at the mean daily displacement (v), as defined above.

$360/v=T$

Substituting v for $360/T$:

$k=224/T$

The length of the sidereal month (T) is dependent on how many cycles (k) the moon has completed in those 224 days. In reality, there are only two possible number of cycles:

$k=7$, $T=32$; $k=8$, $T=28$

Eight cycles is better, because 28 is closer to the true length of the sidereal month. It also aligns better with the text. Applying the 28-day sidereal month to a lunar year with alternating full and hollow months, starting from Nisannu 1st, the moon should complete a whole number of cycles (38 cycles) and return to its original position in relation to the stars after three years and two days, on Nisannu 3rd of the fourth year.

354x3+2=28x38

However, if this month is an intercalary month – an additional Addaru (XII_2) at the end of the third year – the next conjunction of the moon and Pleiades will occur on the first of Nisannu again. The same applies for the second intercalation rule: starting from Araḫasamnu (VIII) 15th, the moon should complete a whole number of cycles (39 cycles) and return to its original location in relation to the stars after three years and 30 days, in Kislimu (IX) 15th, as demonstrated by the following equation: 354*3+30=28*39

The mathematical calculations demonstrate that, unlike Britton's expectations, this intercalation rule does not improve the accuracy of the triennial cycle; the intercalation rule is exactly equivalent to it. Therefore, even though the triennial cycle is more accurate than the ideal 360-day year, the astronomy it uses is still very schematic astronomy. On the other hand, one cannot accept Brown's computation either, and it is evident that the lunar daily displacement in this scheme is not 13° per day.

Interestingly, the ideal position of the moon in relation to the stars was originally calculated with the assumption of the ideal 360-day year. Only the dates of the conjunctions in a leap year assume the triennial cycle. Moreover, a 28-day sidereal month works well with a 364-day year, as a complete number of 13 sidereal months fit into one 364-day year. This supports Horowitz and Albani's opinion that the triennial cycle should be applied to a schematic lunar calendar, and not to the ideal 360-day year.

3. TEXTUAL EVIDENCE: THE BABYLONIAN INTERCALATION SCHEME

Additional evidence for a 28-day sidereal month is found in a scheme for intercalation from 7th-century BCE Babylonia, published by (Hunger and Reiner, 1975).[5] The scheme lists 12 dates for the conjunction of the moon and Stars (Pleiades) in the following pattern: "If in month X on the Yth day you observe the Pleiades and the moon, and they have the same longitude, then this year is normal; if they fall down- it is left behind" (trans. Hunger and Reiner). The list of dates is summarized in Table III below.

Table III. Dates of Babylonian Intercalation Scheme

Month	Day
XII	25
I	23
II	21
III	19
IV	17
V	15
VI	13
VII	11
VIII	9
IX	7
X	5
XI	3

Hunger and Reiner analyzed the pattern of this scheme as follows: $\lambda_p = \lambda_0 + v(a + 26 - 2n)$; when λ_p is the longitude of the Pleiades, λ_0 the longitude of the sun at the preceding conjunction of the sun and moon, v the velocity of the moon in degrees per day (now commonly referred to as daily displacement), n the number of the month in question, and a the time between the conjunction of the sun and moon and the first appearance of the moon following it. Although their computation is accurate, it is unnecessarily complicated, and has nothing to do with the method that the authors of the intercalation scheme used. The authors assumed a 28-day sidereal month, during which the moon returns to the same position in relation to the stars (i.e. in conjunction with the Pleiades).[6]

[5] The edition is based on tablets K. 3923 + 6140 + 83-1-18, 479 (A); K. 9260 (B); and K. 12759 (C).

[6] In addition, later in their paper, Hunger and Reiner compute the dates of the beginning of the year according to the scheme. In their computation they use v=13;10,35 (written sexagesimally). Unfortunately, their computations reveal a discrepancy between the times for the beginning of the new year, as they are implied by each line in the text, and none of these times meet the historical data. They attribute these problems to the schematization of the intercalation rules. However, although the value that they used for lunar velocity is approximately the true value, it contradicts the value of lunar velocity assumed by the scheme. If the

Note that this scheme begins with Addaru, the 12th month. The dates seem to contradict the intercalation scheme of MUL.APIN. However, the contradiction may be attributed to the fact that this scheme comes from Babylonia. According to the Old Babylonian system, the equinoxes and the solstices fall in months XII, III, VI, and IX (van der Waerden, 1951; George, 1991; Al-Rawi and George, 1991/2; Horowitz, 1996). This system reappears in the Late Babylonian period (Britton, 1993). MUL.APIN uses the Neo-Assyrian system, in which the equinoxes and solstices fall in months I, IV, VII, and X. Therefore, an astronomical event that MUL.APIN dictates should occur on the first of Nisannu (I) would occur one month earlier, on the first day of Addaru (XII), if the Old Babylonian system were consulted. If we extrapolate the Babylonian scheme to an additional month beyond the calculations the scheme provides, the next conjunction of the moon and the Pleiades would fall, indeed, on the first day of Addaru. It seems as if the Babylonian scheme simply calculated the conjunctions backward from the beginning of Addaru. Although there is only evidence for this system from the Old and Late Babylonian periods, it does not seem to have been forgotten in the periods in between (although it was likely not practiced during the Neo-Assyrian reign). Indeed, this scheme published by Hunger and Reiner may be evidence of the use of the Babylonian system during the time between the two attested periods.

Another difference between the Babylonian intercalation scheme and the scheme in MUL.APIN is the structure of the rules. Intercalation rules are structured around two types of scenarios: A. If a correct phenomenon occurs on a later date than predicted, or B. An alternate phenomenon occurs on the anticipated date. While the rule in MUL.APIN follows the first structure (the conjunction occurs on Nissanu 3rd instead of Nissanu 1st), the Babylonian scheme follows the second structure (identifying the specific dates on which the moon and Pleiades are still apart, but should not be according to the predictions).

4. TEXTUAL EVIDENCE: ACH SUPPLEMENT II 19, 21-23

A similar intercalation rule concerning the conjunction of the moon and Pleiades, and using the second structure, is found in the unpublished tablet K. 3123, copied by (Virolleaud, 1905-1912) as Ach Suppl. II 19, 21-23 among other texts concerning the moon. It was probably the fourth tablet of the astrological commentary *Sîn ina tāmartišu* (Henceforth SIT 4), composed in the beginning of the first millennium BCE in Babylonia (Frahm 2011). An intercalary rule is found in lines 21-23:

21: If the Pleiad]es is in front, and the moon is behind - this year is right (normal); If the Pleiades is behind and the moon is in front - this year is left behind (intercalated).

22: If in Nisannu, on the third day, you observe the Pleiades and the moon, and they are in conjunction - this is right (normal);

23: If they fall down (not in conjunction) - left behind (intercalated)...

(Schaumberger, 1935) naturally assumed that this rule would agree with the intercalation rules of MUL.APIN, and therefore interchanged the interpretations of the terms *ezbet* (left behind) and *ešret* (right). (Hunger and Reiner, 1975) and (Brown, 2000) agree with each other that the intercalation rule from SIT 4 contradicts entirely the second intercalation rule of MUL.APIN. While in MUL.APIN a conjunction of the moon and the Pleiades indicates that a month should be intercalated, the rule from SIT 4 states the exact opposite, and defines such a year as normal. Here, I offer a different interpretation, reconciling the alleged contradiction without necessitating a reinterpretation of the terms.

All previous scholarly analyses of these lines read line 21 separately from lines 22-23. The rule of line 21 does not refer to a specific date, as the latter two lines do. However, without a date, line 21 has no meaning. The rule states that if the moon is to the west (in front) of the Pleiades, the year should be intercalated. However, every year includes a time at which the moon is to the west of the Pleiades. Schaumberger assumes that this rule refers to the time of the heliacal rising of the Pleiades, although it is not mentioned specifically in the text. However, if line 21 is connected to lines 22-23, then its rule must refer to Nisannu 3rd, as their rules do. Reading the two intercalation rules together, if on Nisannu 3rd the moon is to the west and the Pleiades are to the east, that year is a leap year. If, on the contrary, the moon is to the east and the Pleiades are to the west on Nisannu 3rd, the year is normal. The limit between these two situations is the moment at which the moon and Pleiades are in conjunction. A conjunction can be regarded in this sense as an infinitesimal moment between the different orders of the moon and Pleiades. This means that both MUL.APIN and SIT 4 agree that the day on which the astronomer should examine the relation between the moon and Pleiades is Nisannu 3rd. Both texts agree that the moment at which the moon passes the Pleiades is the determining factor in whether the year should be intercalated. The texts only disagree regarding the

moon passes 360° in 28 days, then its velocity is only approximately 12;51,25 degrees per day. Of course the value assumed by the scheme is not an accurate value, due to the schematic nature of the intercalation rules.

exact moment of the conjunction, a moment which is too short in reality to affect the actual calendar. It seems that the main difference between the two texts is the phrasing of the rule. While MUL.APIN is phrased according to the first structure, which examines the same phenomenon on different dates, SIT 4 is phrased according to the second structure, which examines different positions of the luminaries on the same date. It is not surprising, therefore, that SIT 4 also presents a triennial cycle right after these intercalation rules, as is the case for the second intercalation scheme in MUL.APIN.

5. CONCLUSION

I have demonstrated that the three early intercalation schemes, using the conjunction of the moon and the Pleiades, are in agreement with each other. All three have a schematic nature. Both the second intercalation scheme of MUL.APIN and the Babylonian scheme use the assumption of a 28-day ideal sidereal month to compute their rules. Both MUL.APIN and SIT 4 support a triennial cycle with the intercalation rule, a cycle that adapt an ideal 354-day lunar year with a 364-day approximated solar year. Since these rules are schematic, they do not fit other intercalation rules found in MUL.APIN. Despite the fact that in reality the conjunction of the moon and the Pleiades was used by Assyrian kings and their astronomers to determine if the year required an intercalation (Parpola, 1970-1983, letter n. 325), there is no evidence for the use of a certain attested scheme. Although these intercalation rules are improvements over the 360-day ideal year, they are still schematic in nature, are not based on observations, and do not improve the accuracy of the triennial cycle.

ACKNOWLEDGEMENTS

I thank Dr. Yigal Bloch from the Hebrew University for his assistance with the Akkadian.

REFERENCES

Al-Rawi, Farouk and George, Andrew (1991/92) Enūma Anu Enlil XIV and Other Early Astronomical Tables. Archiv für Oientforschung, Vol. 38/39, 52–73.

Albani, Matthias (1992) Die lunaren Zyklen im 364-Tage-Festkalender von 4Qmischmerot/4QSe. *Mitteilungen und Beiträge der Forschungsstelle Judentum der Kirchlichen Hochschule* (Leipzig), Vol. 4, 3-47.

Albani, Matthias (1994) *Astronomie und Schöpfungsglaube: Untersuchungen zum astronomischen Henochbuch.* Neukirchen-Vluyn: Neukirchener.

Brack-Bernsen, Lis (2005) The "days in excess" from MUL.APIN On the "first intercalation" and "water clock" schemes from MUL.APIN. *Centaurus*, Vol. 47, 1-29.

Brack-Bernse, Lis (2007) The 360-Day Year in Mesopotamia. *Calendars and Years: Astronomy and Time in the Ancient Near East*, Oxford, U.K.

Britton, John (1993) Scientific Astronomy in Pre-Seleucid Babylon. *Morgenländische Studien 3*, 61-76.

Britton, John (2007) Calendar, Intercalations and Year-Lengths in Mesopotamian Astronomy. *Calendars and Years: Astronomy and Time in the Ancient Near East*, Oxford, U.K.

Brown, David (2000) *Mesopotamian Planetary Astronomy-Astrology*. Groningen, STYX publication.

Chadwick, Robert (1992) Calendars, Ziggurats, and the Stars. *The Canadian Society for Mesopotamian Studies Bulletin*, Vol. 24, 7-24.

Englund, Robert K. Administrative Timekeeping in Ancient Mesopotamia. *Journal of Economic and Social History of the Orient*, Vol. 31, 122- 130.

Frahm, Eckart (2011) *Babylonian and Assyrian Text Commentaries. Origins of Interpretation, Vol. 5: Guides to the Mesopotamian Textual Record.* Münster, Ugarit-Verlag.

George, Andrew R. (1991) Review of H. Hunger & D. Pingree, MUL.APIN: An Astronomical Compendium in Cuneiform. *Zeitschrift fur Assyriologie*, Vol. 81, 301–306.

Horowitz, Wayne (1994) Two New *Ziqpu*-Stars Texts and Stellar Circles. *Journal of Cuneiform Studies* Vol. 46, 89-98.

Horowitz, Wayne (1996) The 360 and 364 Day Year in Ancient Mesopotamia. *Journal of the Ancient Near Eastern Society of Columbia University*, Vol. 24, 35–44.

Horowitz, Wayne (1998) *Mesopotamian Cosmic Geography*. Winona Lake, In., Eisenbrauns.

Hunger, Herman and Reiner, Erica (1975) A Scheme for Intercalary Months from Babylonia. *Wiener Zeitschrift für die Kunde des Morgenlandes*, Vol. 67, 21–28.

Hunger, Hermann and Pingree, David (1999) MUL.APIN: An Astronomical Compendium in Cuneiform. *Archiv fuer Orientforschung, Beiheft* 24, Horn, Austria.

Hunger, Hermann and Pingree, David (1999) *Astral Sciences in Mesopotamia* Leiden, Brill.

Koch, Johannes (1996) AO 6478, mul.apin und das 364 Tage-Jahr. *Nouvelles Assyriologiques Brèves et Utilitaires (NABU)*, Vol. 4.111, 97–99.

Koch, Johannes (1996) Kannte man in Mesopotamien das 364 Tage-Jahr wirklich seit dem 7. Jahrhundert v. Chr? *Nouvelles Assyriologiques Brèves et Utilitaires (NABU)*, Vol. 4.119, 109–115.

Parpola, Simo (1970-1983) *Letters from Assyrian Scholars to the Kings Esarhaddon and Ashurbanipal.* Kevelaer, Butzon & Bercker.

Schaumberger, Johann (1935) *Sternkunde und Sterndienst in Babel: 3. Ergänzungsheft zum ersten und zweiten Buch* Münster im Westfalen, Aschendorff.

Van der Waerden, Bartel L. (1951) Babylonian Astronomy. III. The Earliest Astronomical Computations. *Journal of Near Eastern Studies*, Vol. 10, 20–27.

Virolleaud Ch. (1908-1912) *L'Astrologie Chaldeenne: le livre intitulé Enuma Anu Enlil, Second Supplément.* Paris, Paul Geuthner.

Watson, Rita and Horowitz, Wayne (2011) *Writing Science before the Greeks: A Naturalistic Analysis of the Babylonian Astronomical Treatise MUL.APIN.* Leiden, Brill.

14

ANOTHER LOOK AT THE LUNAR SERIES AT DOS PILAS, GUATEMALA

Stanisław Iwaniszewski[1]

[1]Posgrado en Arqueología, Escuela Nacional de Antropología e Historia - Instituto Nacional de Antropología e Historia, Mexico

Corresponding author: siwanisz@yahoo.com

ABSTRACT

Astronomical observations of the ancient Maya consisted of a determination of various calendar cycles commensurating with the synodic cycles of the relevant celestial bodies. This practice of commensuration – of converting distinct synodic cycles into a single calendrical cycle – was a common practice in the life of the Maya day-keepers. By quantifying the multiples of synodic revolutions along a single scale of the Calendar Round combined with the Long Count, the Maya skywatchers created relationships easily represented and compared. The Lunar Series consisted of six glyphs, referred to by letters, E, D, C, X, B, and A and represented the attempts to create the cyclical calendrical structure capable of predicting the synodic period of the Moon. In this paper, I am providing a method of detecting possible intercalations needed to fit the lunar motion and a possibility of the use of a 4784-day period, derived from the Xultun Lunar Table, as a lunar correction cycle. By this means, all Lunar Series from Dos Pilas receive reasonable explanations.

KEYWORDS: Maya Astronomy, Lunar Theory, Lunar Series, Xultun Lunar Table, Dos Pilas monuments.

[1] The paper read at the 23rd SEAC Meeting at Rome was titled "Another Look at the Lunar Series from Dos Pilas and Naranjo". Due to the lack of space, I decided to withdraw the description of the Lunar Series at Naranjo. Instead, I expanded arguments aimed at reconstructing of the lunar correction cycle based on the Xultun Lunar Table. The Naranjo material will be published elsewhere.

1. INTRODUCTION

There is no evidence that the ancient Maya ever had a formal lunar calendar like in the ancient Near East or China. Instead, they devised a complex system of recording lunar months known as the Lunar Series. The Lunar Series is the phrase that contains six glyphs, called E, D, C, X, B and A by Sylvanus Morley (1916) and may briefly be read as

> "n (days after) the Moon (= one of the 18 lunar months) arrived. Glyph X is the (either young or sacred) name of 29 or 30" (see Zender and Skidimore 2012: 8-9; Kinsman n.d.).

This brief statement brings together information about the age of the current lunar month (Glyphs E and D), the number of a completed lunar month grouped in subsets of 6 or 18 differentiated months (Glyphs C), and the record whether the month is 29 or 30 days long (Glyphs A).

This evidence seems to support the idea that each lunar month began at the moment when the lunar crescent was first visible and sighted. Glyphs E and D just record the number of days after the moon has been sighted and we can safely interpret the moon's arrival as referring to her first visibility in the western sky. Changes in Glyphs C, X, and A indicate that the Maya experienced the appearance of the first moon as a sign that they should start a new time unit called "moon" or "month".

As already stated, Glyph A records the length of a lunar month. In each case, it has a coefficient of 9 or 10 while any other coefficients are unattested in the inscriptions. It, therefore, indicates that the length of the month was just either 29 or 30 days. The incorporation of Glyph A into the Lunar Series evidently marks attempts made by the Maya scribes to control the lunar count (Teeple 1931: 63). It may be supposed that their long-term experimentations produced regular and easily predictable sequences, providing strings of alternating 30-day and 29-day lunar months that are fully attested in the Late and Terminal Classic (600-1000 CE) texts, and which could mechanically be predicted in advance (but see below).

As the Maya scribes began to record historical events from an arbitrarily chosen starting point, identified by modern scholars with the year of 3114 BC, their count of days, known as the Long Count, was used both to record the political and religious statements and the movements of the heavenly bodies.

The Lunar Series was attached to dates written in the Long Count, always between the dates expressed by 260-day and 365-day calendars, and often in the company of other calendric-numerical series (the cycles of 7, 9, 63, and 819 days). Together with other dates and calendrical cycles, the Lunar Series functioned as a temporal adverbial part of the whole phrase.

Recorded on public monuments, both strings of differentiating glyphs, those representing the Long Count date and those representing the Lunar Series, visualized the temporal structure of the universe as the sequential flow of events represented by the changing shapes of glyphic forms.

Though during the Postclassic (1000-1542 CE) the major part of Maya chronological notations, including the Lunar Series, was abandoned, a kind of lunar notation was apparently in use at the time of Spanish Conquest. In his Relación de las Cosas de Yucatán Diego de Landa (Tozzer 1941[1566]: 133-134) notices the following

> "They divide it [= year] into two kinds of months, the one kind of thirty days and called *u*, which means "moon", and they counted it from the time at which new moon appeared until it no longer appears. They had the other kind of months of twenty days, and they called *uinal hunekeh*; of these it took eighteen to complete the year, plus five days and six hours."

2. THE PROBLEM

Because the lunar month is determined by a mechanical alternation of 29-day and 30-day units, the lunar record expressed by the Lunar Series continuously shifts about the synodic moon at a rate of about one day in 32.7 lunations. For the Lunar Series to keep up with the lunar synodic period, it was necessary to add (intercalate) a day to one of the 29-day months every 32-33 months. The practice of intercalation of a day to one of the 29-day months is attested both in the Eclipse Table of the Dresden Codex and in the Lunar Table from Xultun where the groups of 6 months sometimes yield 178 days instead of 177. These examples are sufficiently clear to conclude that intercalations consisted of the juxtaposition of 30-day months, eventually causing the interruption in a regular alternation of 30- and 29-day months. While the Dresden Codex intercalations appear to have been quite irregular, those recorded in the early ninth century Xultun Lunar Table were made at regular intervals of 886 days (Iwaniszewski 2014). Naturally, the Classic period Lunar Series recorded hundreds of years before the Xultun Lunar Table, may display irregular or *ad hoc* made patterns or different schemes of intercalation, and it may be difficult to find any regularity in the distribution of intercalary months. Be that as it may, scholars who have studied how these intercalations were made proposed different short-term or long-term solutions (Teeple 1931: 64-69; Beyer 1933, 1935, 1937; Satterthwaite

1947: 86-106; Barthel 1951: 233; Lounsbury 1978: 775-776; Justeson 1989: 88-90; Brauer 2007).

3. METHODOLOGY

In the present paper, I will adopt the emic perspective (i.e. the understanding of the other's ways of life in their own terms). While this point of view is often found in the anthropology of contemporary peoples, in archaeoastronomy we are dealing with past societies and cultures and the emic perspective cannot be directly approached. However, the ancient Maya perception and conceptualization of the lunar cycle may be derived from their written texts and from analogies based on the elements of colonial or contemporary Maya culture. It means that I will not attempt to correlate the recorded lunar dates with the European calendar, nor will I calculate them from a single arbitrarily assumed base. Since my research limits itself to the developments observed at a particular site, I will not compare the lunar data derived from one site with those found at all locations and times.

Throughout the paper, I am using the term of intercalation which applies to the recording of the lunar synodic period. By this, I mean the practice of adding of a day to a 29-day month in a sequence of alternating 29-day and 30-day months.

Finally, I will avoid if possible, modern values of the synodic lunar month, rather, I will employ the average length of 29.5 days as resulting from the regular alternating between 29 and 30 days. The structure of the Lunar Series utilizes a pattern of 30 and 29-day lunar months, and their terms will be used in this paper. For example, in each cycle of 6 lunar months, there are three 30-day and three 29-day months, in total, 177 days. I assume that raising a 29-day month to a 30-day one is the simplest way to intercalate, one intercalation implies there are four 30-day and two 29-day months, in total, 178 days.

My methodology is simple. In most cases during the Late Classic period (600 - 800 CE) the structure of the Lunar Series appears to be fixed, and the sequence of Glyph C head variants followed a predetermined pattern. Since the Lunar Series is always attached to a particular Long Count date, it is easy to determine the number of days separating two Lunar Series. Those intervals are then divided by 29.5 days. If the remainder is zero, it is interpreted that the number of 30- and 29-day months is the same (no intercalations made). If on the other hand, the remainder is found, it is not interpreted regarding mistakes made by the Maya scribes but rather concerning intercalations. The remainder is then analyzed to conclude how many more 30-day months were taken than those of 29 days. This result serves to find a suitable arrangement of the sequence of 30- and 29-day months. For instance, the number of 885 days divided by 29.5 equals to 30 without remainder indicates no intercalations. Now, the number of 886 days divided by 29.5 equals to 30 plus one day revealing that one intercalation was made (i.e. a day is added to one of the 29-day months). Therefore, the number of 886 days is composed of 30 lunar months, but in this case, there are 16 30-day and 14 29-day months (16 x 30 + 14 x 29 = 886). The Lunar Table at Xultun covers 4784 days, which divided by 29.5 produces the remainder of 5, indicating that five intercalations were made (86 x 30 + 76 x 29 = 4784). The Dresden Codex Eclipse Table covers 11960 days, which divided by 29.5 days yields the remainder of 12.5 days suggesting that 12 or 13 intercalations were made (215 x 30 + 190 x 29 = 11960). These numbers are the multiples of the 2392-day period of Teeple (1931), known as the Palenque formula. These numbers mean that the intercalations were made every 956.8 days (2.6196 tropical years, or 2.6214 *haab'* years) and that the synodic period had 29.5308642 days.

This point needs particular emphasis. To some extent, examples mentioned above may be misleading since they all refer to the same moon age. The computing procedure is not as straightforward as it appears to be and must be sensitive to the record of different moon ages (see below).

Finally, it must be said that sometimes the Initial Series date specifies the day when a monument was commissioned; sometimes we may expect that the Lunar Series was calculated by the Maya scribes backward from the date on which this monument was dedicated. Needless to say, this circumstance enables us to reconstruct their cognitive-computational procedures.

4. DOS PILAS

By the mid-seventh century, Dos Pilas was emerging as a regional capital extending its hegemony over the Pasión River Valley to control its trade routes. It rose under the guise of B'ajlaj Chan K'awiil who was born at Tikal in 625 CE but at the age of four (in 629 CE) was sent and installed on Dos Pilas throne to establish presumably the Tikal hegemony in the region and to substitute the local ruling families. Dos Pilas was never powerful enough to become an entirely independent polity but always was able to produce balanced and nuanced links with Tikal and Calakmul, the two Maya superpowers. The Lunar Series at Dos Pilas is recorded on six monuments erected within the period between 682 and 735 CE, covering the time span of 53 years (see Table 1). Five records of the Lunar Series are found on stelae and two on Hieroglyphic Stairway 4. Three Lunar Series texts are attached to the IS dates that commemorate period endings (stelae 14, 15, and 2) while four are associated with non-period-ending dates. Basic information is provided in Table 1.

During that time Dos Pilas was ruled by three lords: B'ajlaj Chan K'awiil (Ruler 1), Kokaaj K'awiil (formerly Itzamnaaj K'awiil, Ruler 2), Vega Villalobos 2012) and Ucha'an K'in B'alam (Ruler 3).

The earliest known Lunar Series at Dos Pilas is known from the Hieroglyphic Stairway 4 (see Table I). This monument was dedicated in 682 by B'ajlaj Chan K'awiil (Ruler 1) to commemorate half Period Ending (Step II: 9.12.10.0.0 9 Ajaw 18 Sotz' or 8.05.682). About two and a half years later another step was added (Step I) to celebrate his third k'atun anniversary (see Table 2; Symonds et al. 1992: 215; Guenter 2003).

Table 1. Dos Pilas summary. IS = Initial Series, LS = Lunar Series, DD = Dedicatory Dates, HS = Hieroglyphic Stairway, PE = Period Ending; Entries: 1, 2 – according to Guenter 2003: 34-36; 5 – Mathews 2001 [1979]. The letters E, D, C, X, B, and A stay for Glyphs E, D, C, X, B, and A respectively. Combined with Glyph C, are its lunar patrons denoting, s- skull (God A), m - Young Maize (Tonsured) God, and j - Jaguar God of the Underworld. Numerical coefficients with Glyphs E, D, and A stay for the number of days, numerical coefficients with Glyph C denote numbers of lunar months. Finally, the nomenclature for Glyph X is: Roman number designate three main classes associated with three lunar patrons of Glyph C while their numerical coefficients are linked to those of Glyph C.

No	Monument	IS Date	LS	Event	Protagonist	DD
1	HS4 Step II	9.12.10.0.0 9 Ajaw 18 Sotz'	2ED 3Cm II.3 A10	half-PE, Stairway 4 construction	B'ajlaj Chan K'awiil	9.12.10.0.0
2	HS4 Step I	9.12.12.11.2 2 Ik' 10 Muwan	20ED 5Cs I.5 A10	60th tuun birthday, dance	B'ajlaj Chan K'awiil	9.12.12.11.2
5	Stela 8	9.12.[0.10].11 13 Chuwen 19 Kayab [9.12.6.15.11 11 Chuen 4 Sip]	3D 3Cs	Birth of Kokaaj K'awiil	Ucha'an K'in B'alam "Master of Sun Jaguar"	9.14.15.5.15
3	Stela 14	9.14.0.0.0 6 Ajaw 13 Muwan	16D 3Cm	PE, "stone binding" *k'altuun* ritual	Kokaaj K'awiil	9.14.5.3.14
4	Stela 15	9.14.10.0.0 5 Ajaw 3 Mak	17D 5Cj	half-PE casting incense ceremony	Kokaaj K'awiil	9.14.10.4.0
6	Stela 5	9.15.0.0.0 4 Ajaw 13 Yax	11D 1Cm II.1 A10	PE-celebration	Uch'an K'in B'alam	9.15.0.0.0
7	Stela 2	9.15.4.6.4 8 K'an 17 Muwan	10D 1Cm II.1 A10	War against Seibal "star-over-Seibal"	Uch'an K'in B'alam	9.15.5.0.0

Both dates are separated by 942 days and the same number of days is observed between two Lunar Series. The sequence of Glyph C head variants combined with Moon Age records (Glyphs D and E) indicates that the lunar months should have followed a regular order of 30 and 29 days. It is easy to compute that 942: 29.5 = 31 plus 27.5 days indicating the Maya scribes calculated 16 months of 29 days and 15 months of 30 days plus 28 days. There is no place for an intercalation. The number of 942 days may be structurally represented as 531 + 177 + 177 or 885 + 57 = 942 days.

Table 2. Monuments with the Lunar Series commissioned by B'ajlaj Chan K'awiil

Entry	Monument	IS Date	Days	Difference (in days)	LS recorded	Computed LS
1	HS4 Step II	9.12.10.0.0	1,386,000	942	22 3Cm 30	22 3Cf 30 + 942 = 20 5Cs 30
2	HS4 Step I	9.12.12.11.2	1,386,942		20 5Cs 30	

Kokaaj K'awiil (formerly known as Iztamnaaj K'awiil, Ruler 2, 673 – 726 CE, Vega Villalobos 2012), was the son of B'ajlaj Chan K'awiil and ascended to the throne in 698. My first step in analysis is to compare the last LS record of B'ajlaj Chan K'awiil (HS4 Step I (V)) with the first record of the LS of Kokaaj K'awiil (Stela 14, see Table 3). The time interval between two IS dates yields 9,858 days, indicating there were 334 months of 29.5 days plus five days, or 167 x 30 + 167 x 29 days plus 5 days. However, the difference between the LS computed and recorded amounts to 9 days (25D computed against 16D recorded, see Table 3) implying that nine intercalations were made. For better clarity, this may be provisionally interpreted either as (9 x 532 + 9 x 531 + 177 + 89 + 25) or (9 x 886 + 2 x 885 + 89 + 25).

Table 3. The last rectord of B'ajlaj Chan K'awiil (HS4 Step I) compared with the first record of Kokaaj K'awiil (Stela 14)

Entry	Monument	IS Date	Days	Difference (days)	LS recorded	Computed LS
2	HS4 Step I	9.12.12.11.2	1,386,942	9,858	20 5Cs 30	20 5Cs 30 + 9,858 = 25 3Cj 30
3	Stela 14	9.14.0.0.0	1,396,800		16 3Cj	

Table 4. The Lunar Series recorded during the reign of Kokaaj K'awiil

Entry	Monument	IS Date	Days	Difference (in days)	LS	Computed LS
3	Stela 14	9.14.0.0.0	1,396,800	3600	16 3Cj	16 3Cj + 3600 = 17 5Cm
4	Stela 15	9.14.10.0.0	1,400,400		17? 5Cm	

Kokaaj K'awiil commissioned stelae 14 and 15 with the Lunar Series to commemorate period intervals and to describe his military achievements and document famous shell-star wars (Houston 1993: 111). In first years of his reign, he defeated Tikal, however, for unknown reasons, the Maya scribes recorded the abbreviated LS resembling the Early Classic Tikal system of registering the lunar data. The time elapsed between the IS dates and Lunar Series is in agreement (3,600 days) supporting the reading of the Moon Age on Stela 15 as 17 days (see Table 4). These data are sufficient to deduce that no effort was made to add intercalary days within this period.

After his death in CE 727 at Dos Pilas throne was installed Ruler 3 (Ucha'an K'in B'alam, possibly as a regent for K'awiil Chan K'inich (Ruler 4). This ruler left the Lunar Series on three monuments.

Following my previous procedure first I compare the latest Lunar Series registered during the reign of Ruler 2 with the earliest Lunar Series of Ruler 3. Both rulers commissioned monuments commemorating period endings (Stelae 15 and 5). My computations indicate that between those dates seven intercalations occurred (see Table 5).

Table 5. The latest LS of Ruler 2 compared with the earliest LS of Ruler 3

Entry	Monument	IS Date	Days	Difference (days)	LS recorded	Computed LS
4	Stela 15	9.14.10.0.0	1,400,400	3,600	17 5Cm	17D 5Cm + 3600 = 18D 1Cm A30
6	Stela 5	9.15.0.0.0	1,404,000		11 1Cm A30	

Now, the magnificent Stela 8 shows that Ruler 3 was enthroned only 76 days after the death of Ruler 2. This monument describes important dates in the life of Ruler 2, including his birth and death, and probably ends with the celebration of the 15th k'atun ending (at 9.15.0.0.0, AD 731, Mathews 1979[2001]: 404) under the auspices of Ruler 3. The text on the back of Stela 8 begins with the IS date referring to the birth of Kokaaj K'awiil to which the abbreviated LS is added (that is, 9.12.0.10.11, or AD 673). This date is recorded as 9.12.6.15.11, but corrected as 9.12.0.10.11. This correction is made on calendrical grounds (see for details Mathews 2001[1979]: 397-398, Martin and Grube 2008: 59; Vega Villalobos 2012 58-59). The lunar date is probably backward computed from the monument's last presumed date, 9.15.0.0.0 (see Table 6). If we assume that the last date on Stela 8 marked the period ending at 9.15.0.0.0, then the Lunar Series would be computed backward from this last date, covering the time span of more than 58 years (21,389 days). Fortunately, the lunar data for the date 9.15.0.0.0 are given by Stela 5 dedicated by Ruler 3 to commemorate k´atun ending, and I think it is reasonable to suppose that the same information was known to the scribes who made computations on Stela 8 (see Table 6). The number of days (21,389) divided by 29.5 yields 725 with the remainder of 1.5 days. The Lunar Series counted backwards (362 x 30 + 362 x 29 + 31) arrive at 9D 2 Cs, whereas the recorded data yield 3D 3Cs.

The difference between these two records amounts to 23 days, indicating that between 9.15.0.0.0 and 9.12.0.10.11 the Maya scribes made 23 intercalations (see Table 6).

Table 6. The comparison of the Lunar Series on Stelae 8 and 5

Entry	Monument	IS date	Days	Difference (days)	LS recorded	Computed LS
5	Stela 8	9.12.0.10.11	1,382,611	21,389	3 3Cs	11 1Cm 10
6	Stela 5	9.15.0.0.0	1,404,000		11 1Cm A30	- 21,389 = 9 2Cs 9

The first conclusion that can be drawn from this analysis is that Rulers 2 and 3 made intercalations with very irregular frequency. Columns 2 and 3 in Table 8 clearly show that periods without intercalations were followed by periods when intercalations were often made. It seems that after their accession both rulers were tempted to add days to avoid shifts in relation to the synodic month, but later the lunar data followed this shifting again. These data suggest there was no regular pattern at all. It is interesting to observe that number of intercalations between a first date and the last one is the same as displayed in columns 3 and 5. While intercalations in column 3 were presumably made in irregular fashion, those in column 5 seem to follow an unknown numerological model.

A few years after the celebration of the 15th k'atun, Ruler 3 commissioned Stela 2 which begins with the IS date recording 9.15.4.6.4 (AD 735). The IS date on Stela 2 falls 1,564 days after the last Period Ending at 9.15.0.0.0 recorded on Stela 5 (see Table 7), whereas the Lunar Series calculable from the same date is wrong. The Lunar Series of 9.15.4.6.4 would have been 11D 6Cs A29 if the scribes reckoned the lunar months without intercalations. Instead, Stela 2 carries the Lunar Series read as 10D 1Cm A10, which indicates that the lunar count has the Long Count position of 9.15.4.7.12 or 1,592 days after the Period Ending at 9.15.0.0.0. What happens is that the IS date refers to the interval of 53 lunar months (18+18+17), but the LS event relates to the period of 54 lunar months (18+18+18). In other words, the 54th lunar month is being ascribed to the IS date of the 53rd lunar month and one lunar month is suppressed (consult Table 7). One of the units of 6 or 18 lunar groups is shortened to contain only 5 or 17 such months. Now, the interval 1564 days equals to 27x30 + 26x29.

Table 7. The comparison of the Lunar Series of Stelae 5 and 2

Entry	Monument	IS Date	Days	LS recorded	Computed IS	Computed LS
6	Stela 5	9.15.0.0.0	1,404,000	11D 1Cm 30	11 1Cm 30	11 1Cm 30
7	Stela 2	9.15.4.6.4	1,405,564	10D 1Cm 30	+ 1564 = 11 6Cs 29	+ 1592 = 10 1Cm 30

Though Dütting (1986:124) defined this interval as 8 x 6 plus 1 x 5 lunations, counted from 9.15.0.0.0, other explanations suggest an eclipse interest. The distance of 18, 18, and 17-month (together 53 lunar months) intervals implies that the 17-month period consists of subsets of 177, 177 and 148 days, and the presence of a 148-day interval may involve an eclipse computation. A rough confirmation of this possible explanation is found in the Dresden Codex Eclipse Table (Dresden 55b) where one of the intervals recording eclipses yields 1565 days. According to Aveni (1980: 80, Table 5), this number is useful to commensurate 53 lunar months with 57.5 draconic months (of 27.21222 days each) that mark the average interval between successive passages of the moon through the node: 57.5 x 27.21222 = 1564.70265 days. Here, the cycle of 57.5 draconic months implies that the moon is (was) observed at the opposite node. However, we do not know whether the number of 1564 days from Stela 2 has the same significance as the interval of 1565 days recorded in the Dresden Codex (but see below). It may or may not be interpreted as concerning eclipse watching.

The IS date refers to the event which is 28 days (roughly one lunar month) less than the date implied by the recorded Lunar Series. The moon age recorded on Stela 2 will be correct if we add 28 days more to the restored Lunar Series to arrive at 9.15.4.7.12, that is, 1592 days from the starting point at 9.15.0.0.0. It also means that the Maya did not reckon any intercalation during the period that started with the Period Ending at 9.15.0.0.0, and the date recorded on Stela 2.

Stela 2 commemorates the victory of Ruler 3 over Seibal and is known for the substitution of the Dos Pilas Emblem Glyph, which probably originated at Tikal, for its new version (Houston 1993: 98-99, Fig. 4-1; Martin and Grube 2008: 63). The conquest of

Seibal may, therefore, mirror attempts made by Ruler 3 to establish a more distinctive identity for Dos Pilas. The lunar data appear to herald significant changes affecting the structure of the Lunar Series, signaling its possible use to track the eclipses.

The "star over Seibal" war event mentioned on Dos Pilas Stela 2 is also recorded on Aguateca Stela 2. Unfortunately, the Aguateca monument does not include a Lunar Series. Both monuments also state that one day later, on 9.15.4.6.5 9 Chicchan 18 Muwan, Ruler 3 captured Yich'aaak B'alam, the ruler of Seibal. This date is separated by 1565 days from 9.15.0.0.0.

Together with many similar statements, the "star over Seibal" Emblem Glyph motif on Stela 2 gave rise to hypotheses that the Maya timed battles or other warlike events according to the stations of Venus (e.g. Lounsbury 1982; Schele and Freidel 1990). This date falls roughly one month after the superior conjunction and coincides with the day of the first appearance of Venus as Evening Star (Lounsbury 1982: 152-53; Aveni and Hotaling 1994: S27 Table 1). As already recognized by Kelley (1977: 64 Table 5.1, 65.) the Evening Star event may also be deduced from the Dresden Codex Venus Table base at 9.9.9.16.0. Supposing that the base date denoted the first appearance of Venus as Morning Star, its first evening appearance is marked as the 326-day of the whole cycle of 584 days (see Dresden 46). In other words, we have the following computing:

9.9.9.16.0 + 326 = 9.9.10.14.6 2 Kimi 19 Muwan + 70 x 584 = 9.15.4.6.6 10 Kimi 19 Muwan

But even, in this case, there is no obvious patterning of these dates. Both monuments, Dos Pilas Stela 2 and Aguateca Stela 2, do not record the date of 9.15.4.6.6 indicating that any reference to this event was not intended, or that the "star over Seibal" glyph referred to the observed phenomenon, not the computed one. While the planet Venus was believed to act a kind of patron of war, there is very limited evidence that Venus stations had ever been utilized to schedule military events (Aldana 2005; Bricker and Bricker 2011:245-248). The study of this topic is beyond the scope of the present paper.

5. INTERCALATIONS

Table 8 lists all the lunar dates. The first column gives the recorded long count date while the second determines intervals separating two successive dates. The third column gives the estimated number of intercalations and below the average number of days between those intercalations (rounded off to the nearest full day). In general, within the span of 22,953 days, the Maya scribes presumably made 23 intercalations which the average of one intercalation per 998 days. The record registered on Stela 5 and given in the fourth column, which relates the number of 23 intercalations made within the span of 21,389 days offers an opportunity to infer the sequence of intercalations. We can guess that from 9.15.0.0.0 at Dos Pilas, the intercalation was fixed through the adoption of a 4784-day cycle (comprising 5 intercalations). At least, this cycle was used to perform backward lunar computations:

[9.15.0.0.0] 2D 3Cs 30 - 21, 389 days (= 4 x 4784 + 2 x 886 + 481) = [9.12.0.10.11] 2D 3Cs

This calculation predicts 22 intercalations: 5 with-in the each cycle of 4784 days and 1 within the cycle of 886 days. It is observed, however, that the final result varies 1 day from the recorded moon age: it has 2 days instead of 3 days displayed on Stela 5. One day difference may indicate that the Maya scribes already performed the 23rd intercalation, or that the lunar month (not indicated by Glyph A) has 29 days in-stead of 30 days advocated by the 4784-day cycle.

Table 8. Hypothetical intercalations at Dos Pilas.

IS Date	Duration (days) between neighbor dates		Intercalations made between neighbor dates /below are average intercalations per days		Duration between the dates on telae Stelae 8 and 5 days)	Intercalations made/average intercalations per day
9.12.0.10.11	3389		7		21389	23
9.12.10.0.0		942	484	0		930
9.12.12.11.2	9858		9			
9.14.0.0.0		3600	1095	0		
9.14.10.0.0	3600		7			
9.15.0.0.0		1564	514	0		
9.15.4.6.4						
total	22953		23 998			

It is also observed that all other intercalations inferred, those between Stelae 14 and HS Step I and between Stelae 15 and 5 cannot utilize the 4784-day intercalary mechanism. It may suggest that the Maya scribes could not afford a reasonably accurate model of intercalation or decisions to determinate intercalations were made *ad hoc*, at the will of the rulers or elite skywatchers.

6. CONCLUDING REMARKS

The study of celestial motions recorded by the Maya scribes has too often depended on a kind of correlation of the Long Count with the European calendars. The lack of references to the Julian Calendar dates or the concept of the synodic lunar month enabled my research to attempt to infer the rules of the Maya scribes for perceiving, recording and explaining the motion of the Moon. Such an approach helped to explain all variations observed in the moon age data, within the reasonable framework of hypothetical intercalations and confirmed the non-altered structure of the Lunar Series. It also allowed for the application of an intercalary scheme indicated by the Xultun Table. Finally, my study advocates the idea that most Late Classic lunar data were computed not observed.

ACKNOWLEDGEMENTS

The study of the Lunar Series at Dos Pilas and Naranjo was supported by the Instituto Nacional de Antropología e Historia (project 296 "La teoría lunar en astronomía maya"). I thank an anonymous reviewer who has found errors in my draft.

REFERENCES

Aldana, G. (2005) Agency and the "Star War" Glyph: A historical reassessment of Classic Maya astrology and warfare. *Ancient Mesoamerica*, Vol. 16, 305-320.

Aveni, A. F. (1980) *Skywatchers of Ancient Mexico*. Austin, University of Texas Press.

Aveni, A.F and Hotaling, L.D. (1994) Monumental Inscriptions and the Observational Basis of Maya Planetary Astronomy. *Archaeoastronomy*, Vol. 19 (supplement to the *Journal of the History Astronomy*, Vol. 25), S21-S54.

Barthel, T. S. (1951) Maya-Astronomie: Lunare Inschriften aus dem Südreich. *Zeitschrift für Ethnologie*, Vol. 76, 216-238.

Beyer, H. (1933) Zur Konkordanzfrage der Mayadaten mit denen der christlichen Zeitrechnung. *Zeitschrift für Ethnologie*, Vol. 65, 75-80.

Beyer, H. (1935) On the Correlation between Maya and Christian Chronology. *Maya Research*, Vol. 2, No. 1, 64-72.

Beyer, H. (1937) Lunar Glyphs of the Supplementary Series at Piedras Negras. *El Mexico Antiguo*, Vol. 4, No. 3-4, 75-81.

Brauer, T. (2007) A New Approach for Systemizing the Dates in the Maya Lunar Series. *Mexicon*, Vol. 29, No. 4, 99-102.

Bricker, H.M. and Bricker V.R. (2011) *Astronomy in the Maya Codices*. [Memoirs of the American Philosophical Society, Vol. 265), Philadelphia, The American Philosophical Society.

Dütting, D. (1986) Lunar Periods and the Quest for Rebirth in the Mayan Hieroglyphic Inscriptions. *Estudios de Cultura Maya*, Vol. 16, 113-147.

Guenter, S. P. (2003) The Inscriptions of Dos Pilas associated with B'ajlaj Chan K'awiil. Mesoweb:<www.mesoweb.com/features/guenter/DosPilas.html.>

Houston, S. D. (1993) *Hieroglyphs and History at Dos Pilas: Dynastic Politics of the Classic Maya*. Austin, University of Texas Press.

Iwaniszewski, S. (2012) Los ciclos lunares y el calendario maya. *Arqueología mexicana*, Vol. 19, No. 118, 38-42.

Iwaniszewski, S. (2015) Counting Lunar Phase Cycles in Mesoamerica. In *Handbook of Archaeoastronomy and Ethnoastronomy*, Ruggless, C.L. N. (ed., New York, Springer Reference, 709-714

Justeson, J. S. (1989) Ancient Maya ethnoastronomy: an overview of hieroglyphic sources. In *World Archaeoastronomy*, Aveni, A.F. (ed.) Cambridge, Cambridge University Press, 76-129.

Kelley, D. H. (1977) Maya astronomical tables and inscriptions. In *Native American Astronomy*, Aveni, A.F. (ed.), Austin, University of Texas Press, 57-73.

Kinsman, H. (n.d.) Grammar in the Script: Nouns and Statives. Crabs and Glyphs "Maya Glyph Decipherment including Notes on Ancient Maya Astronomy. www.crabsandglyphs.com.org , text acceded 22.06.2015.

Lounsbury, F. G. (1978) Maya Numeration, Computation, and Calendrical Astronomy. In *Dictionary of Scientific Biography*, vol. 15, suppl. 1, Gillespie, C. C. (ed.), New York, Charles Scribner's Sons, 757-818.

Lounsbury, F.G. (1982) Astronomical Knowledge and Its Uses at Bonampak, Mexico. In *Archaeoastronomy in the New World*, Aveni, A.F. (ed.), Cambridge, Cambridge University Press, Cambridge, 143-168.

Martin, S., and Grube, N. (2008) *Chronicle of the Maya Kings and Queens. Deciphering the Dynasties of the Ancient Maya*. New York, Thames & Hudson.

Mathews, P. 2001[1979] Notes on the Inscriptions on the Back of Dos Pilas Stela 8. In *The Decipherment of Ancient Maya Writing,* Houston, S.D., Chinchilla Mazariegos, O. F., and Stuart, D. (eds.), Norman, University of Oklahoma Press, 394-415.

Morley, S. G. (1916) The Supplementary Series in the Maya inscriptions. In *Anthropological Essays. Holmes Anniversary Volume*, Washinton, J.W.Bryan Press, 366-396.

Satterthwaite, L., Jr. (1947) *Concepts and Structures of Maya Calendrical Arithmetics*. [Museum of the University of Pennsylvania and the Philadelphia Anthropological Society Joint Publication, 3].Philadelphia, University Museum.

Schele, L. and Freidel, D. 1990 *A Forest of Kings: the Untold Story of the Ancient Maya,* New York, William Morrow.

Symonds, S. Houston, S.D., and Stuart D. (1992) Implicaciones del descubrimiento de una nueva escalinata Jeroglífica en Dos Pilas, Petén. In *Simposio de Investigaciones Arqueológicas en Guatemala,1990,* J. P. Laporte, J.P., Escobedo, H. and Brady, S. (eds.), Guatemala, Museo Nacional de Arqueología y Etnología, 214-217.

Teeple, J. (1931) *Maya Astronomy*. [Publication 403], Washington, Carnegie Institution of Washington.

Tozzer, A. M. (1941) *Landa's Relación de las cosas de Yucatan*. Translated and edited by Alfred M. Tozzer. [Papers of the Peabody Museum of American Archaeology and Ethnology, Harvard University, vol. 18], Cambridge, Mass.

Vega Villalobos, M. E. (2012) Los dioses, el gobernante y la comunidad. Las estrategias político-religiosas del Gobernante 2 de Dos Pilas, Guatemala. *Estudios de Cultura Maya,* Vol. 40, 51-78.

Zender, M. and Skidmore, J. (2012) Unearthing the Heavens: Classic Maya Murals and Astronomical Tables at Xultun, Guatemala. *Mesoweb Reports*. www.mesoweb.com/reports/Xultun.pdf, electronic text consulted 1.12.2012.

15

ARA PACIS AVGVSTAE: AN ASTRO-ARCHAEOLOGICAL ANALYSIS

Vance R. Tiede

Astro-Archaeology Surveys

Corresponding author: ***vance.tiede@aya.yale.edu***

ABSTRACT

The *Ara Pacis Augustae* (Altar of Augustan Peace) is examined as a shrine of light consistent with the astro-orientation principles of Roman architect Vitruvius (c.25 BC). Italian archaeologists excavated (1937-38) and relocated the altar along Rome's Tiber River – rotating it 75° counter-clockwise from the original orientation. While its magnificent marble sculpture may be viewed at the *Museo dell'Ara Pacis,* a key astronomical component of the altar's architectural design has until now remained encoded on a paper site map. The author calculated the horizon declination (+11.3° decl.) of the altar's original northeast opening with Program STONEHENGE (Hawkins 1983, 328-330), Guglielmo Gatti's 1938 site map, and Google Earth satellite imagery: True Azimuth ≈ 75°; Latitude/Longitude = 41° 54′ 10.47″ N/12° 28′ 44.60″ E; Elevation ≈ 10 meters ASL; and Horizon Altitude ≈ +1.1°. The *Ara Pacis* was dedicated in 9 BC (Moretti 1947,7). Analysis shows that the axis of the *Ara Pacis* enclosure's northeast opening originally was oriented to the rising Sun (+11.0° decl.) on both April 21st and August 27th. However, the iconography of the *Ara Pacis* supports an interpretation that only the April 21st sunrise orientation was intended, because each sculptural element of the *Tellus* frieze (young woman, flowing amphora, pair of fishes, ram, bull, infant twins) uniquely mirrors the astronomy: Venus (Morning Star phase) and the Zodiac constellations Aquarius-Pisces-Aries-Taurus-Gemini at dawn on the festival of *Par ilia* (Pales, goddess of shepherds), 21 April 9 BC. This month and day coincide with the founding of Rome on 21 April 1 *ab urbe condita* (*a.u.c*) as recorded by Roman historian Marcus Terentius Varro. The dedication year of 9 BC coincides with the 93rd iteration of the eight-year Earth-Venus Synodic Period when Venus also appeared as Morning Star on the first *Dies Natalis Romae*, 21 April 753 BC.

KEYWORDS: Ara Pacis, archaeoastronomy, Augustus, Roman architecture, Roman religion, Venus, Vitruvius, Tellus

1. INTRODUCTION

1.1 Background

The Roman Senate voted on 4 July 13 BC to honor Emperor Gaius Julius Caesar Octavianus Augustus (b. Equinox, 23 September 63 BC, Imperator 31 BC, d. 19 August 14 AD) with the *Ara Pacis Augustae* (Altar of Augustan Peace) in recognition of his pacification of the enemies of Rome, legal reforms (*leges Iuliae*), and revival of traditional Roman religion (prompted by newly arrived Mystery Cults from the East):

> *When I returned from Spain and Gaul after successfully settling the affairs of those provinces, in the consulship of Tiberius Nero and Publius Quintilius [13 BC], the Senate decreed that the Altar of Pax Augusta should be consecrated for my return near the Campus Martius, and ordered that the magistrates, priests, and the Vestal Virgins should there make an annual sacrifice. (Augustus 14 AD, XII.ii)*[1]

The *Ara Pacis Augustae* was dedicated on 30 January 9 BC, about 83m (≈ 280 Roman Feet) northeast of the contemporary *Horologium Augusti* obelisk whose shadow is claimed to have pointed towards or even touched the *Ara Pacis* steps every Equinox (Buchner 1982; 2007; IDIA Lab 2013; Frischer et al. 2014; cf. Heslin 2007, 10-15). Apart from the modern discussion regarding an *Horologium – Ara Pacis* alignment on the birth date of Caesar Augustus, Vitruvius (c. 75-15 BC) makes clear what he understood as the geometry of gnomon shadow length at the Equinox:

> *...we must ourselves explain the principles which govern the shortening and lengthening of the day. When the sun is at the equinoxes, that is, when passing through Aries and Libra, he makes the gnomon cast a shadow equal to eight ninths of its own length, in the latitude of Rome. (Vitruvius, De Architectura, IX. Vii. 1)*

Italian archaeologists excavated (1937-38) the *Ara Pacis* buried some 10 meters under Rome's *Via in Lucina* no.16 and reassembled it 455 meters northwest along the left bank of the Tiber River (Moretti 1947, 6; Andersen 2003, 5). Ignoring Vitruvius' astro-orientation principles, they rotated the altar 75° counter-clockwise from its original orientation. While the altar's marble sculpture, digitally reconstructed in polychrome by art historians (Foresta 2011; Rossini 2015), may be seen at the *Museo dell'Ara Pacis* in Rome, until now the significance of the altar's original orientation has remained encoded on a paper map (Gatti 1938). However, Vitruvius makes clear that the astronomical orientation of Roman public architecture was intended as a metaphor written in stone linking the Emperor's civil authority on Earth below with that of the universal order (*kosmos*) of the Immortal gods above:

> *But when I saw that you [Caesar Augustus] were giving your attention...to the providing of public buildings...so that the greatness of its [SPQR] power might likewise be attended with distinguished authority in its public buildings, I thought that I ought to take the first opportunity of lay before you my writings on this theme.* (Vitruvius, I.preface.2)

Not only did Vitruvius write "that one who professes himself an architect should be ...acquainted with astronomy and the theory of the heavens" (Vitruvius, I:i:iii), but two of his astro-architectural principles which were incorporated into the design of the *Ara Pacis*:

> *Altars should face the east, and should always be placed on a lower level than are the statues in the temples, so that those who are praying and sacrificing may look upwards towards the divinity.* (Vitruvius, IV.ix.1)

> *The quarter toward which temples of the immortal gods ought to face is to be determined by the principle that....the temple and statute place in the cella should face the western quarter of the sky. This will enable those who approach the altar with offerings or sacrifices to face the direction of the sunrise in facing the statue in the temple, and thus those who are undertaking vows look toward the quarter from which the sun comes forth and likewise the statues themselves appear to be coming forth out of the east to look upon them as they pray and sacrifice.* (Vitruvius, IV.v.1)

1.2 Description

The marble altar of the *Ara Pacis* is located inside a rectangular walled enclosure (Figure 1). The architectural inspiration of twin opposing openings of the *Ara Pacis* has been attributed to both the (as yet unlocated) Shrine of Janus Geminus (TPQ 753 BC) in Rome, and to the Altar of the Twelve Gods (c. 450 BC) in the Agora of Athens (Müller 1943, 437; Thompson 1952, 79-82; Kleiner 1998, 90; Mauzy 2006, 9,90-91). Archaeological excavation confirms that the opposing enclosure openings of the Greek Altar of the Twelve Gods and of the Roman *Ara Pacis* are both oriented SW-NE (Figure 1; Tiede 2015, 36).

[1] Caesar Octavianus Augustus *Res Gestae Divi Augusti*, http://en.wikisource.org/wiki/Res_Gestae_%28Anon%29 ; Tiede 2015, 2.

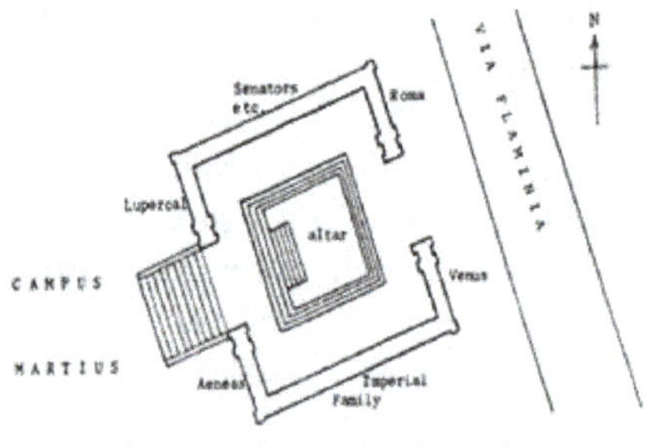

Figure 1. Original Orientation of the Ara Pacis (Toynbee 1953, 56; Gatti 1938 in Moretti 1947, 17)

Unlike its Athenian antecedent, the sculpture of the Roman *Ara Pacis* is remarkably well preserved. The exterior wall panels on the north face depict Roman senators in procession from the *Via Flaminia* (modern *Via del Corso*) to the western stairway, while the southern panel mirrors the processional motif with the imperial family and entourage. The southwest stairway is flanked by twin friezes depicting scenes of the mythological founding of Rome. The northeastern raised opening is flanked by twin friezes: the goddess *Roma*; and the altar's largest figure, a goddess variously identified as *Dea Tellus, Terra Mater, Italia, Pax Romana,* or *Venus Genetrix* (Figures: 1, 3 top, and 4 bottom).

2. METHODOLOGY

2.1 Research Objective

Although scholarship on the *Ara Pacis Augustae* has focused primarily on its magnificent Luna marble sculpture, our research objective is to determine if the *Ara Pacis* itself functioned as a shrine of light lending "distinguished authority" to the Emperor Augustus by means of the astro-architectural principles of Vitruvius.

2.2 Research Design

The present investigation is grounded in the inter-discipline of Astro-Archaeology, a term coined by astronomer Gerald S. Hawkins, to describe the hybrid study of the prehistories of architecture and astrononomy:

> *...I wish that many, or all, of the hundreds of Neolithic and Bronze Age stone monuments still standing could be accurately surveyed and examined for astronomical orientation. Much information fruitful for archaeologists, anthropologists, historians and others might result. If any university or foundation is casting about for promising fields for exploration and research, let it consider astro-archaeology!* (Hawkins 1965, 121)

In order to discover any astronomical significance to the original orientation of the *Ara Pacis*, it was necessary to determine the True Azimuths of the sides of the structure when *in situ* under *Via in Lucina* no. 16. The site map by Guglielmo Gatti 1938 (in Stampini 1970, 36; cf. Buchner 1982, 338-341; Heslin 2007, 14) shows a 75° Azimuth for the SW to NE axis (Figure 2; Tiede 2015, 9). The next step was to determine if the site plan were oriented to *Magnetic* North or *True* North, as the 1938 magnetic declination was 4° 27′ West (NOAA). Although, Gatti's map does not specify which azimuth was used, and his field notes remain unpublished at Rome's *Archivio Centrale dello Stato*, the issue was resolved by layering the Gatti map over a Google Earth satellite image and using the 1938 map's street corners as Ground Control Points (GCPs). Figure 2 (top) confirms the Gatti map to be oriented to True North (cf. Tiede 2015,9-11 for larger images; Ur 2009).

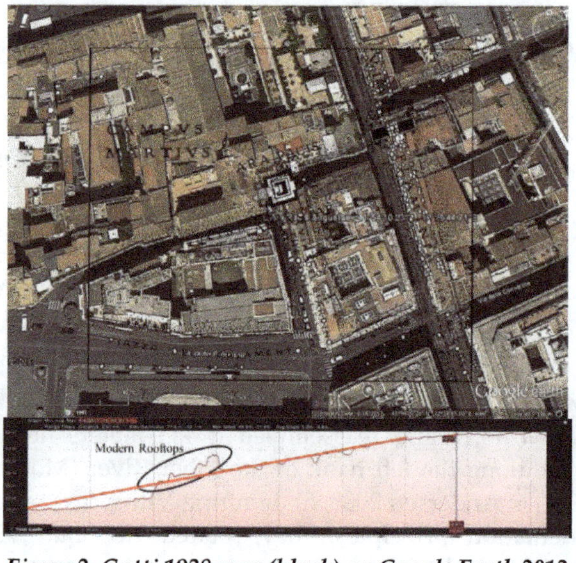

Figure 2. Gatti 1938 map (black) on Google Earth 2013 image & 75° True Azimuth Elevation Profile

The remaining input data was entered into an Excel version of Program STONEHENGE (Hawkins 1983, 328-330) to calculate the declination for the horizon intercept point of the 75° True Azimuth. The eye level of an observer over the *Ara Pacis* altar is estimated at 10 meters (m) Above Sea Level (ASL), based on Google Earth's 30m Digital Elevation Model (DEM) spot elevation for the *Via in Lucina* of 16m, less 10m soil below street level of the 1937–38 excavation (Moretti 1947, 6), and plus 4m for an observer standing on the altar podium. Google Earth locates the original position of the *Ara Pacis* at Latitude: 41° 54′ 10.47″ North, Longitude: 12° 28′ 44.60″ East. From there Google Earth's Elevation Profile tool shows that a 75° True Azimuth has a horizon Altitude of +0.13° (Elevation/Distance = 580m/30,100m). The modern rooftops may be ignored as they did not yet exist on the Campus Martius (Figure 2 bottom). The *Ara Pacis Augustae* was completed and dedicated in 9 BC (Moretti 1947, 7), i.e., the year to test for any hypothetical astro-target. Program STONEHENGE calculated the horizon declination for the SW-NE axis (75° True Azimuth) of the *Ara Pacis.* Solar declinations were generated with the planetarium software Starry Night Pro Plus 6 and compared with the *Ara Pacis'* northeast axis intercept point with the local horizon (skyline) expressed as declination.

2.3 *Findings*

Program STONEHENGE shows that the original axis of the *Ara Pacis Augustae* northeast enclosure opening (75° True Azimuth) met the local horizon at +11.3° declination. Starry Night Pro Plus 6 shows that the first gleam of the rising Sun was +11.0° declination on both 21 April and 27 August 9 BC (Figure 3). Because the difference in declination between the sun (+11.0° decl.) and the horizon vector (+11.3 ° decl.) is less than the angular diameter of the solar disk (0.5°), we may infer that the sun rose on the northeast axis of the *Ara Pacis* on both dates.

2.3 *Analysis*

Vitruvius prescribed that Roman altars should face east towards sunrise, rather than west towards sunset. Figures 1, 2 and 3 show that the *Ara Pacis* conforms to Vitruvius' sunrise orientation principle. To choose between the two solar alignment dates as either April 21st or August 27th, one needs either historical or archaeological confirmatory evidence. Unfortunately, the *Res Gestae* is silent regarding the month and day; stating only that "the Vestal Virgins should there make an annual sacrifice." (Augustus, 14 AD, XII:ii) However, the marble iconography of the *Ara Pacis* contains clues in the motifs found in Art History, Astrology and Classics literature (Gundel 1992; Galinsky 1992; Herbert-Brown 2002, 79; Pasco-Pranger 2006, 141) that may resolve the choice of date. For example, if the sculpted motifs on the much disputed *Tellus* frieze correspond to the respective Zodiac constellations and planets appearing on the local horizon at sunrise on 21 April versus 27 August 9 BC, then the iconography would confirm simultaneously both (1) the date of the annual sacrifice, and (2) that the altar's orientation was intentional rather than random.

Figure 3. Ara Pacis Augustae at dawn 21 April & 27 August 9 BC (Altair Multimedia 2014; Starry Night Pro Plus 6; Tumbler)

Upon inspection of the *Tellus* frieze, one sees twin *velificantes* – "typical of the deities of the atmosphere 'over whom vaults the firmament' (Galinsky 1966, 229) framing the polychrome iconography. Moreover, all the other figures are consistent with an astrological motif (cf. Gundel 1992), viz.: swan (Cygnus); flowing amphora (Aquarius); sea monster (Cetus); fish head and tail (Pisces); ram (Aries); bull (Taurus); twin infants (Gemini); seated woman in blue with exposed right shoulder (Venus), and all above a tetra-gamma meander (*Ursa minor* at each of the Four Seasons at midnight) serving as a horizon separating the verdant earthly lower panel from the celestial motifs above (Figure 4 bottom; Tiede 2015, 33-34). Figure 4 (top) shows the same Zodiac constellations (Aquarius-Pisces-Aries-Taurus-Gemini) appearing adjacent to the helically rising Venus above the horizon on 21 April 9 BC.

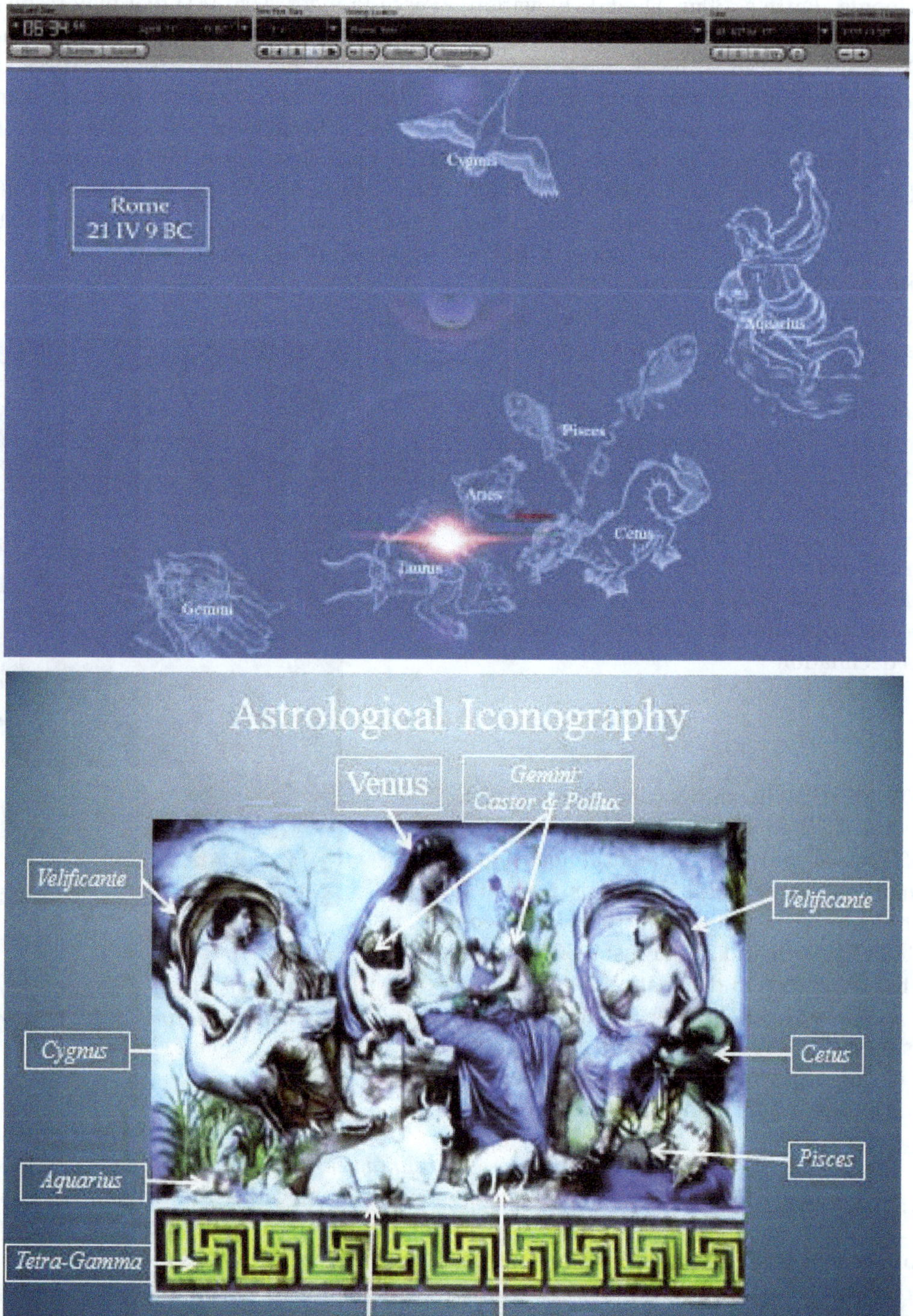

Figure 4. Sun, Venus (Morning Star) in Zodiac, 21 April 9 BC (top), and Astrological Iconography of the NE Frieze, Ara Pacis Augustae (bottom)

Conversely, Venus was invisible *below* the horizon when the Zodiac constellations Cancer-Leo-Virgo rose with the sun at dawn on 27 August 9 BC; i.e., *the astronomy does not match* the iconography of the *Tellus* frieze. Therefore, the sculptures on the *Tellus* frieze bear silent witness that they were intended to mirror their celestial counterparts at sunrise on the festival of *Par ilia* (Pales, goddess of shepherds), 21 April 9 BC. This day was the anniversary of the founding of Rome (*Dies Natalis Urbis*) in 1 *a.u.c.* (753 BC) according to the Roman historian Marcus Terentius Varro. Moreover, another solar-calendric alignment of politico-religious significance occurs on April 21st, viz., the oculus-to-entrance solar-noon meridian transit alignment of the Pantheon (completed by Emperor Hadrian c.125 AD) (Hannah & Magli 2011).

Art historians have long speculated on the identity of the seated lady in Figure 4. She has been interpreted variously as the goddess *Italia* (Van Buren 1913, 134 ff; Sellers-Strong 1937, 114 ff); *Terra Mater* or *Dea Tellus* (Momigliano 1942, 228-231; Moretti 1947,9); *Pax Romana* (Weinstock 1960); *Venus Genetrix* (Booth 1966; Galinsky 1966, 1992); or even as "a synthetic symbol of all of these personifications and divinities" (Kleiner 1992, 96). However, Astro-Archaeology provides new direct physical evidence supporting the identification of the goddess as *Venus Genetrix*, viz., the planet Venus actually rose heliacally at its maximum brightness as Morning Star (magnitude -4.4) on the *Ara Pacis* axis at dawn 21 April 9 BC, the year recorded in *the Res Gestae*. Moreover, because of the eight-year Venus-Earth Synodic Period, Venus also appeared as Morning Star in Aries on the first *Dies Natalis Urbis Romae,* 21 April 753 BC (Tiede 2015, 22). Therefore, the dedication of the Ara Pacis in 9 BC coincided with the 93rd iteration of Venus appearing as Morning Star at the founding of Rome (753- 9 BC = 744 years, and 744/8 = 93; cf. Fiorenza 2012). Significantly, ancient Roman writers as well as modern scholars have remarked on the connection of April to Venus and as Morning Star:

> *The second was April: so called of the name Aphrodite, that is to say Venus, unto whom they make open sacrifice in this moneth. And the first day of the same, women do wash themselves, having a garland of myrtle upon their heads.*
> (Plutarch, 262-3)

> *This fact may best be recognized from Venus. When she is following the sun, she makes her appearance in the sky after his setting, and is then the Evening Star, shining most brilliantly. At other times she precedes him, rising before daybreak, and is named the Morning Star.*
> (Vitruvius, IX.i.7)

> *...others have considered the swan as the personification of Venus, the morning star.*[81] *The latter interpretation would give us, of course, an additional connection between scenes of the Primaporta armor and the Ara Pacis since the Veneris astrum on the armor is represented carrying off Aurora.* (Galinsky 1966, 230)

Recent digital reconstructions by art historians of the polychrome sculpture of the *Ara Pacis* (Foresta 2011; Rossini 2015, 134-141) offers another line of evidence for identifying Venus as the central figure in the northeast frieze. Inspection of eight frescos from Pompeii (TAQ 79 AD) suggests that there was an artistic convention regarding the color of clothing for Mars and Venus in imperial Roman iconography such that Mars wore red, while Venus wore blue (Tiede 2015, 26-28).

The meander of the *Ara Pacis* is formed by 118 tetra-gamma girding the precinct enclosure wall exterior. The origin of the symbol is uncertain but is seen in Etruscan, Mycenaean and Greek iconography (D'Alviella 1894). The tetra-gamma is formed by four Greek "gamma" (Γ) upper-case letters set at right angles. One suggestion is that the tetra-gamma is a solar symbol and therefore appropriate for an Augustan altar, as Augustus "envisioned ...Apollo as his personal protector." (Green 1991, 46-49; Kleiner 1992, 62) An alternate interpretation is that the 118 tetra-gamma symbolizes the eternal order of the Cosmos by its depiction of the constellation *Ursa Minor* with the North Pole Star Polaris at the tip of its tail at midnight on each of the Four Seasons of the solar year (Tiede 2015, 33-34). Either interpretation is consistent with astronomical iconography promoting the economic, political and social stability of the *Pax Augusta.*

3. CONCLUSIONS

Making explicit the astro-architectural principles of Vitruvius as seen in *the Ara Pacis Augustae* serves as a reminder of the limitations of scholarly interpretations informed solely by art motifs and mythology. One has only to cite Stoic philosophers to realize the significance of astronomy in Roman architecture, e.g., "*Omnis ars naturae imitatio est*" ("All art is but an imitation of nature") (Senaca 65). The juxtaposition of the frieze pairs of the Imperial family and Roman Senate versus those the mythical past and Immortal gods is emblematic of the Stoic philosophy of the Augustan age, e.g.,

> *The Stoic doctrine of universal 'sympathy" between the microcosm (man) and the macrocosm (the universe) - expressed in the dictum used later when Stoicism no longer existed, 'as above, so below' - and astrology seemed made for each other.*
> (Christie-Murray 1990, 115)

Our analysis of the *Ara Pacis Augustae* demonstrates that astronomy played an integral role in the architectural design memorializing the politico-religious bond between Imperium and Cosmos (cf. Rehak 2009). Therefore, what could be a more fitting venue than the *Museo dell'Ara Pacis* for exhibiting a virtual 3D animation (cf. Altair4 Multimedia 2014) depicting an annual dawn sacrifice by Vestal Virgins which intertwined the *Pax Augusta* of the SPQR with the Cosmos of the Immortal Gods on April 21st, the *Dies Natalis Romae*?

ACKNOWLEDGEMENTS

We thank Orietta Rossini and the anonymous reviewer for their constructive comments.

REFERENCES

Altair4 Multimedia (2014) Ara Pacis, https://www.youtube.com/watch?v=xcnz3QQSaPo

Andersen, W. (2003), *The Ara Pacis of Augustus and Mussolini*, Boston: Editions Fabriart.

Augustus (14) *Res Gestae Divi Augusti*, http://en.wikisource.org/wiki/Res_Gestae_%28Anon%29 .

Bertarione, S. V. and G. Magli (2015) Under Augustus' sign: the role of Astronomy in the foundation of *Augusta Praetoria Salassorum*, *Cambridge Archaeological Journal*, Vol. 25, pp. 1-15. https://www.academia.edu/5904495/Under_Augustus_sign_the_role_of_Astronomy_in_the_foundation_of_Augusta_Praetoria_Salassorum

Buchner, E. (1982) *Sonnenuhr des Augustus: Nachdruck aus RM 1976 u.1980 u. Nachtrag über die Ausgrabun 1980/*1981, Mainz a. Rhein: Verlag Philipp von Zaber.

Booth, A. (1966) Venus on the Ara Pacis, *Latomus*, Vol. 25, pp. 873-879.

Christie-Murray, D. (1990) *The Practical Astrologer*, New York: St. Martin's Press.

D'Alviella, G. (1894) *The Migration of Symbols*, London: A. Constable and Co.

Fiorenza, N. A. (2012) http://www.lunarplanner.com/HCpages/Venus.html

Foresta, S. (2011) La policromia dell'Ara Pacis e I colori del Campo Marzio settentrionale, https://www.academia.edu/945654/La_policromia_dell_Ara_Pacis_e_i_colori_del_Campo_Marzio_settentrionale

Frischer, B., P. A. Auber, and O. Rossini (2014) The Obelisk, Meridian, and Ara Pacis of Augustus, https://www.youtube.com/watch?v=GUXZ0d0sxpo

Galinsky, K. (1966) Venus in a Relief of the Ara Pacis Augustae, *American Journal of Archaeology*, Vol. 70, pp. 223-243.

(1992) Venus, Polysemy, and the Ara Pacis Augustae," *American Journal of Archaeology*, Vol. 96, pp. 457-475.

Green, M. (1991) *The Sun-Gods of Ancient Europe*, London: B.T. Batsford.

Gundel, H. G. (1992) *Zodiakos : Tierkreisbilder im Altertum : kosmische Bezüge und Jenseitsvorstellungen im antiken Alltagsleben*, Mainz a. Rhein: Verlag Philipp von Zabern.

Hannah, R. (2009) *Time in Antiquity*, London: Routledge.

(2005) *Greek and Roman Calendars: Constructions of Time in the Classical World*, London: Duckworth.

Hannah, R. and G. Magli (2011) The Role of the Sun in the Pantheon's Design & Meaning, *Numen* Vol. 58, pp. 4, 486-513.

Hawkins, G. S. (1983) *Mindsteps to the Cosmos*, New York: Harper & Row.

(1965) *Stonehenge Decoded*, New York: Doubleday.

Herbert-Brown, G. (2002) *Ovid's 'Fasti': Historical Readings at its Bimillennium*, Oxford: Oxford University Press.

Heslin, P. (2007) Augustus, Domitian and the So-called Horologium Augusti, *Journal of Roman Studies*, Vol. 97, pp. 1-20.

IDIA Lab, Ball State University (2013) https://www.youtube.com/watch?list=UUtQrLF1JPprEvP4AO-iSNvw&v=-o2Crtam3R8#t=13 .

Kleiner, D. E. E. (1992) *Roman Sculpture*, New Haven: Yale University Press.

Moretti, G. (1947) *L'Ara Pacis Augustae*, Rome: Istituto Poligrafico dello Stato.

Müller, V. (1943) The Shrine of Janus Geminius in Rome, *American Journal of Archaeology*, Vol. 47. pp.4, 437-440.

National Oceanic and Atmospheric Administration (NOAA), http://www.ngdc.noaa.gov/geomag-web/ .

Pasco-Pranger, M. (2006) *Founding the Year: Ovid's 'Fasti' and the Poetics of the Roman Calendar*, Boston: Brill.

Plutarch (c.100), Numa Pompilius, *Plutarch's Lives* (T. North, trans.), Vol. I, London: J. M. Dent & Co., 1898.

Rehak, P. (2009) *Imperium and Cosmos: Augustus and the Northern Campus Martius*, Madison: University of Wisconsin Press.

Rhyne, C. (2011) http://cdm.reed.edu/ara-pacis/contents.php .

Rossini, O. (2015), *Ara Pacis guide*, Milan: Mondadori Eleta S.p.A.

Seneca (65) *L. ANNAEI SENECAE EPISTULARUM MORALIUM AD LUCILIUM LIBER SEPTIMUS LXV. SENECA LUCILIO SUO SALUTEM*, 3, http://www.thelatinlibrary.com/sen/seneca.ep7.shtml.

Stampini, P. (1970) Ara Pacis Augustae, *Dibattiti Rotariani*, III:5-6, Roma.

Thompson, H. A. (1952) The Altar of Pity in the Athenian Agora," *Hereperia: The Journal of the American School of Classical Studies at Athens*, Vol. 21, pp. 1, 47-82.

Tiede, V. (2015), https://www.academia.edu/19159136/ARA_PACIS_AVGVSTAE_An_AstroArchaeologi cal _Analysis_PPT_slides

Toynbee, J.M.C. (1953) *Ara Pacis Reconsidered and Historical Art in Roman Italy*, London: Geoffrey Cumberledge.

Tumbler, http://mappaemundi.tumblr.com/post/87884514617/lara-pacis-augustae-ovvero-laltare-della-pace

Ur, J. (2009) Lab 4: Georeferencing Imagery, CPAS Archaeological GIS Workshop. http://www.fas.harvard.edu/chinaarch/Downloads/gis/L04_images/L04%20Georef%20Imagery %20English.pd

Van Buren, A. W. (1913) The Ara Pacis Augustae," *The Journal of Roman Studies*, Vol.3, pp. 1, 134-141.

Vitruvius (c. 25 BC) *De Architectura Libri Decem* (M.H. Morgan, trans.), New York: Dover Press, 1960.

Zanker, P. (1990) *The Power of Images in the Age of Augustus*, Ann Arbor: University of Michigan Press.

16

ORIENTATIO AD SIDERA (OAS): HIGHLIGHTS OF A DECADE OF ARCHAEOASTRONOMICAL RESEARCH IN THE MEDITERRANEAN REGION AND BEYOND

Juan Antonio Belmonte[1], A. César González-García[2], Andrea Polcaro[3], Andrea Rodríguez-Antón[1] and Mosalam Shaltout[4]

[1]Instituto de Astrofísica de Canarias and Universidad de La Laguna, Tenerife, Spain
[2]Instituto de Ciencias del patrimonio, Incipit-CSIC, Santiago, Spain
[3]Università degli Studi di Perugia, Italy
[4]N.R.I.A.G., Helwan, Cairo, Egypt

Corresponding author: Juan A. Belmonte (jba@iac.es)

ABSTRACT

The OAS Project has been run with the support of the Spanish research funding agencies during the last dozen years. Within its framework, research on cultural astronomy has been developed for a series of ancient civilizations, from the Atlantic Islands to the Arabian Peninsula and beyond, with the Meditterranean Sea as the principal axis of action of the project. A catalogue of studies has been performed in a set of cultures such as the Megalithic Phenomenon (González-García and Belmonte, 2010; Belmonte; González-García and Polcaro, 2013a), ancient Egypt (Belmonte and Shaltout, 2009; Belmonte, 2012), Middle East Bronze Age (notably on the Hittites: González-García and Belmonte, 2011) and Iron Age (notably on the Nabataeans: Belmonte, González-García and Polcaro, 2013b) civilizations and the Roman World (e.g. González-García, Rodríguez-Antón and Belmonte, 2014), among many others. In this essay a general scope of the project and a series of most interesting highlights will be presented. The evolutionary ties of the megalithic monuments of the Iberian Peninsula and elsewhere, the pattern of orientation of Egyptian temples and skyscaping practices within the Hittite or Nabataean cultures, among others, will be briefly explored; including a comprehensive, statistical and comparative study of the orientation patterns of thousands of ancient monuments of the Mediterranean region (González-García and Belmonte, 2014). Finally, a sketch of our most recent, still ongoing, research will be a compelling and promising closing remark of our analysis.

KEYWORDS: Cultural astronomy, OAS Project, Göbekli Tepe, ancient Egypt, Nemrud Dag, Magan

1. INTRODUCTION

Astronomy has played a major cultural role in human societies. On the one hand, it has been the principal tool, if not the only one, to obtain an appropriate orientation in time and space for generations. On the other hand, sky watching and state of the art explanations of the cosmos have been the main generators of metaphysics in human mentality in such a way that, for instance, as astronomy progressed in the Western World, the space for religious speculation was consequently reduced. Celestial objects, returning once and again, provided a sense of security, allowing up the possibility of transcendence of death, inferred by similarity. For this reason, people mapped the firmament from very early dates in an attempt to find the order where, only in appearance, chaos reigned, and constructed their sacred buildings in such a variety of aspects, forms and realities seeking for a reflection of their cultural worldview.

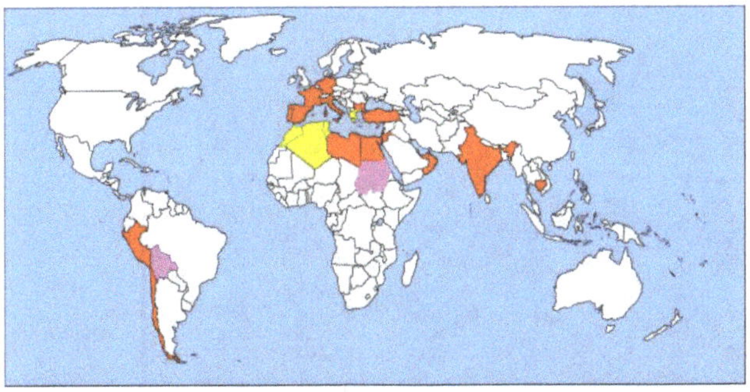

Figure 1. The international projection of the Orientatio ad Sidera research project on Cultural Astronomy. The twenty countries where fieldwork has been made during the dozen years of three successive projects of the same name are coloured red. Pink colour marks those countries where research has been done using high-resolution satellite images (Bolivia and Sudan). Yellow stands for countries whose data, obtained in previous campaigns by our team or by other researchers (as in the case of Algeria and Greece), have been used in analyses developed under OAS framework. The nucleus of our work has been the Mediterranean Region.

Pursuing ancient people mind, and how this was reflected in art, iconography, architecture and epigraphy, our research team promoted a scientific project in the early 2000s where cultural astronomy studies may be situated within an appropriate funding and institutional framework. This would precisely be "Orientatio ad Sidera". Since then, and on a three/four years period basis, three consecutive versions of the project have been approved and have permitted to follow up scientific objectives in five "continents" and more than twenty countries (see Figure 1). There have been lights and shadows in our work but the highlights have been inspiring and have proven beyond any doubt that sky- and landscape relationships have been more the rule that the exception in the way humans have connected to the cosmos. Let us show some examples.

2. SELECTED HIGHLIGHTS

2.1. *Before writing: Göbekli Tepe*

Until very recent times, the megalithic monuments in Europe, and in particular Stonehenge, were the archaeological remains earning all the credit on any potential astronomical knowledge of our earliest ancestors. However, a discovery in the steppes of southeast Anatolia has disrupted previous ideas. There, on a barren isolated hill called Göbekli Tepe a team of German and Turkish archaeologists (Schmidt, 2006) are excavating a cluster of suggestive stone monuments erected with large, megalithic pillars in the form of a T within a series of dry-stone enclosures. They started to be built by a completely unknown hunter-gatherer society more than 11,000 years ago. These series of sanctuaries, built presumably one after – and even upon – the other, would have remained in use for centuries, perhaps millennia, but were deliberately buried by their own constructors for unknown reasons, a singular fact that contributed to their excellent state of preservation despite of their great antiquity.

These monuments are mostly ellipsoidal in form and had megalithic accesses mostly open to the S-SE that might define a favourite orientation (see Figure 2). A few contradictory proposals have been put on the table. For example, Collins (2014) defends an orientation to the northern skies, notably to Deneb, offering certain clues for the importance of the Cygnus constellation in Prehistoric times. This would be highly dependent on the original landscape and orography of the site which is not easy to reconstruct. On the contrary, Magli (2015) defends a southern orientation centred on Sirius, the brightest star of the sky that would be visible in that region precisely in that epoch after have been out of sight for centuries due to precession.

However, what is undebatable is that between the series of monumental structures, there is one with nearly rectangular walls which were almost perfectly aligned according to the cardinal points (see Fig. 2). This circumstance alone would force us to think that we are faced with a society that had a look at the sky and used it as a guide to find appropriate ways of orientation in space and, almost certainly, also in time. Within this context, we could perform additional exercises, analysing the profuse decoration of the T-pillars where we may already find atavistic astronomical representations such as the Crescent and the star, so common in later cultures of the Middle East and beyond, or even totemic representations of animals which, allowing a little speculation, would remind us constellations, such as Leo, Taurus or Scorpius, that we can recognize in the skies of other evolved cultures in the region several centuries later. Interestingly, one of the pillars of the cardinally orientated hall, which was framing an altar on the eastern side of the structure, has a representation of a lion; and Leo was rising with the sun at the Spring Equinox precisely at east in the epoch of construction of this particular shrine (see Fig. 2). A moment in time when Taurus and Scorpius where marking other annual cycle milestones such as the solstices.

Unfortunately, when the data of Göbekli Tepe is analysed, a problem is faced: its singularity. There is no other preserved monument of the same culture where data can be contrasted and hence it will always be extremely difficult to further develop any hypothesis on the site relationship with skyscaping and to prove it beyond any reasonable doubt.

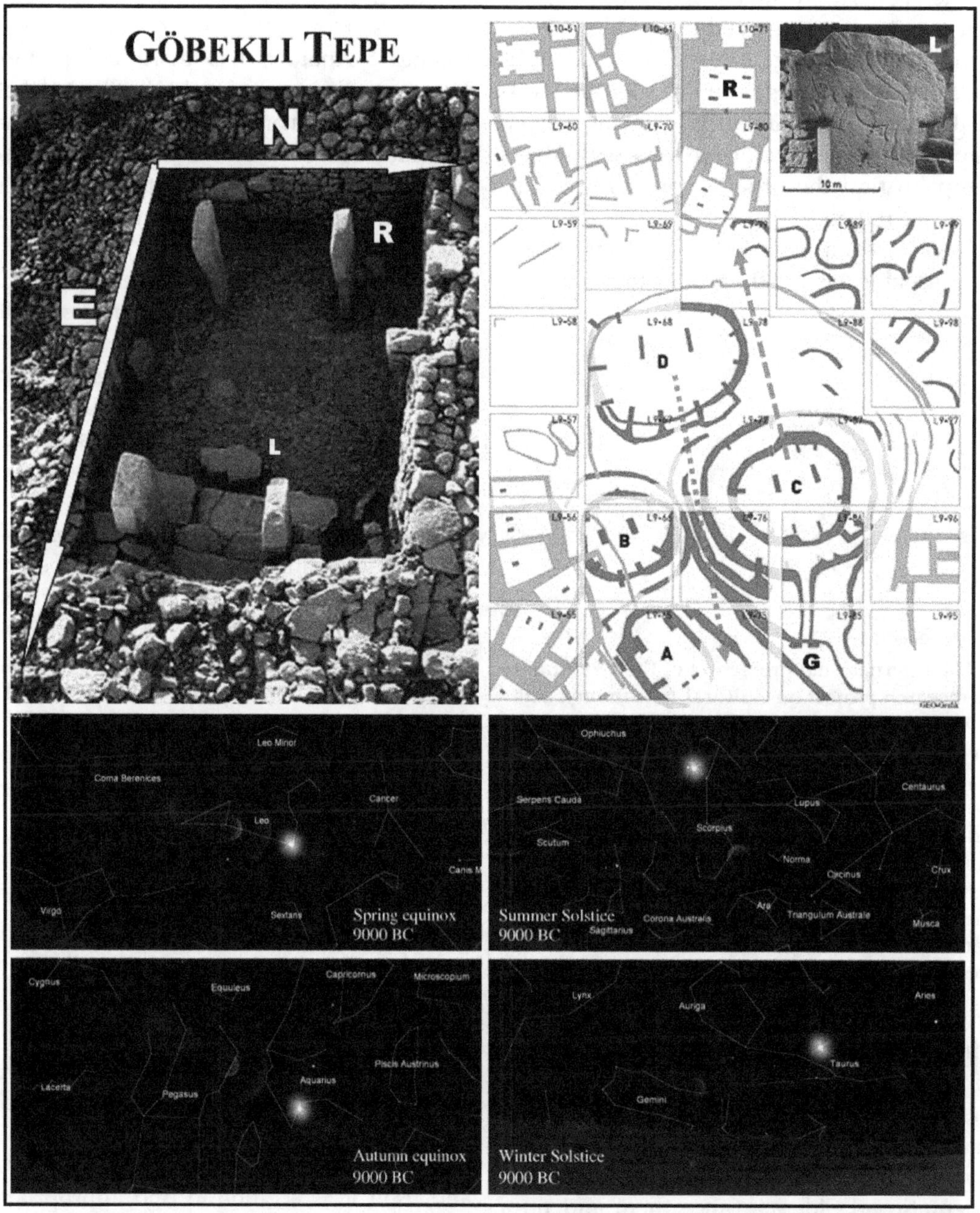

Figure 2. Composite diagram of some astronomical proposals for Göbekli Tepe. The walls of the rectangular structure (R) built c. 8500 B.C. in the upper sector of the site are perhaps the first orientated in the cardinal directions by humankind discovered so far. One of the pillars (L) was decorated with a image of a lion. Either by chance or design, the equinoctial sun was easterly rising in conjunction with our Leo in that epoch (other seasonal constellations of the epoch: Scorpius for the summer solstice and Taurus for the winter one had also his graphic counterparts in the pillar decoration on the site). The orientation of the main axes of the older larger enclosures (notably C and D) has been alternatively related to Sirius (dot-line; Magli, 2015) or to Deneb (dashed-line; Collins, 2014) See the text for more details. Diagram by the authors with images adapted from photographs by courtesy of M. Sanz de Lara and Stellarium.

However, statistics can be a good approach to perform cultural astronomy studies when dealing with illiterate extinct cultures, or cultural phenomena, of the past. In this sense, the earliest unmistakable evidence of an interest in the local landscape, including the skyscape, can be found throughout the megalithic monuments of the Iberian Peninsula. This is especially true in the southwest where Neolithic settlers built a series of monuments with a very similar architectural structure – they almost are clones of each other – known as Alentejan antas, for the name of the Portuguese region where they are a majority. They have been dated in the early fourth millennium B.C. Most important, they show a pattern of orientation that clearly speaks to us of an unmistakable astronomical interest, because all the studied exemplars, without an exception (Hoskin, 2001), are pointing at sunrise – or moonrise in a naïve alternative – at a certain moment in the annual cycle. We think that this very simple pattern of orientations was undistinguishable of megalithic monument design in the Iberian Peninsula in the 4th millennium B.C., if not earlier, and that construction methods and overall orientation patterns lately evolved and changed when the megalithic tradition moved across new extended territories (González-García and Belmonte, 2010).

2.2. *Reading the Bronze Age*

For various reasons, archaeoastronomy has not been one of the favourite disciplines of Egyptologists in the past. Hence, important questions such as the orientation of Egyptian temples and the relevance of astronomy in this respect had never been afforded with the requisite seriousness and depth. The "Egyptian-Spanish Mission for the Archaeoastronomy of Ancient Egypt" (Belmonte and Shaltout, 2009) funded by the successive OAS Projects has contributed to the solution of this problem. In order to achieve this, the Mission measured in five years the orientation of more than 350 temples in the Valley, the Delta, the Oases and the Sinai (reaching 400 if we include satellite image data from Sudan; Belmonte, 2012). The aim was to find a correct and almost definitive answer to the question of whether the ancient Egyptian sacred constructions were astronomically aligned or not.

The data seem to answer this question in the affirmative (see Figure 3). In addition, they offer a very interesting new perspective both in chronology and in the field of landscape archaeology, a new discipline in which few have engaged so far in Egypt, and in which landscape, dominated by the Nile, and skyscape, dominated by the sun and the stars, combine in order to permit the establishment of Ma'at, the Cosmic Order, on Earth (Belmonte, Shaltout and Fekri, 2009).

Hieroglyph sources, such as astronomical papyri, star-clocks and celestial diagrams were scrutinized in an effort to gain as much information as possible on how astronomy and its archaeological implications could help to better understand the ancient Egyptian civilization. Within the same framework, we should also highlight the identification of a series of Egyptian stars and constellations obtained on the analysis of the Senenmut's astronomical ceiling (the oldest complete symbolic representation of Egyptian skies) and the well-known Zodiac of Dandara. The analysis drives, under very simple assumptions, to the potential identification of the stars, asterisms and constellations that populated the skies of ancient Egypt, and their use as star-clocks within the framework of a sophisticated astral eschatology (Lull and Belmonte, 2009).

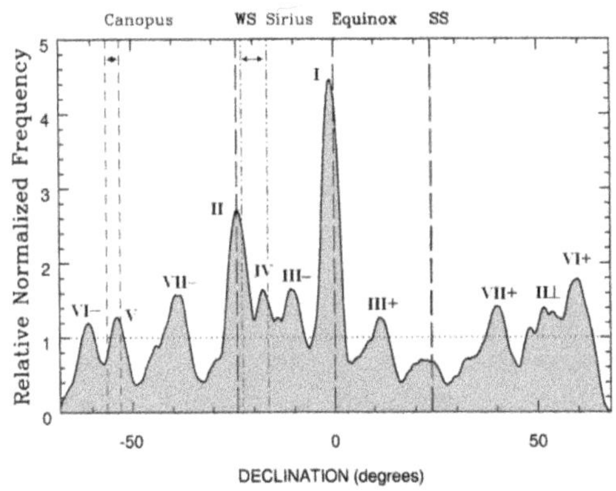

Figure 3. The seven (I to VII) families of astronomical orientations of the temples of ancient Egypt. Three of them have a solar character: eastern (I), solstitial (II) and seasonal (III). The other four (five) have a stellar character: eastern (I, in the perpendicular direction), Sothic (IV), Canopus' (V), meridian (VI) and quarter cardinal (VII). The cardinal families (I, VI and VII) would probably be related to alignments towards the constellation of Meskhetyu (the Plough, in Ursa Maior). Adapted from Belmonte and Shaltout (2009).

The last effort that our project has devoted to the ancient Egyptian civilization has dealt with the earlier steps of Egyptian monumental construction and the role astronomy played within (Belmonte and Magli, 2015). The two pyramids built during the Old Kingdom by the 4th Dynasty King Sneferu at Dahshur are usually considered as two consecutive projects: the second - that of the Red Pyramid - being generated by a presumably failure of the first, the Bent Pyramid. It has been shown that the archaeological proofs of such a scenario are far from obvious and that, on the contrary, a series of architectural, topographical, epigraphic and astronomical hints point to a unitary project probably conceived from the very beginning in terms of the two pyramids and their annexes. Similar conclusions could be reached for other royal tombs of the 4th Dynasty, notably at Giza, Abu Rowash and Zawiyet el Aryan. The pyramids of the period all-together are thus shown to form a conceptual, sacred landscape related to the power of the King and his afterlife within the context of a sophisticated astral eschatology.

But ancient Egypt was not isolated. Within the framework of our project, the geographical, historical and anthropological context of a possible interest on the heavens by the ancient Hittites has also been scrutinized. Contemporaneous with the Egyptian New Kingdom, the inhabitants of the Hittite Empire produced a most sophisticated society, heir to a long Anatolian cultural tradition lasting several millennia (see section 2.1).

A review on ancient Hittite religion was performed, looking for those aspects of the cult that could be of interest to our project. These included an analysis of the solar, and other astral, divinities, the sacred space and its administrators, the calendar of festivals – i.e. Hittite sacred time – and, finally, the dead cult and related spaces and topography. This was the first serious insight dedicated to cultural astronomy studies of the Hittite civilization (González-García and Belmonte, 2011).

The importance of solar cults has been clearly emphasized and certain hints on the ancient Hittite sacred time were established with a certain degree of certitude. This was useful for a late comparison with the data provided by the archaeoastronomical research. The results showed that the analysis of a statistically significant sample of Hittite sacred structures allowed to affirm, beyond a reasonable doubt, that ancient Hittite monuments were not randomly orientated as had been previously argued. On the contrary, there were well defined patterns of orientation that could be interpreted within the context of Hittite culture and religion.

Our team was able to obtain data for nearly a hundred sacred structures erected by ancient Anatolian civilizations, such as the Hittite and the Phrygians, as a result of an extensive field prospective campaign. The analysis of the Hittite data has proven extremely fruitful. For example, our data confirm the textual evidence and have shown the relevance of solstitial and "equinoctial" orientations that could be explained within the context of ancient Hittite solar cult necessities. Indeed, Hattusha, the capital, has shown a striking and highly interesting astronomical and topographical landscape, where potential relationships between astronomical phenomena, built structures such as temples and monumental gates, and different elements of the local landscape have been illustrated.

2.3. *Commemorating the past: Nemrud Dag*

There are some occasions in life when one has the impression of being envisaging something unique. This was our case a few months ago when an extraordinary planetary conjunction took place in the constellation of Leo (see Figure 4). However, the most fascinating aspect of this view was that we were actually evocating a completely similar situation which happened in the sky more than 2000 years ago in the northern banks of the Euphrates River.

The Kingdom of Commagene was a small country between the upper course of the Euphrates and the mountains of Anti-Taurus in the south-east of Anatolia. Commagene played, despite her tiny size, a relevant role in the history of the Middle East during the late Hellenistic and early Roman periods as a buffer state between the powerful Seleucid (later Roman) and Parthian Empires. Antiochos I Theos (c. 69-36 B.C.) arguably was the most important of her kings, governing for more than 30 years in one of the most challenging periods of the region's history.

The world heritage site of the "hierothesion" (burial monument) of Antiochos I at Nemrud Dag certainly constitutes one of the most fascinating historical enigmas in human culture worldwide. The monument includes a stone slab with the famous lion "horoscope". On the slab, a lion with stars on his body, likely the constellation of Leo, is represented together with a crescent moon on his chest and three planets, identified in Greek as Pyroeis of Heracles, Stilbon of Apollon and Phaeton of Zeus, standing for Mars, Mercury and Jupiter, respectively (see Fig. 4). The possibility that the slab depicts a real or schematic astronomical scene or an astrological image introduced the idea of dating the monument and interpreting its nature since the earliest archaeological studies of the site. The most accepted conclusion so far had the support of Neugebauer and Van Hoessen (1959) who argued that the scene might represent a sort of horoscope for the date July 7 62 B.C. at the beginning of the reign of Antiochos I.

One of the main highlights of our work in the OAS Project has been a new approach in the analysis of the burial monument at Nemrud Dag, precisely (Belmonte and González-García, 2010), reaching different conclusions. After a visit to the monument, with on-site observations, in the summer solstice of 2009, our team proposed an alternative and more substantiated explanation which deals not only with the lion slab but also, and most important, with the orientation of the eastern and western terraces of the hierothesion and the religious tradition of the country as stressed in on the local inscriptions known as *nomos*.

Our main results can be summarized as follows. On the one hand, the five cyclopean statues of the eastern terrace would have been facing sunrise followed by the rising (obscured by the solar glare) of their celestial manifestations (the planets) in the constellation Leo on July 23 49 B.C., commemorating Antiochos' ascent to the throne as explicitly mentioned in the inscriptions on site (*nomos*). On the other hand, a few months later, their equivalents of the western terrace would have been facing sunset on December 23 49 B.C. in commemoration of the king's birthday. Consequently, according to our proposal, the main elements of the eastern and western terraces of the hierothesion should have been deliberately aligned to sunrise of Loios 11 and sunset of Audnayios 16 (the two festival dates mentioned in the *nomos*), respectively, in the year 49 B.C. The lion's horoscope could consequently be assigned a new date to July 12 49 B.C. (see Fig. 4).

The conclusion of our team was that Antiochos' monument reflected the situation of the skies at exclusive moments of the year 49 B.C. as confirmed by the local monumental inscription. This represents a paradigmatic change to well established theories on the interpretation of Nemrud Dag. It is indeed appealing that we could commemorate this fact with a completely similar phenomenology twenty centuries later at dawn on October 9 2015.

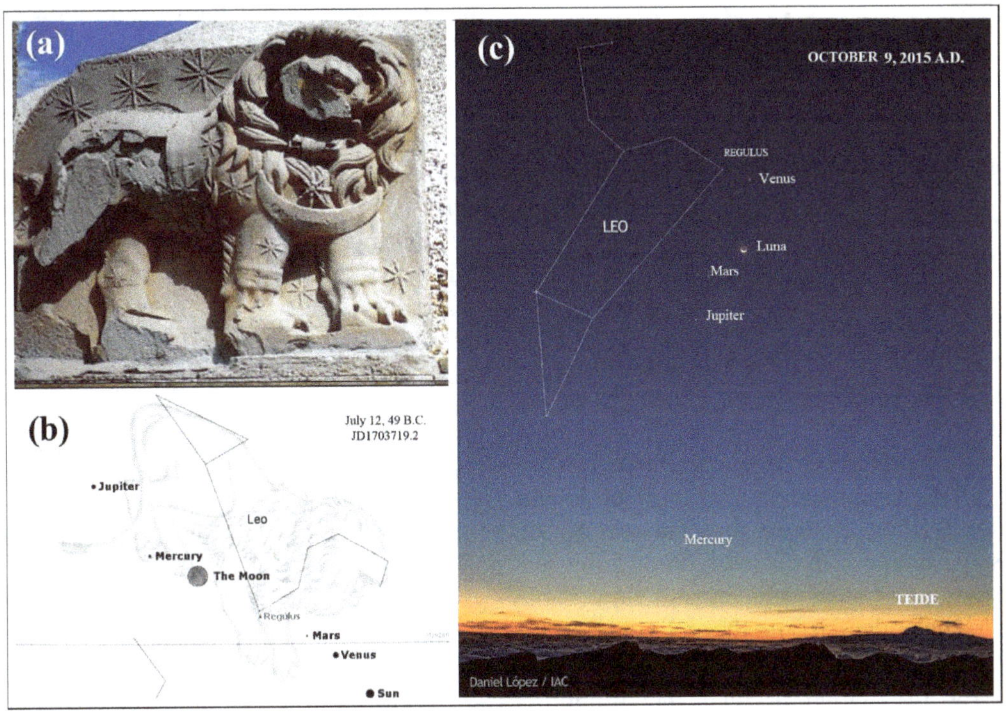

Figure 4. The famous "Lion Slab" at the eastern terrace of Nemrud Dag (a), and the corresponding astronomical diagram, corresponding to Leo's setting after sunset for July 12 49 B.C. in the Gregorian proleptic calendar. This date has suggestive connections with those yielded by the orientation of the main architectural elements of the hierothesion and the inscriptions on site. This kind of splendid celestial planet conjunction with such an important constellation as Leo are seldom repeated in the sky as the one that the authors could enjoy above Tenerife in October 9 2015 from the summit of La Palma, in the Canary Islands (c). Diagram by the authors, adapted from Belmonte and González-García (2010) and from a sky image by courtesy of Daniel López.

2.4. *Present and future*

The Arabian Peninsula is for obvious reasons one of the last virgin territories for cultural astronomy studies in the Middle East. However, its relevance as a cultural crossroad at the frontier of the Fertile Crescent can hardly be discussed. This is for example the case of the Haffit and Umm an Nar cultures that dominated the northeast of the Arabian Peninsula during the Chalcolithic and the Early Bronze Age (Third Millennium BC), in the so-called land of Magan (present day Oman and the Emirates). In a field campaign in January 2012 several ancient EBA necropoleis of the region were visited and (when possible) measured in an attempt to shed some light on the orientation customs of these ancient populations. The sample, of c. 70 monuments, consists of two different types of tombs: earlier (c. 3000 BC) dry-stone cairns of the so-called Haffit type and later (c. 2500 BC) megalithic structures of the so-called Umm an-Nar type. This was the first systematic archaeoastronomical approach ever conducted in this area of the Arabian Peninsula (Belmonte and González-García, 2014). However, the work is uncompleted and there still are several options for the future. For example, an area we did not visit in our campaign was the Eastern Hajjar Range in Oman where, as archaeology proves, the research can still offer interesting surprises (see Figure 5).

Figure 5. Orientation histogram and images of the monumental cyclopean tombs of Al Jailah in the Eastern Hajjar Mountains of Oman. It was obtained using the archaeological data collected by Yule and Weisgerber (1998). The two most significant peaks may correspond to equinoctial and winter solstice alignments. Images by courtesy of Jorge Notivoli.

The other extreme of the Arabian Peninsula was dominated at the term of the era by the powerful Nabataeans. This Arab lineage people built several monuments in Petra and elsewhere displaying a decoration with a preference for astronomical motifs, possibly as a reflection of their religion. Due to the lack of direct written accounts and the scarcity of inscriptions we do not have a clear knowledge on the precise nature of such believes and how these reflected on the calendar or the religious timekeeping system of this ancient society. A statistical analysis of the orientation of their sacred monuments (Belmonte, González-García and Polcaro, 2013b) demonstrated that astronomical orientations were often part of an elaborated plan and possibly a trace of the astral nature of Nabataean religion. Petra and other monuments in the ancient Nabataean kingdom have proven to be marvellous laboratories of the interaction between landscape features and astronomical events showing impressive hierophanies on particular monuments related to cultic times and festivals. Our work has demonstrated that the sky was a substantial element on Nabataean religion and revealed new evidence for cultic worship centred on the celestial sphere in this fascinating region of the Mediterranean contour (for further details see e.g. Polcaro et al. 2013 and González-García, Belmonte and Polcaro, 2016, in this same volume).

The most recent effort of the OAS project has been the one devoted to the study of the pattern of orientations within the ancient Roman world, notably in the Iberian Peninsula (González-García, Rodríguez-Antón and Belmonte, 2014). This is the nucleus of the first PhD Thesis undertaken within the project and that is being done by Andrea Rodríguez-Antón under the supervision of the first two authors (see also Rodríguez-Antón, Belmonte and González-García, 2016, this volume). The research includes the in depth analysis of certain substantial Roman sites. A good example is Cartagena, the ancient Punic Qart Hadašt, or Roman Carthago-Nova. Archaeological excavations have manifested the existence of ritual, urban and topographical elements that could be analysed from the perspective of cultural astronomy.

Therefore, in October 2013, an interdisciplinary team of astronomers and archaeologists conducted a field campaign of the main topographic and archaeological landmarks of the Punic and Roman periods of the city. Methodologically, a basic guide criterion was established for each particular element, measuring its corresponding azimuth(s). The data demonstrated the relevance, within the ancient city, of a series of orientations towards sunrise and sunset at the summer solstice, whose significance could be fully integrated within the context of the Punic ritual (González García et al., 2015). This skyscaping was merged and reinterpreted in the framework of the subsequent Roman appropriation of the city landscape, including their successive urban and architectural programs, particularly that of the period of Augustus, when certain astronomical orientations could serve to strengthen the image of Rome and the *Princeps* as restorers of peace and guarantees of a new order based in cosmological elements. There are plans to repeat similar approaches in other ancient Roman cities.

3. CONCLUSION

In our opinion, the OAS Project has made a substantial and essential contribution to the advance of cultural astronomy research in the Mediterranean region and elsewhere in the last decade. This is reflected in the recently published "Handbook of Archaeoastronomy and Ethnoastronomy" (Ruggles, 2015), where the participation of members of the project (and immediate collaborators) has been notable, contributing with a total of 17 essays. Additionally, the IP of the OAS project was one of the seven Section Editors of what hopefully will become a reference volume in our discipline.

OAS is a living project! Progressing, growing and in permanent reenovation (an IVth extension to 2019 has just been approved). We now expect that it will become an important partner in the promotion and development of the Astronomy and World Heritage initiative (Ruggles and Cotte, 2010).

ACKNOWLEDGEMENTS

Substantial parts of the fieldwork and data interpretation during these dozen years of hard work were carried out with the help of a series of persons whose collaboration was fundamental for the success of the project. We would like to especially acknowledge astronomers Lourdes Costa-Ferrer and Giulio Magli, anthropologist Margarita Sanz de Lara, Egyptologists José Lull, Miguel Ángel Molinero and Magdi Fekri, and archaeologists María Antonia Perera and Marco García Quintela for their ongoing permanent support. This work is partially financed under the framework of the projects AYA2011-66787-P "Orientatio ad Sidera IV" of the Spanish MINECO (and earlier versions of the same project), and P310793 "Arqueoastronomía" of the IAC. ACGG is a Ramon y Cajal Fellow of the Spanish MINECO.

REFERENCES

Belmonte, J.A. (2012) *Pirámides, templos y estrellas: astronomía y arqueología en el Egipto antiguo*, Barcelona, Crítica.

Belmonte, J.A. and Shaltout M. (eds.) (2009) *In Search of Cosmic Order: selected essays on Egyptian Archaeoastronomy*, Cairo, Supreme Council of Antiquities Press.

Belmonte, J.A., Shaltout, M. and Fekri, M. (2009) Astronomy, landscape and symbolism: a study of the orientation of ancient Egyptian temples. In *In Search of Cosmic Order: selected essays on Egyptian Archaeoastronomy*, J.A. Belmonte and M. Shaltout (eds.), Cairo, Supreme Council of Antiquities Press, 213-285.

Belmonte, J.A., and Gonzalez Garcia, C. (2010) Antiochos's hierothesion at Nemrud Dag revisited: adjusting the date in the light of astronomical evidence, *Journal for the History of Astronomy*, Vol. 41, 469-481.

Belmonte, J.A., González García, A.C. and Polcaro, A. (2013a) On the orientation of megalithic monuments of the Transjordan Plateau: new clues for an astronomical interpretation, *Journal for the History of Astronomy*, Vol. 44, 429-455.

Belmonte, J.A., González García, A.C. and Polcaro A. (2013b) Unveiling Petra: astronomy and landscape in the Nabataean lands, *Nexus Network Journal*, Vol. 15, 497-502.

Belmonte, J.A. and González García A.C. (2014) On the orientation of Early Bronze Age tombs in ancient Magan. *Mediterranean Archaeology and Archaeometry*, Vol. 14, 233-246.

Belmonte, J. A. and Magli, G. (2015) Astronomy, architecture, and symbolism: the global project of Sneferu at Dahshur, *Journal for the History of Astronomy*, Vol. 46, 173– 205.

Collins, A. (2014) *Göbekli Tepe: Genesis of the Gods*, Rochester, Bear & Co.

González Garcia, C. and Belmonte, J.A. (2010) Statistical analysis of megalithic tomb orientations in the Iberian Peninsula, *Journal for the History of Astronomy*, Vol. 41, 225-238.

González Garcia, A.C. and Belmonte, J.A. (2011) Thinking Hattusha: astronomy and landscape in the Hittite lands, *Journal for the History of Astronomy*, Vol. 42, 461-494.

González García, A.C. and Belmonte, J.A. (2014) Sacred architecture orientation across the Mediterranean: a comparative statistical analysis. *Mediterranean Archaeology and Archaeometry*, Vol. 14, 95-113.

González García, A.C.; Rodríguez Antón, A. and Belmonte J.A. (2014) On the orientation of Roman cities in Hispania: preliminary results. *Mediterranean Archaeology and Archaeometry*, Vol. 14 (2014), 3, 107-119.

González García, A.C. and Belmonte, J.A. (2015) The Orientation of Pre-Romanesque Churches in the Iberian Peninsula *Nexus Network Journal*, Vol. 17, 353-377.

González-García, A. C., Noguera Celdrán, J. M., Belmonte Avilés, J. A., Rodríguez Antón, A., Ruiz Valderas, E., Madrid Balanza, M. J., Zamora, E. and Bonnet Casciaro, J. (2015) Orientatio ad sidera: astronomy and landscape in Qart Hadašt/Carthago Nova, *Zephyrus*, Vol. 75, 139-160.

González-Garcia, A.C., Belmonte, J.A. and Polcaro, A. (2016) A diachronic analysis of orientation of sacred precincts across Jordan, *Mediterranean Archaeology and Archaeometry*, this volume.

Hoskin, M. (2001) *Temples, tombs and their orientations*, Bognor Regis, Ocarina Books.

Lull, J. and Belmonte, J.A. (2009) The constellations of ancient Egypt, In *In Search of Cosmic Order: selected essays on Egyptian Archaeoastronomy*, J.A. Belmonte and M. Shaltout (eds.), Cairo, Supreme Council of Antiquities Press, 155-195.

Magli, G. (2015) Sirius and the project of the megalithic enclosures at Göbekli Tepe, *Nexus Network Journal*, in press. http://link.springer.com/article/10.1007%2Fs00004-015-0277-1

Neugebauer, O. and Van Hoessen, H.B. (1959) Greek Horoscopes, *American Philosophic Soc. Memories* Vol. 48, 14-16.

Polcaro, A., González-Garcia, A.C. and Belmonte, J.A. (2013) The orientation of the Bronze Age Temple of Pella, Jordan: the dying god Baal and the Rituals of the Summer Solstice, *Anthropological Notebooks*, Vol. 19, 481-492.

Rodríguez-Antón, A., Belmonte J.A. and González García, A.C. (2016) Romans in Near East: Orientation of Roman settlements in modern Jordan , *Mediterranean Archaeology and Archaeometry*, this volume.

Ruggles, C.L (ed.) (2015) *Handbook of Archaeoastronomy and Ethnoastronomy*, Heidelberg, Springer.

Ruggles, C.L.N. and Cotte, M. (eds.) (2010), *Heritage sites of astronomy and archaeoastronomy in the context of the UNESCO World Heritage Convention*, Paris, ICOMOS.

Schmidt, K. (2006) *Sie bauten die Ersten Temple*, Munich, Verlag C.H. Beck.

Yule, P. and Weisgerber, G. (1998) Prehistoric Tower Tombs at Shir/Jaylah, Sultanate of Oman, *Beiträge zur allgemeinen und vergleichenden Archäologie*, Vol. 18, 183-241.

17

SKY BEAR RESEARCH: IMPLICATIONS FOR "CULTURAL ASTRONOMY"

Roslyn M. Frank

University of Iowa, Department of Spanish and Portuguese, Iowa City, Iowa, USA

Corresponding author: Roslyn M. Frank (rozfrank14@yahoo.com)

ABSTRACT

This chapter examines research carried out to date on the Sky Bear and seeks to demonstrate the implications of this line of research for "cultural astronomy". It begins by reviewing research that has been done cross-culturally on bear ceremonialism, focusing on the role of circumpolar stars, Dipper stars and stars adjacent to them, and how they came to be integrated, cognitively, into an over-arching cosmology shared by different hunter-gatherer populations across the N. Hemisphere. Next, using three mutually reinforcing ethnographic datasets, the manner in which, specifically in Europe, this archaic worldview, characterized by embodied reciprocity, allowed humans, animals and nature to be bound together. The cosmology, grounded in the belief that humans descended from bears, integrated skyscape and landscape into a single interlocking reality. Furthermore, the worldview can be understood as embodying a "relational epistemology" or "relational ecology". This more culturally-informed approach coincides with the goals of "cultural astronomy" as well as the methodology and goals of the emerging field of archaeological ethnography. The reasons that these particular sky resources were chosen to project this set of spirtual beliefs skyward are also addressed. In the final section I suggest that the skyscape acts a kind of mnemonic device. As such, it is a cognitive resource, readily available to the social collective in question, which can act as a repository for past beliefs. Anchoring key components of a cosmology in the stars above allows the resulting skyscape to act as an enduring "memory bank". In short, the datasets analyzed facilitate the reconstruction of a European-wide ethno-cultural substrate that points to an archaic relational cosmovision and the belief that humans descended from bears as well as providing evidence for the way that skyscape and landscape were integrated into this cosmology.

KEYWORDS: Ursa major, bear ceremonialism, carnival, relational ecology, Bear's Son tale, archaeological ethnography, Good-Luck Visits, circumpolar stars.

1. INTRODUCTION

Over the past hundred years, significant research has been carried out cross-culturally on bear ceremonialism as well as on how the Dipper stars and stars adjacent to them came to be integrated, cognitively, into an over-arching cosmovision shared by different hunter-gatherer populations across the N. Hemisphere (Hagar, 1900; Hallowell, 1926; Speck, 1945). At the same time, there have been repeated attempts to define an entity called "cultural astronomy" (Iwaniszewski, 1991, 1995; Rappenglück, 2009; Ruggles & Saunders, 1993).

Furthermore, in recent years research on bear ceremonialism has charted the way-faring and time-telling strategies attached to these stars which have ended up being` passed on from generation to generation, often woven into sacred narratives shared by the members of the same hunter-gatherer populations (Berezkin, 2005; d'Huy, 2012; Frank, 2014a, 2014b). For example, among circumpolar peoples, there is substantial evidence for an entrenched belief in an ursine ancestor. We find this bear ancestor represented as a kind of divinity whose dwelling place is located in the heavens above, often in the same amorphous manner that Christian believers conceptualize God: as residing in the sky above, but without specifying exactly where his abode actually is. At other times the ursine cosmology is associated with hunting narratives and projected onto specific sets of circumpolar stars.

Moreover, there is increasing evidence that a similar cosmology was at work in Europe which evolved from a similar set of environmental circumstances existing in times past. At the core of this cosmology is the belief that humans descended from bears. Until recently investigations into bear ceremonialism and the projection of its associated cosmology skyward have focused on N. Hemisphere manifestations of this belief system, particularly among populations whose mode of subsistence is, at least in many respects, still similar to that of hunter-gatherers of times past who inhabited the same geographical zones.

Now that attention has turned to reconstructing the cognitive and cultural factors that led Europeans to create a skyscape with a bear at its center, these investigations have become a source of analogical reference. Similarly, in the case of Europe the belief in a bear ancestor would appear to harken back to a forager mentality, at least to one that was not compatible with a pastoral-agro mode of subsistence. In short, this European-wide ethno-cultural substrate is one that points to an archaic relational ontology founded on the belief that humans descended from bears. As will be shown, it opens a window onto the cognitive prehistory of European peoples.

2. EMBODIED RECIPROCITY

In contrast to this culturally informed interpretive approach, many attempts have been made to project, quite literally, the image of a bear onto the stars of Ursa Major as if the resulting putative resemblance was sufficient to explain how a bear came to be associated with these circumpolar stars, especially the Dipper stars.

Figure 1. Ursa Major as depicted in Urania's Mirror, a set of constellation cards published in London c.1825.

As we will see, the older worldview was characterized by embodied reciprocity where humans, animals and nature were bound together. The resulting cosmology is one that researchers have referred to as a kind of "relational ecology" (Betts et. al., 2015), "relational epistemology" (Bird-David, 1999), or "relational ontology" (Hill, 2011). In this respect, Betts, et al. (2015) and Watts (2013) also speak of the need to pursue a "relational archaeology".

In line with these proposals and the increasing interest in the relationship between "landscape" and "skyscape", I suggest we place more emphasis on the relational nature of human interactions with the *environment*, understood in its broadest distributed sense and that includes engaging with other-than-human animals, like bears, which in times past not only inhabited essentially all of Europe, in most zones they were actually more numerous than their human animal offspring.

3. THREE DATASETS: INTERCONNECTED AND MUTALLY REINFORCING

Three basic types of interlocking datasets are found across Europe relating to the belief that humans descended from bears. They are:

1. The Bear's Son tales, the most wide-spread motif in European folklore
2. The identification of the "King of Carnival" with the Bear's Son

3. Analogies with historically documented as well as extant beliefs concerning bear ceremonialism and the Celestial Bear among circumpolar peoples, especially Finno-Ugric peoples, e.g., the Khanty of western Siberia

3.1. *Bear's Son Tales: Orally transmitted `narratives*

Of late, research has focused on a figure who is half-human, half-bear, his mother is a human female while his father is a bear. The figure of this intermediary being is a central component of this much earlier worldview grounded on the belief in an ursine ancestry. At the same time, the story of the birth, life and exploits of this character is found in a myriad of European folktales, known collectively as the "Bear's Son Tale".

Moreover, in the Pyrenean zone, the "Bear Fests" celebrated each year[1] still incorporate elements taken from the Bear's Son Tale, reenacting, for instance, the initial encounter between his mother and father and his subsequent birth in the bear cave (Alford, 1930; Lajoux, 1996; Pauvet, 2012). While this motif represents the most widely disseminated European folktale, until the belief that humans descended from bears was plugged into the interpretative frame of these tales, they were not viewed as particularly significant.

3.2. *European Carnival: Performance transmitted narratives*

More recently, evidence has surfaced which links this "Jesus-Bear"—as one of my informants called him—to the character who dies, is mourned and is resurrected during Carnival. More specifically there is evidence that in Europe the so-called "King of Carnival" is actually an embodiment of this intermediary being, a topic treated in depth in a recently published monograph (Frank, 2015). This also suggests that the older layer of pre-Christian belief in which Carnival is embedded comes from this ursine informed ethno-cultural substrate. It also suggests that Carnival itself, as well as pre-Christian beliefs concerning February 2, known as "Bear Day" or Candlemas, should be traced back to this same archaic cosmology (Lajoux, 1996; Pauvert, 2012, 2014).

"Good Luck Visits" conducted with a retinue that regularly includes a human dressed as a bear also take place all across Europe throughout the Winter and Spring carnival periods (Bertolotti, 1992; Frank, 2008a, 2008b, 2009). These house visits are still part of traditional social practice in many locations where they often take place on December 5th and 6th.[2] They are still popularly believed to insure good health, promote fertility in animals and humans alike as well as guarantee the fruitfulness of crops. In short, their purpose has been viewed as prophylactic: they are intended to bring about the well-being of those visited.

Figure 2. The "bear" of Saint Laurent de Cerdans, Pyrenees, France. Fréger (2012: 226).

Many other remarkable examples of contemporary European Bear performers, dressed in straw, fur and plant materials, who take part in European Winter and Spring carnivals have been photographed by Charles Fréger (2012) and are available for viewing online at http://www.charlesfreger.com/portfolio/wildermann/. Meanwhile it is important to note that Fréger, in his photographic study of the European performers, writes the following concerning the figure of the "Wild Man", known in German as *Wilder Mann* and in French as *L'Homme sauvage*. His comments are based on the pan-European folktale called the Bear's Son, who is

> according to legend a son born of the union of a bear and a woman, whether by consent or not. Belonging to two worlds and knowing the intricacies of both, this mythological being is considered a 'superheroe' [...]. In most cases the Wild Man wears a costume made of natural materials or animal skins, his face is rendered unrecognizable, either by a mask or by a costume that covers him entirely, or by a black make-up. An accessory —a stick, a club or something similar— and one or more bells complete his outfit. (Fréger, 2012: 243)

Figure 3. An Austrian Krampus. Fréger (2012: 21).

The double-souled nature of this figure, half bear and half human, allowed it to be conceived as a kind of intermediary being who bridged the connection between humans and their bear ancestors. This has a parallel in Christianity where a similar role is played by Jesus Christ, born of a human female but whose father was the Holy Spirit, in short, an intermediary, partly human and partly divine.

Figure 4. Sourvakari. Bulgaria. Fréger (2012: 166).

However, there is a major difference in that the older ursine cosmology appears to be one in which real human beings, properly dressed, fulfilled this role, the role of intermediary, watching over the community, both in a physical and spiritual sense. Hence, Good Luck Visits were performances that were intended to insure the well being of the members of each community. In these performances in which the humans, dressed as bear-humans, that is, donning special clothing, as if they, too, were dual-natured beings, carried out their duties vis-à-vis the community. However, far from being lost in mists of time, Good Luck Visits continue to be celebrated in many parts of Europe.[3]

Figure 5. Ursul (Bear), Romania. Fréger (2012: 71).

These survivals harken back to the view that bears were ancestors of humans. While the struggle of the medieval Church against the deeply entrenched belief in this ursine intermediary is well documented, efforts to eradicate the old belief system were not totally successful. As Pastoureau explains, even after Christianization, the old beliefs continued to circulate:

> Indeed, in a large part of non-Mediterranean Europe in the Carolingian period, the bear continued to be seen as a divine figure, an ancestral god whose worship took on various forms but remained solidly rooted, impeding the conversion of pagan peoples. Almost everywhere, from the Alps to the Baltic, the bear stood as a rival to Christ. (Pastoureau, 2011: 3)

3.3. *Sources of analogy: Research on the role of the Sky Bear in Finno-Ugric cultures*

Over the past fifty years the ursine relational cosmology along with its various celestial projections has been investigated in depth, most extensively among the Saami, Finnish and Khanty with the latter still practicing Bear Fests (Pentikäinen, 2007; Sarmela, 2006; Schmidt, 1989).

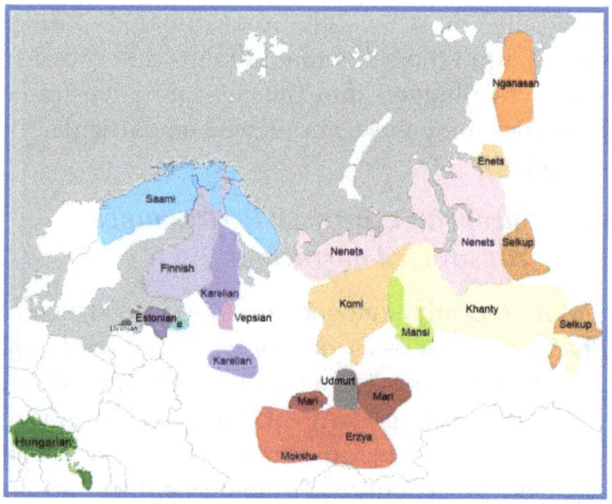

Figure 5. Finno-Ugric Languages. https://en.wikipedia.org/wiki/Finno-Ugric_languages.

There are indications that a similar ursine belief was present among other circumpolar peoples across N. Eurasia (Anisimov, 1963; Lushnikova, 2002; Sokolova, 2000). Moreover, a number of insights can be drawn based on well attested cognitive parallels associated with circumpolar bear ceremonialism. For instance, in the ursine cosmology documented for the Khanty people of western Siberia, the sacred narrative speaks of a celestial Sky Bear ancestor who dwells in the heavens above. It is the son (or daughter) of the Sky Bear who comes down to Earth explicitly to teach humans how to behave, treat each other and nature properly and carry out the necessary "sending home" ceremony performed for the souls of individual earthly bears slain by hunters (Shepard, 2007; Shepard & Sanders, 1992; Schmidt, 1989).

The "sending home" ceremony was needed so that the earthly bear could report back to the Sky Bear concerning the behavior of its human offspring. (Frank, in press; Frank & Silva, 2012). In this respect, given that the figure of the European Bear's Son has also been identified with the "King of Carnival" who dies, is mourned and is resurrected, there is reason to believe that the pan-European version of this ursine belief system contained a similar narrative motif (Bertolotti, 1992, 1994; Frank, 2015; Pauvert, 2012).

4. FURTHER INSIGHTS INTO THE SKY BEAR COSMOLOGY

But we are still left with a question: why associate a divinity in the form of a bear ancestor with circumpolar stars such as those of the Big Dipper? The well-known astronomer Ed Krupp who is also one of the foremost historians of astral myths, wrote the following about Ursa Major, however, without any awareness of the belief in an ursine ancestor:

> Because real bears hibernate, they are symbols of renewal and seasonal change. They represent the immortal spirit of life. We don't have a bear in the northern sky because the stars there look like a bear. The celestial bear prowls around the pole, performs a seasonal death and returns to life when the world does. It is a celestial power-house, and the energy churned out by this revolving dynamo in the sky seems to activate the entire world. The celestial bear performs another one of those necessary sacrifices that ensure the perpetuation of life—this time in the northern sky. By dying, the celestial bear lives forever. (Krupp, 1991: 239-240)

Then we have the well-known nature writer Paul Shepard (1995) who made these cogent observations about the association of a bear with these circumpolar stars:

> The motion of the stellar hemisphere, which to later, agricultural civilizations suggested grindstones and other wheels, turned by some invisible hand, may have been very differently interpreted by earlier peoples, as the heavenly, unitary spectacle of the dynamic process or energy system that animates life itself, of which in diurnal experience we see only fragments. ... The 'cult of the sacred bear' seems to be much larger than a local theriophany, indeed, to be a complex, multiform, ancient metaphysics.

In this sense we might view the skyscape as a kind of mnemonic device. As such, it is a cognitive resource, readily available to the social collective in question, which can act as a repository for archaic beliefs. Anchoring key components of a cosmology in the stars above allows the resulting skyscape to act as an enduring "memory bank". In this way traces of past beliefs and social practices are perpetuated across time. It would appear that the circumpolar stars, especially those of the Big Dipper, have functioned in a similar fashion.

5. CONCLUSIONS

What is most remarkable to me is the discovery of the staying power, that is, the cognitive resilience of beliefs and social practices attached to the Sky Bear, bear ceremonialism and its associated relational ontology. All across the N. Hemisphere these beliefs have served to bind together humans and bears, landscape and skyscape, utilizing the stars of the Big Dipper and stars nearby to illustrate and hence help insure the transmission of this aspect of the ursine belief system from one generation to the next (Frank, 2014a, 2014b, in press).[4] And this is also true of Europe. As we have seen. these conceptual linkages are particularly striking, even today, once we begin to investigate the European ethnographic materials, both past and present. In short, the cosmovision in question has left an indelible mark on the culture of Europeans.

Finally, if we are to take the term "cultural astronomy" seriously we need to pay attention to the role of culture in the creation, understanding and projection of skyscapes, and in this instance, in the choices that led people to conceptualize their bear ancestor as dwelling in the sky above them. Also, as we have seen, the worldview in question was characterized by embodied reciprocity: humans, animals and nature were bound together. Consequently, the cosmology is one that we can define as epitomizing a kind of "relational ecology" which, in turn, speaks to us about the need to pursue a "relational archaeology". At the same time this more culturally-informed approach, one that blends data from the past and present, coincides with the methodology and goals of the emerging field of archaeological ethnography (Hamilakis & Anagnostopoulos, 2009).

In summary, cultural astronomy provides a more cognitively-oriented approach that in this case allows for the recognition that among hunter-gatherer cultures the landscape and the skyscape were interlocking realities, each of which provided a way to structure and project the fundamental beliefs of the cultural group in question, beliefs and social practices that quite remarkably have survived, albeit in a fragmented form, into the 21st century.

ACKNOWLEDGEMENTS

I would like to express my thanks to Dr Paul Shepherd who years ago helped me come to appreciate the importance of the skyscape in this ursine cosmology and to Dr Dominique Pauvert for his constructive comments on the current research project.

REFERENCES

Alford, V. (1930) The Springtime Bear in the Pyrenees. *Folklore*, XLI, 266-279.

Anisimov, A.F. (1963) Cosmological concepts of the peoples of the north. In *Studies in Siberian Shamanism*, H. N. Michael, (ed.). Arctic Institute of the North, University of Toronto Press, Toronto, 157–229.

Barbeau, M., (1946) Bear Mother. *Journal of American Folklore*, 59 (231), 1-12.

Berezkin, Y. (2005) The cosmic hunt. *Folklore: Electronic Journal of Folklore*, Vol. 31, 79-100. http://www.folklore.ee/folklore/vol31/berezkin.pdf.

Betts, M., Hardenberg, M., and Stirling, I. (2015) How animals create human history: Relational ecology and the Dorset-Polar Bear connection. *American Antiquity*, Vol. 80 (1), 89-112.

Bird-David, N. (1999) Animism revisited: Personhood, environment, and relational epistemology. *Current Anthropology*, Vol. 40, 67-91.

Bertolotti, M. (1992) *Carnevale di Massa 1950*. Torino: Guilio Einaudi Editore.

Bertolotti, M. (1994) La fiaba del figlio dell'orso e le culture siberiane dell'orso. *Quaderni di Semantica*, 15 (1), 39-56. [The tale of the bear's son and the Siberian culture of the bear].

d'Huy, J. (2012) Un ours dans les étoiles: Recherche phylogénétique sur un mythe préhistorique. *Bulletin Préhistoire du Sud-Ouest*, 20, 91-106. [The bear in the stars: Phylogenetic research concerning a prehistoric myth]. http://tinyurl.com/d-huy-un-ours.

Fréger, C. (2012) *Wilder Mann: The Image of the Savage*. Stockport, England: Dewi Lewis Publishing.

Frank, R.M. (2008a) Evidence in Favor of the Palaeolithic Continuity Refugium Theory (PCRT): *Hamalau* and its linguistic and cultural relatives. Part 1. *Insula: Quaderno di Cultura Sarda*, Vol. 4, 91-131. http://tinyurl.com/hamalau14.

Frank, R.M. (2008b) Recovering European ritual bear hunts: A comparative study of Basque and Sardinian ursine carnival performances. *Insula: Quaderno di Cultura Sarda* (Cagliari, Sardinia), Vol. 3, 41-97. http://tinyurl.com/hamalau.

Frank, R.M. (2009) Evidence in Favor of the Palaeolithic Continuity Refugium Theory (PCRT): *Hamalau* and its linguistic and cultural relatives. Part 2. *Insula: Quaderno di Cultura Sarda*, Vol. 5, 89-133. http://tinyurl.com/hamalau14.

Frank, R.M. (2014a) The origins of 'Western' constellations. In *The Handbook of Archaeoastronomy and Ethnoastronomy*. C.L.N. Ruggles (ed.). Berlin: Springer Publishing Company, 147-163.

Frank, R.M. (2014b) The skylore of the indigenous peoples of Eurasia. In *The Handbook of Archaeoastronomy and Ethnoastronomy*, C.L.N. Ruggles (ed.). Berlin: Springer Publishing Company, 1679-1686.

Frank, R.M. (2015) Bear Ceremonialism in relation to three ritual healers: The Basque *salutariyua*, the French *marcou* and the Italian *maramao*. In *Uomini e Orsi: Morfologia del Selvaggio*, E. Comba & D. Ormezzano (eds.). Torino: Accedemia University Press, 41-122.

Frank, R.M. in press. A status report: A review of research on the origins and diffusion of the belief in a Sky Bear. *The Materiality of the Sky*, J. K. Malville, F. Silva, F. Ventura and T. Lomsdalen (eds.). University of Wales Trinity Saint David, Wales: The Sophia Centre Press.

Frank, R. M., and Silva, F. (2012) European folklore in the *longue durée:* Palaeolithic Continuity and the European ursine genealogy. Paper presented at the Folklore & Archaeology Conference, October 13-14, 2012. University College London, London. http://tinyurl.com/folklore-longueduree.

Hallowell, A. I. (1926) Bear Ceremonialism in the Northern Hemisphere. *American Anthropologist*, 28, 1-175.

Hamilakis, Y. and Anagnostopoulos, A. (2009) What is archaeological ethnography? In *Archaeological Ethnographies: A Special Issue of Public Archaeology*, Y. Hamilakis and A. Anagnostopoulos (eds.), Vol. 8 (2-3), 65-87). Maney Publishing. http://dx.doi.org/10.1179/175355309X457150.

Hill, E. (2011) Animals as agents: Hunting ritual and relational ontologies in Prehistoric Alaska and Chukotka. *Cambridge Archaeological Journal*, 21 (3), 407-426.

Iwaniszewski, S. (1991) Astronomy as a cultural system. *Interdisciplinarni izsledvaniya, Vol. 18,* 282-288.

Iwaniszewski, S. (1995) Archaeoastronomy and Cultural Astronomy: Methodological issues. *Atti dei Convegni Lincei 121. Convegno Internazionale sul tema: Archaeologia e astronomia: esperienze e prospettive future (Roma, 26 novembre 1994).* Roma: Accademia Nazionale dei Lincei, 17-26.

Krupp, E.C. (1991) *Beyond the Blue Horizon: Myths and Legends of the Sun, Moon, Stars, and Planets*. New York: HarperCollins Publishers.

Lajoux, J.-D. (1996) *L'homme et l'ours*. Grenoble: Glénat.

Lanciano, N. and Tutino, M. (2005) Orse del cielo, orse della terra *Rivista italiana di Archeoastronomia, 3*, 107-118. [Bear of the sky, bear of the earth].

Lushnikova, A. (2002) Early notions of Ursa Major in Eurasia. In *Astronomy of Ancient Societies*, T. M. Potyomkina and V. Obridko (eds.). Moscow: Nauka, 254-261.

Newcomb, F. J. (1967) *Navajo Folk Tales*. Santa Fe, N. M.: Museum of Navaho Ceremonial Art.

Pastoureau, M. (2011) *The Bear: A History of a Fallen King*. Cambridge: Cambridge University Press.

Pauvert, D. (2012) *La religion carnavalesque*. Meuzac: Chamin de Sent-Jaume. [*Carnavalesque Religion*]

Pauvert, D. (2014) Le rituel de l'ours des Pyrénées aux steppes. In *Traditions en devenir (coutumes et croyances d'Europe et d'Asie face au monde moderne), EURASIE N°. 2*, Société des Études euro-asiatiques (ed.) (pp. 17-51). Paris: Hartmattan, 17-51. [The ritual of the bear from the Pyrenees to the steppes]. https://www.academia.edu/13070219/Les_rituels_de_lours_des_Pyr%C3%A9n%C3%A9es_aux_steppes.

Pentikäinen, J. (2007) *Golden King of the Forest: The Lore of the Northern Bear*. Helsinki: Etnika.

Rappenglück, M. (2009) Constructing worlds: Cosmovisions as integral parts of human ecosystems. In *Cosmology across Cultures*, J. A. Ribiño-Martín, et. al. (eds.). San Francisco, CA: Astronomical Society of the Pacific, 107-115.

Ruggles, C.L.N. and Saunders, N.J. (1993) The study of cultural astronomy. In *Astronomies and Cultures*, C.L.N. Ruggles and N. J. Saunders (eds.). Niwot: University of Colorado Press, 1-31.

Sarmela, M. (2006) The Bear in the Finnish Environment: Discontinuity of Cultural Existence. Translated by Annira Silver (2005). Appendix: Ritva Boom (1982). Helsinki. http://www.kotikone.fi/matti.sarmela/bear.html.

Schmidt, É. (1989) Bear cult and mythology of the Northern Ob-Ugrians. In *Uralic Mythology and Folklore*, M. Hoppál and J. Pentikäinen (eds). Budapest: Ethnographic Institute of the Hungarian Academy of Sciences/Helsinki: Finnish Literature Society, 187-232.

Shepard, P. (1995) A bear essay (unpublished manuscript).

Shepard, P. (2007) The biological bases of bear mythology and ceremonialism. *The Trumpeter: Journal of Ecosophy*, 23 (2), 74-79.

Shepard, P. and Sanders, B. (1992) *The Sacred Paw: The Bear in Nature, Myth and Literature*. New York, NY: Arkana.

Sokolova, Z. P. (2000) The bear cult. *Archaeology, Ethnology & Anthropology of Eurasia*, 2 (2), 121-130.

Speck, F. C. (1945) *The Celestial Bear Comes Down to Earth: The Bear Ceremony of the Munsee-Mahican in Canada as Related by Nekatcit*. In collaboration with Jesse Moses, Delaware Nation. Ohsweken: Reading Public Museum and Art Gallery.

Watts, C. (2013) *Relational Archaeologies: Humans, Animals, Things*. London/New York: Routledge.

---

[1] Perhaps the most remarkable Bear Fests still celebrated today are found in three remote Pyrenean villages. Although in centuries past, their existence was rarely noted, today modern technology makes them readily available to the researcher. The villages are: 1) Prats de Mollo, "Enquête d'Ailleurs : La fête de l'ours" ["A survey from elsewhere: The bear festival"] (filmed in 2015), https://www.youtube.com/watch?v=bs0xfwYx4lE&app=desktop; 2) Arles sur Tech, "La fête de l'ours" ["The bear festival"] (filmed in 2010), https://www.youtube.com/watch?v=RHlttofhBIE; 3) Saint Laurent de Cerdans, "Fête de l'Ours" ["The bear festival"] (filmed in 2015), https://www.youtube.com/watch?v=xi_xGbliKLQ.

[2] The amalgam of pre-Christian and Christian elements is clear in the traditional outings of "straw-bears" and their retinues, a commonplace scene in Bavaria on December 5th and 6th, as shown in these videos. Bavaria, especially Berchtesgaden, is a location where conducting visits to the houses of the village is still a central component of the performance: "Alpendämonen - Maskenlauf in Berchtesgaden" ["Alpine Demons: Mask Running in Berchtesgaden"] (filmed in 2012), https://www.youtube.com/watch?v=XXq073okUIl, and "Der Nikolaus mit seinen Buttnmandl in Bischofswiesen" (2015 12 05) ["Nicholaus with his Buttnmandl (Straw Bears) in Bischoswiesen"]. https://www.youtube.com/watch?v=uggy9KE6cP4https://www.youtube.com/watch?v=YV4pt6NZWdQ. A more urban setting is found in this video entitled "Buttnmandllaufen Berchtesgaden [Buttnmandl und Kramperl]" (filmed in 2015), dedicated to the running of the Buttnmandl straw bears and their handlers, the Kramperl, in Berchtesgaden, https://www.youtube.com/watch?v=joaZfZLnFCQ.

[3] Good Luck Visits are also popular at this time of year in the Frisian Waddeneilanden, islands off the coast of The Netherlands, for example, in the carnival of Sunderum in the municipality of Terschelling: "Sunderum Terschelling 6 december 2012" ["The Festival of Sunderum, Terschelling"], https://www.youtube.com/watch?v=nHqkUvmj2OY.

---

[4] Although outside the scope of this study, another avenue that is still open to investigation is the degree to which the Sky Bear ancestor was perceived by the social collective in a gendered fashion, that is, as a male or female. Shepard (1985) has suggested that in the oldest stratum of this cosmology the ancestral figure may have been conceptualized as a Mother Bear, a belief that could have had its continuation in the gendering of Ursa Major (Lanciano & Tutino (2005). However, given the range of variations found in N. Hemisphere bear ceremonialism, it is still too early for us to reach a definitive conclusion in this respect. Indeed, whereas a mythical Mother Bear is common in narrative and iconic representations, it is far less common to find circumpolar stars themselves associated with her as opposed to their utilization to portray action scenes with motifs related to hunting (Barbeau, 1946; Berezkin, 2005; Lushnikova, 2000).

18

FROM HAGIOGRAPHY TO CELTIC COSMOLOGY: ARCHAEOASTRONOMY AND CHRISTIAN LANDSCAPE IN OURENSE (NW SPAIN)

A. César González García[1], Marco V. García Quintela[2]

[1] *Institute of Heritage Sciences, Incipit-CSIC*
[2] *University of Santiago de Compostela*

Corresponding author: a.cesar.gonzalez-garcia@incipit.csic.es

ABSTRACT

The cult of Santa Mariña is very popular in the region of Galicia, in northwest Spain. According to tradition, she was born and martyred in two sites at the heart of the modern-day province of Ourense: she was born and grew up around the lagoon of Antela, and was martyred in the parish of Santa Mariña de Augas Santas, where she performed miracles and where her tomb is still preserved. Both places are located in the territory of the *Limici*, a pre-Roman Celtic tribe, and contain a remarkable amount of archaeological material from the Iron Age and Roman times. An archaeoastronomical study has revealed that the most important archaeological sites have a number of significant solar and lunar relationships attributable to the Celtic tradition (lunistices, Celtic mid-season festivals, the cosmos divided into three levels). Christianity preserved these structures through the feast dates of the saints worshipped in different parishes and other places, and their arrangement in different local landscapes. Episodes of Mariña's life and her places of worship are important because they coincide with significant points in the astronomical alignments that have been detected. Hydatius of Chaves (c. 400-469) a *Limici* scholar, bishop and author of a Chronicle, is considered responsible for introducing the cult of Santa Mariña, and as the driving force behind the Christianization of a landscape/skyscape that was previously defined by a Celtic worldview.

KEYWORDS: Galicia, Celtic cosmology, Romanization, hagiography, Christianization, landscape building, archaeoastronomy, Hydatius of Chaves

1. INTRODUCTION

Santa Mariña is a martyr who is widely worshipped in Galicia, in northwest Spain. Together with saints such as St. Eulalia, St. Mamede, or St. Quitera, she defines a very primitive stage of Christianization of the Galician countryside, probably dating from the sixth century on average[1]. It is interesting to note that parishes and places of worship dedicated to this saint are often associated with visible archaeological remains dating from different times. The popularity of the saint and her hagiography appear frequently in the literature from the sixteenth century onwards, coinciding with the spread of printing and the compilation of pious traditions and grassroots hagiographies, albeit with shaky historical roots.

The religious context of the Reformation and Counter-Reformation affected debates on the authenticity of the hagiography of Mariña, resulting in the appearance of two schools of thought that have continued to the present day.

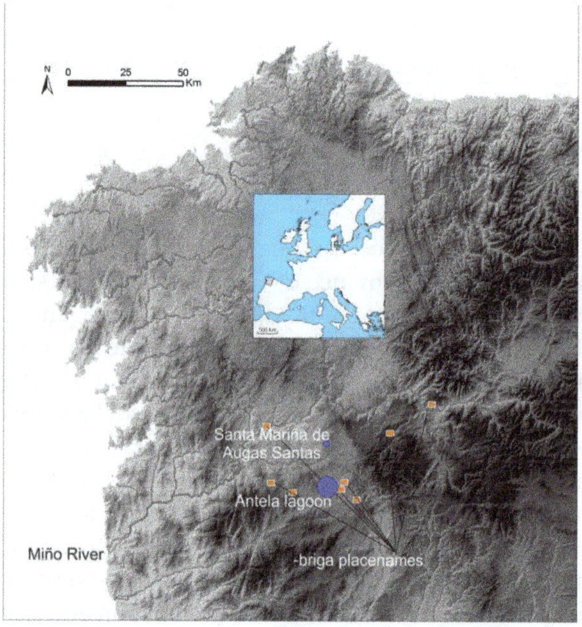

Figure 1. Location of the main places referred to in this article.

The first sceptical or learned school of thought compares the hagiography of Mariña with traditions about her namesake Marina, Pelagia in Greek, or Margaret, a saint from Pisidian Antioch in Anatolia. As the stories attributed to Mariña in Galicia were identical to those attributed to the martyr of Anatolia, the conclusion was that the historicity of the Galician saint was questionable, as she would have been an anonymous saint onto whom the life and martyrdom of the Anatolian were imprinted. The main proponents of this idea are non-Galician, Catholic scholars.

The second school of thought defends the "archaeological truth" of the saint, supported by Galicians who defend the veracity of the tradition of Mariña from episodes of her life attached to specific places, buildings, rocks or trees, in the geography at the heart of the modern-day province of Ourense. How can a saint be false who gave Christian meaning to the local landscape for the peasant community? This school believes that the idea of the "myth" of Mariña is equivalent to saying that the local peasant communities were mistaken about the world they lived in for centuries (García Quintela and González-García, submitted).

This idea is based on the fact that Mariña's biography is located in two places (Fig. 1). The first is her birthplace and where she spent her younger years, until she was taken prisoner by the evil Roman Olibrius. This is the area around the lagoon of Antela, in the upper reaches of the River Limia, which dried out in the late 1950s. In addition to the legend of Mariña, the lake has a rich folklore and archaeological sites from different eras abound in its surroundings (http://www.altalimia.org/). The second location is where she was imprisoned and martyred, in the modern-day parish of Santa Mariña de Augas Santas, whose rich archaeological heritage is inventoried, so to speak, by the episodes from the martyrdom of the saint (http://santamarinadeaugassantas.com/).

Our study involves these locations, all of which are archaeological sites. The question is, therefore, to find out the role of the saint in the transformation of a pre-existing skyscape. In other words, our aim was to identify how a pre-Christian skyscape, involving different places and monuments, served as a vector for the spread of Christianity, by attaching the local topography to a narrative that came all the way from Anatolia. This process resulted in the construction of the present-day Christian landscape.

We first present the archaeoastronomical observations of both areas with some explanations to help understand its regional scope. Then we introduce the pre-Christian Celtic and Roman context, and how well they coincide with the observations about Mariña. Finally, we argue that the chronicler Hydatius of Chaves (c. 400-469), who was born in Xinzo de Limia, close to the lagoon of Antela, may have been the architect behind the religious engineering that led to the creation of this Christian landscape.

[1] J.C. Sánchez Pardo, personal communication on the C14 dating of martyr's churches in several parts of Galicia, data to be published.

2. ARCHAEOASTRONOMY AND ASTRONOMY IN CULTURAL OBSERVATIONS

2.1. *Around the lagoon of Antela (Fig. 2., table I)*

Three sites in the vicinity of the lagoon of Antela are connected with Saint Mariña.

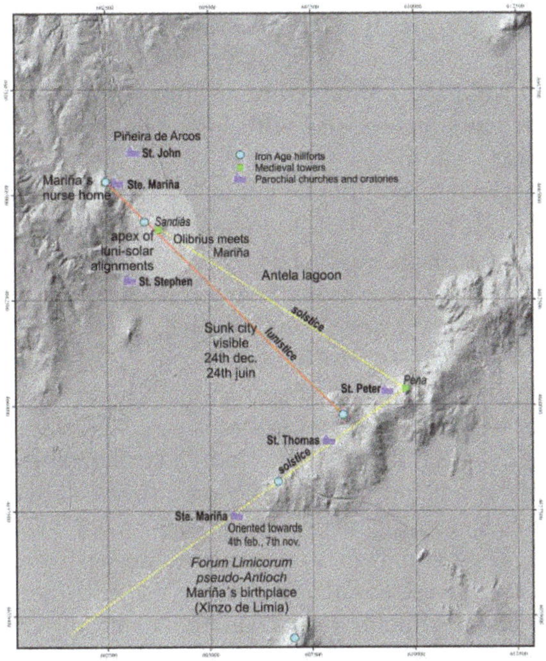

Figure 2. Archaeoastronomical observations around the Lagoon of Antela.

Table I. Astronomical orientation of the elements indicated in the text around the lagoon of Antela (latitude 42° 7'). The columns show the site or structure measured, the azimuth (A) and horizon altitude (h) measured with a precision compass (error estimated to be ½°), and the calculated astronomical declination (δ). The last column gives the possible astronomical event or date. For further details, see the text.

Structure	A	h	δ	Event
St. Mariña Church (Xinzo)	112¼	0¾	−16¼	4th Feb/7th Nov
Sandias to Pena	123½	0½	−23¾	Winter Solstice
St. Mariña to Pena	57¾	1	24	Summer Solstice
Porqueira to Pena	60	0	22	Summer Solstice
Pendón to Cidá	132	0½	−29¾	Lunistice

1) The Saint was born in Xinzo de Limia, traditionally called "Antioch" due to a false etymology of "Antela" (evoking the Anatolian origin of the reused hagiographic tradition). Xinzo church, dedicated to St. Mariña, is built on top of a Roman building whose orientation has survived successive renovations, and differs from the standard orientation of the pre-Romanesque churches in the region (Xusto, 2000, González-García, 2015).

2) The home of the wet nurse who cared for the child and raised her as a Christian is in the hillfort of Pendón de Santa Mariña (Piñeira de Arcos). On the slopes of this hillfort there is a 17th century church dedicated to St. Mariña. This hillfort and the hillfort of A Cidá, on the opposite shore of the lagoon, are the only ones that directly face towards Antela and have a relationship compatible with the major southern lunasticice.

3) At the base of the tower of Sandiás in the NW sector of the Lagoon, the Roman Olibrius tried to seduce Mariña, and after her rejection, she was captured and led to Augas Santas. This place has a long cultural stratigraphy: at its peak there is a rock shrine from the Iron Age; in Roman times there was a post on the XVIII road bordering the lagoon of Antela to the West, and in the late Middle Ages a tower was built at the top. This tower has a solstitial relationship with the tower of Pena, its functional equivalent on the eastern side of the Lagoon (Fig. 3).

Figure 3. The landscape of the lagoon of Antela today as seen from the Pendón de Santa Mariña hillfort. The arrows indicate the different landmarks, from left to right: Pena medieval tower, Sandias tower and A Cidá hillfort. Upper left inset: sunrise at the winter solstice of 2015 over the Tower of Pena, seen from the Tower of Sandiás. Upper right: the lagoon from the Tower of Sandias, with the rock altar in the foreground.

We can add another three observations from cultural astronomy to these archaeoastronomical observations.

4) One of the folk tales connected to the lagoon refers to the existence of a sunken city. It was also called Antioch, and was sunk because of the sins of its inhabitants who, when tested by Jesus, were found to be uncharitable. The city could be perceived "on the morning of St. John's Day, when the first ray of sunlight shines on the lagoon, deep, deep down the church tower can be seen. And on Christmas Day, at midnight, the bells can be heard ".

5) The parishes bordering the Lagoon form a dual interplay of solstitial dates: the eastern parishes are dedicated to St. Peter (celebrated on 29th July) and St. Thomas (21st December), while those to the west are dedicated to Saint John (23rd June) and Saint Stephen (26th December).

6) To the south of the Limia valley, away from the Lagoon, in Porqueira, a third medieval tower has a solstitial relationship with the tower of Pena, in the eastern part of the valley. Also, the solstice line connecting both towers passes over the church of Santa Mariña in Xinzo de Limia.

2.2. *The parish of Santa Mariña de Augas Santas (Fig. 4, table II).*

The summit of the Os Canteiros hill occupies the centre of the territory of the parish of Santa Mariña de Augas Santas.

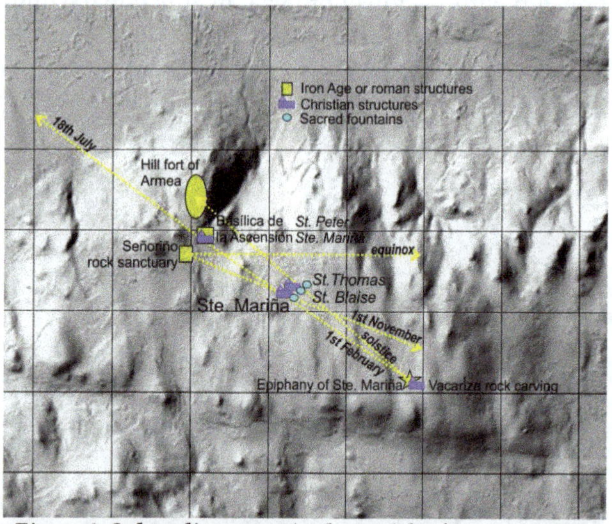

Figure 4. Solar alignments in the parish of Santa Mariña de Augas Santas.

Table II. Astronomical orientation of the different elements indicated in the text around Augas Santas (latitude 42° 14'). The last two lines include data from the hillfort of Avión and the rock art site of A Ferradura, mentioned in the text. Columns are as in Table I.

Structure	A	h	δ	Event
Basilica-Crypt	119¾	6½	−16¾	2nd Feb/9th Nov
Basilica-Nave (west)	297½	1½	20.5	26th May/18th July
Armea hillfort to Vacariza	126¼	3	−23¾	Winter Solstice
Señoriño rock altar	89¾	2	1½	Equinox
Señoriño basin	114	4½	−14¼	10th Feb/1st Nov
Avión basin	109	0¼	−14	11th Feb/1st Nov
A Ferradura	236	1	−23¾	Winter Solstice
	245	1	−17½	1st Feb/13th Nov

The main enclave in honour of the saint today is the village of Augas Santas (= Holy Springs), so called because according to tradition, this was where she was beheaded. Her head hit the ground in three places, where three holy springs emerged. Also, the saint's tomb is in the Romanesque parish church (celebrated on July 18th). St. Thomas (December 21st) and St. Blaise (February 3rd) are also worshipped in the church. This place became an important centre of population after the spread of Christianity. Previously, the habitat was located in the hillfort of Armea and its surroundings (1 km to the NW of the church), while the petroglyph of A Vacariza is located at the top of Os Canteiros.

Around the hillfort there are three places of religious significance, and with interesting solar alignments.

1) The "Outeiro dos Pendóns" is the highest point of Armea hillfort. This hilltop was surrounded by a wall that encircled three large rocks that seem to point towards the present church and the mountain top where the petroglyph is located. The sun rises at the winter solstice in this direction. Also, the most miraculous of the three springs is in a chapel dedicated to St. Thomas (December 21st).

2) On the hill of "O Señoriño," a structure was recently excavated that forms a rock shrine similar to others in the region known to date from the Iron Age. The main structures are oriented eastward. In addition, a rectangular basin carved into the rock and eccentric with respect to the main structures, faces diagonally towards the sunrise on November 1st in coincidence with the parish church. A similar phenomenon is also observed in the Avión hillfort (Ourense), without any known justification for its function.

3) The most complicated structure is the unfinished church known as Church of the Ascension or *Forno da Santa*. The two denominations feature aspects of the complex "biography" of the building (Blanco et al., 2015).

The original building is an underground sauna adjacent to the Iron Age hillfort (García Quintela and Santos-Estévez, 2015; García Quintela et al., 2014). Traditionally, this is where Mariña was said to have been martyred by fire. However, she did not die, as she was rescued by St. Peter, who turned her into steam, allowing her to escape up the chimney. Thus, the hagiography accurately evokes the original function of the building: Mariña, water, in contact with the fire, becomes steam.

The sauna suffered an initial Christian transformation in the sixth century (548 ± 45), probably becoming a chapel dedicated to the cult of the martyr. Then the building underwent two enlargements and renovations until the 13th century, when it became the crypt of a larger church, built over it, which remained unfinished, and commemorates the Ascension (Blanco et al., 2015). The entire structure features an interesting array of solar alignments.

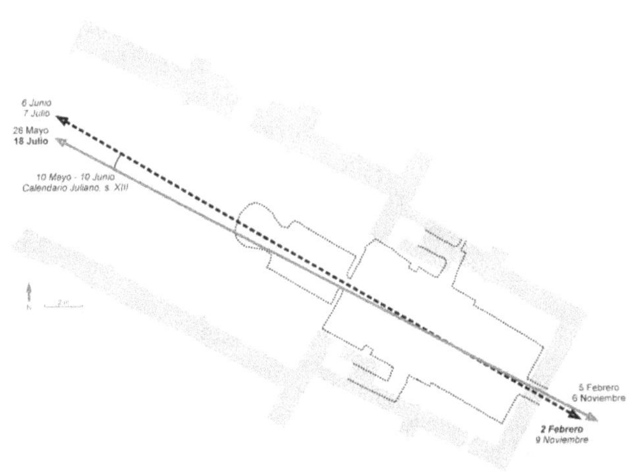

Figure 5. Solar alignments at the Forno da Santa / Church of the Ascension sanctuary in the parish of Augas Santas. The grey shaded sections show the plan of the upper basilica. The crypt and Iron Age sauna are indicated by the dotted line. The dashed line indicates the orientation of the crypt and apse, the grey solid line indicates the orientation of the nave. It is interesting to note that the angle between the two orientations towards west coincides with the range of dates for the festivity of the Ascension in the middle ages.

The early sauna and different phases of the Christian crypt are consistently oriented towards the solar sunrise on February 2nd (coinciding with the local celebration of St. Blaise). On the surface, the unfinished building of the basilica has an apse with the same orientation as the underground part. To the west, the orientation of the nave is towards the sunset on July 18th, the feast day of Santa Mariña (Fig. 5).

Furthermore, the beginning of February is important in one of the petroglyphs in A Ferradura, a rock art site in the vicinity of Augas Santas, where the orientation of the site is similar to the church of Santa Mariña in Xinzo de Limia (supra) and the feast day is Candlemas (February 2nd) (García Quintela et al., 2015).

3. PRE-CHRISTIAN CULTURAL CONTEXT

The protohistoric cultural context of the region is Celtic. This is suggested by names ending in *-briga*, a standard criteria in defining areas with speakers of Celtic languages. The latest tally puts eight of the eighty known cases in the Iberian Peninsula in the modern-day province of Ourense (Guerra, 2005) therefore the province, accounting for 1.25% of the territory of the Peninsula, concentrates 10% of these names. This finding is corroborated by the Celtic characterization of god's names and pre-Roman people attested in Latin inscriptions of the *Limici*, the tribe that inhabited the area around the River Limia (Lujan, 2009).

The Roman footprint in this cultural horizon from the year 26 BC onwards is important, but discontinuous. Around Antela, the *Via XVIII* route was of great importance, connecting *Bracara Augusta* (now Braga, in Portugal) with *Asturica Augusta* (Astorga, in Spain), built from 80 AD onwards. Its route along the west edge of the Lagoon is shown by a large number of Roman milestones. Also, the site of the tower of Sandiás probably coincided with the *mansio Gemina* that appears in the Antonine Itinerary (Rodriguez et al., 2004). The *Forum Limicorum* (Ptolemy II 5, 44) was probably to the south east of the lagoon, coinciding with the current town of Xinzo de Limia, with important Roman remains under the current parish church of Santa Mariña (Xusto, 1999, 2000). If we add the Pendón de Santa Marina hillfort, we have all the points where the hagiography locates relevant episodes of Mariña's life that were important in antiquity.

Turning to Augas Santas, the hillfort and the Vacariza petroglyph are pre-Roman (García Quintela and Seoane-Veiga, 2011), but the hillfort was strongly influenced by the Romans, and the surrounding area was also the backdrop to scenes in the martyrdom of Mariña.

Moreover, the Celtic linguistic features that have been identified reveal the existence of a Celtic worldview between the pre-Roman inhabitants of the region, which remained active during Roman rule.

However, the Celtic worldview is not well defined, and its relevance in all of the areas with speakers of Celtic languages is disputed. To avoid a circular argument, we will explore the basic arguments, and how an archaeoastronomical observation supports or disproves its ethno-cultural Celtic definition. We will basically highlight three features:

1. The Celtic cosmos was divided into three levels. A marine or aquatic underworld, an intermediate space inhabited by men, and a celestial and divine world. The Irish mythological texts and linguistic analysis of the Gaulic "kings of the world" agree in this respect (Sayers, 1996; Delamarre, 1999; Mac Mathúna, 1999).
2. The moon and night-time governing the account of time established by the Druids, as evidenced by Caesar, the Gaulish Coligny calendar and ethnographic and linguistic uses in Celtic-speaking countries (Caesar, *Gallic Wars* VI, 14.6, 18.2; Loth, 1904). We have observed the relevance of the lunastices in the orientation of public monuments with Celtic cultural influence (García Quintela and González-García 2014; García et al., 2014; García Quintela, 2016).
3. The four festivals marking the start of the season attested in Ireland, on November 1st, February 1st, May 1st and August 1st (Le Roux

and Guyonvarc'h, 1995). Their pan-Celtic nature has been disputed, although our observations seem to support such character (González-García and García Quintela, 2014, 2016 this issue on *Lugdunum Conuenarum*).

On this background occurs the Christianization of the celestial landscape.

4. THE CHRISTIAN TRANSFORMATION OF A CELTIC LANDSCAPE.

We have seen a predominance of relationships connecting the solstice with Roman and medieval structures in the vicinity of the lagoon of Antela. However, the pre-Roman hillforts seem to be related to the lunistice. Moreover, in Antela the solstice relationships are present in three levels: underwater, indicated by the dates when the sunken city of "Antioch" can be seen; in the middle plane of the life of rural communities flanking the Lagoon, with parish churches dedicated to solstitial saints; and in the upper plane, by the towers of the medieval lords that dominate the territory. This situation can evoke a triple-levelled, local Celtic worldview that was reinterpreted and transformed into Christian. In addition, the church of Xinzo and the previous Roman building are oriented towards the sunrise at the beginning of season dates.

Turning to Augas Santas, we have found orientations towards the solstice, Christianized by St. Thomas, and beginning of season dates in February and November in the sauna and the rock shrine. It is important that the orientation of the sauna is similar to that of the Romanesque church dedicated to Santa Mariña in Xinzo: two pre-Christian buildings of similar orientation in the area were Christianized through the worship of Santa Mariña, becoming the places where her birth and martyrdom are commemorated. We have also already commented on the orientation on February 1st and November 1st for other structures from the Iron Age in this area.

These observations lead us to suggest a Christian reinterpretation of a culturally Celtic skyscape, using the hagiography of St. Mariña as a vector for the transformation of religious and cultural semantics of the places that supported the previous conceptions.

5. CONCLUSION: A WORK OF RELIGIOUS ENGINEERING

In conclusion, we believe that we are able to identify the author of this transformation: Hydatius of Chaves (400-469).

He was an important regional figure, known to have been the bishop of *Aquae Flaviae* (Chaves, in Portugal), and author of a historical chronicle. In the preface of his work he is presented as "*Limicus,*" meaning he came from the territory in question. In his youth he travelled to the Holy Land, and throughout his chronicle mentions events in the eastern Mediterranean (Burgess, 1993). One such item mentions an earthquake in Antioch of Isauria in the year 462, referring to it as God's punishment for failing to listen to the message of Salvation. He writes:

> "Greater Antioch in Isauria, paying no heed to the warnings for its salvation, was swallowed up when the earth split open, and all that remained above the ground was the tops of the towers. Only the bishop of this city was saved from death, along with a number of people who followed him in obedience to the fear of the Lord." (Translation by Burgess, 1993)

This news is very similar to the legend of the sunken city in the lagoon of Antela: the name of the city, its sins, its punishment, and even the details of the towers of the churches are the same. In addition, the wordplay of Antela/Antioch present in the traditions of Mariña, is typical of the early Middle Ages (Amsler, 1989). It is interesting to note that Hydactius lists 5 solar and 2 lunar eclipses in his chronicle. These are difficult to study (Tabov, 2013), although they do reflect his interest in celestial phenomena as omens. Finally, datings of mortar from the first transformation of the sauna using Optically Stimulated Luminescence has provided a date of 548 ± 45 (Blanco et al., 2015), close to the lifetime of Hydatius or his successors.

In short, Hydatius has all of the skills and the political and religious knowledge to be considered as the person responsible for introducing the history of Margaret of Antioch into the pagan celestial landscape of his homeland. In this way, he sought to transform the religious orientation of the landscape and, with it, the temporal and spatial coordinates of its inhabitants.

His work can be summed up in four ideas:

1. The three-layered Celtic cosmos becomes Christian. There is a striking similarity between the stories of the sunken cities of Antioch in Asia Minor (Isauria in Hydatius) and Antela (popularly known as "Antioch"). If we also remember that the history of Margarita is located in Antioch of Pisidia, the similarities could not be any closer.
2. To Christianize this cosmic structure, a previously lunisolar landscape was solarized. Hydatius therefore followed the pattern of the Julian solar calendar spread through Christianization.

3. He provides Christian references for monuments with solar orientations related to mid-season festivals; and in particular:
4. He systematically uses the different episodes of the hagiography of Mariña as stories that give new religious and cultural meaning to the existing monuments, landmarks and landscape.

By studying the celestial landscape linked to the hagiography of Santa Mariña, we have studied one case among many of a local process of Christianization. We have also shown that archaeoastronomical observations are essential in order to correctly understand the process. Finally we found support in "negative", through the footprint in a Christian culture, for the idea of the community of celestial conceptions of the Celts over very large spaces and very long periods.

ACKNOWLEDGEMENTS

Our work in Santa Mariña has depended greatly of the joint work of an interdisciplinary team, from whose work we have only offered the aspects relating to archaeoastronomy and Celtic studies. However, our in-depth knowledge of the landscape and its transformations depends of their different professional expertise. We cite in alphabetical order: Rebeca Blanco Rotea, Cruz Ferro Vázquez, Sonia García Rodríguez, Ana Lois Gómez, José Carlos Sánchez Pardo, Jorge Sanjurjo Sánchez, Yolanda Seoane-Veiga. ACGG is a Ramon y Cajal fellow of the Spanish Ministry of the Economy and Competitiveness.

REFERENCES

Amsler, M. (1989) *Etymology and Grammatical Discourse in Late Antiquity and Early Middle Ages*. Amsterdam and Philadelphia, Johns Benjamins Publising Company.

Blanco-Rotea, R., García Rodríguez, S., Mato-Fresán, C. and Sanjurjo-Sánchez, J. (2015) La Basílica da Ascensión y Os Fornos (Allariz, Ourense) y la cristianización de la arquitectura en la Antigüedad Tardía. *Estudos do Quaternário*, Vol. 12, 111-132.

Burgess, R.W. (1993) *The Chronicle of Hydatius. And the Consularia Constantinopolitana: two contemporary accounts of the final years of the Roman Empire*. Oxford, Clarendon Press.

Delamarre, X. (1999) Cosmologie indo-européenne, 'Rois du Monde' celtiques et le nom des Druides. *Historische Sprachforschung*, Vol. 112, 32-38.

García Quintela, M.V. (2016) Los druidas y la luna. *Hommage B. Sergent*. In press.

García Quintela, M.V. and González-García, A.C. (2014) Le 1er août à Lugdunum sous l'Empire Romain : bilans et nouvelles perspectives. *Revue Archaeologique de l'Est*, Vol. 63, 157-177.

García Quintela, M.V. and González-García, A.C. (2016) De Bibracte à Augustodunum: observations archéoastronomiques. *Revue Archaeologique de l'Est*, Vol. 65, in press.

García Quintela, M.V. and González-García, A.C. (submited) The Cult of Ste. Mariña and the Making of a Christian Landscape in Antela Lagoon (Ourense, Spain). *History of Religions.*

García Quintela, M.V. and Santos Estévez, M. (2015) The Iron Age Saunas of Northern Portugal: State of the Art and Research Perspectives. *Oxford Journal of Archaeology*, Vol. 34/1, 67-95.

García Quintela, M.V. and Seoane-Veiga Y. (2011) La larga vida de dos rocas ourensanas. *Archivo Español de Arqueología* Vol. 84, 243-266.

García Quintela, M.V., González-García, A.C. and Seoane-Veiga Y. (2014) The Iron Age Saunas of NW Iberian Peninsula: An Archaeoastronomical Perspective. *Mediterranean Archaeology and Archaeometry*, Vol. 14/3, 133-141.

García Quintela, M.V., González-García, A.C. and Seoane-Veiga Y. (2015) From Hillforts to Saints: Sun Tropoi and Patterns of Christianization in NW Spain. F. Pimenta, N. Ribeiro, F. Silva, N. Campion, A. Joaquinito, L. Tirapicos (eds.), *SEAC 2011. Stars and Stones: Voyages in Archaeoastronomy and Cultural Astronomy*. Oxford, Archaeopress, 240-245.

González-García, A.C. (2015) La Orientación de las Iglesias Prerrománicas de Galicia: análisis y resultados preliminares. *Estudos de Quaternário*, Vol. 12, 133-142.

Guerra, A. (2005) Povos, cultura e língua no ocidente peninsular: uma perspectiva, a partir da toponomástica. *Acta Palaeohispanica* IX, *Palaeohispanica*, Vol. 5, 793-822.

Le Roux., F. and Guyonvarc'h, Ch.J. (1995) *Les fêtes celtiques*. Rennes, Ouest-France.

Loth, J. (1904) L'année celtique d'après les textes irlandais gallois bretons et le calendrier de Coligny. *Revue Celtique*, Vol. 25, 113-162.

Luján, E.R. (2009) Pueblos celtas y no celtas de la Galicia Antigua: fuentes literarias frente a fuentes epigráficas. *XXII Seminario de Lenguas y Epigrafía Antiguas.* Valencia, 235-237.

Mac Mathúna, L. (1999) Irish perceptions of the cosmos. *Celtica* 23, 174-176.

Rodríguez Colmenero, A., Ferrer Sierra, S. and Álvarez Asorey, R.D. (2004) *Miliarios e outras inscricións viarias romanas do Noroeste Hispánico (Conventos Bracarense, Lucense e Asturicense).* Santiago, Consello da Cultura Galega.

Sayers, W. (1996) Tripartition in Early Ireland, Cosmic or Social Structure? E.C. Polomé (éd.) *Indo-European Religion after Dumézil.* Washington: Institute for the Study of Man, 168-173.

Tabov, J. (2013) Model for astronomical Dating of the *Chronicle* of Hydatius. *International Journal "Information Models and Analysis"*, Vol. 2/1, 23-36.

Xusto Rodríguez, M. (1999) O núcleo urbano de Xinzo de Limia e as súas orixes romanas. A. Rodríguez Colmenero (ed.) *Los orígenes de la ciudad en el Noroeste Hispano.* Lugo, Deputación, 1291-1304.

Xusto Rodríguez, M. (2000) Iglesia Vieja de Xinzo de Limia (Ourense) Intervención arqueológica y lectura histórica. *Restauración & Rehabilitación*, Vol. 37, 26-33.

19

ANCIENT AND MEDIEVAL COSMOLOGY IN ARMENIAN HIGHLAND

Sona V. Farmanyan*[1], Areg M. Mickaelian[2]

[1]*M. Abeghyan Institute of Literature, National Academy of Sciences, 15 Grigor Lusavorich Str., Yerevan 0015, Armenia*
[2]*Byurakan Astrophysical Observatory (BAO), National Academy of Sciences, Byurakan 0213, Aragatzotn Province, Armenia*

Corresponding author: Sona V. Farmanyan (sona.farmanyan@mail.ru)

ABSTRACT

Humankind has always sought to recognize the nature of various sky related phenomena and tried to give them explanations. It is especially vivid in ancient cultures, many of which are related to the Middle East. The purpose of this study is to identify ancient Armenian's pantheistic and cosmological perceptions, world view, notions and beliefs. By this study we answer the question *"How did the Universe work in Ancient Armenian Highland?"* The paper focuses on the structure of the Universe and many phenomena of nature that have always had major influence on ancient Armenians thinking. Here we weave together astronomy, anthropology and mythology of Armenia, and scientific thinking revealed in local astronomy traditions. The initial review of the study covers Moses of Khoren, Yeznik of Koghb, Anania Shirakatsi and other 5th-7th centuries historians' and scientists' records about the Universe related superstitious beliefs and cosmological understanding. By discussing and comparing Universe structure in various regional traditions, myths, folk songs and phraseological units we very often came across "seven worlds", "seven earths" and "seven layers" concepts. We draw parallels between scientific and mythological Earth and Heaven and thus find similar number of layers on both of the ancient and modern thinking. In the article we also give some details about the tripartite structure of the Universe and how these parts are connected with axis. This axis is either a column or a Cosmic Tree (*Kenatz Tsar*). In Armenian culture the preliminary meanings of the Kenatz Tsar are more vivid in folk songs (*Jan gyulums*), plays, epic, and so on, which was subsequently mixed with religious and spiritual views. We conclude that the perception of the Universe structure and celestial objects had a significant impact on culture and worldview of the people of the Armenian Highland; particularly it was one of the bases of the regional cultural diversity.

KEYWORDS: Cultural Astronomy, Ancient Cosmology, Universe, Earth, Heaven, Cosmic Tree.

1. INTRODUCTION

The roots of Armenian astronomy go back to several thousands of years. Still ancient Armenians were quite far from the scientific recognition of the nature: that is why they attributed their birth and death, strength and power, fortune and misfortune mostly to planets and constellations. There are ancient Armenian names of planets, stars and constellations, mostly related to the mythological origin. Before the creation of the Armenian alphabet, Armenians expressed their astronomical knowledge through the carvings on rocks and stone structures. Since the ancient times, astronomical notions contributed and formed the development of natural philosophical thought, through the frame of the mythological perceptions (Danielyan, 1987). The initial review covers Moses Khorenatsi, Eghishe, Eznik Koghbatsi, Davit Anhaght, Anania Shirakatsi and other 5th-7th centuries historians' and scientists' records about sky related superstitious beliefs and cosmological understanding. The above-mentioned authors had recorded not only the celestial perceptions of their time, but also transferred the cosmic views of Pagan Armenians. By drawing parallels between Pagan and Christian cosmic views, they proved that astronomy of the time was heavily tied to the religious and spiritual outlook of the humankind. However, some of the Armenian Christian historians very often avoided giving many details of exact sciences, and they concentrated on theological works. Still we can see these historians' scientific perceptions in the interpretations of the Bible.

It is noteworthy to mention about the 7th century famous Armenian historian, philosopher, mathematician, astronomer and geographer Anania Shirakatsi's thoughts that sometimes is not far even from modern astronomical knowledge. Shirakatsi was the first Armenian author who distinguished the branches of sciences, and considered that those branches should be developed separately, thus Shirakatsi became the founder of the natural sciences in Armenia. Shirakatsi authored several works related to the calendar and Cosmography, a 48-chapter work that discusses astronomy, meteorology, and physical geography (Anania Shirakatsi, 1962; Hewsen, 2001). Shiaraktsi's basic astronomical views are presented in this work, which is based on the ideas of ancient Greek astronomers and philosophers, and his further modifications based on his own observations. He described the world as *"being like an egg with a spherical yolk (the globe) surrounded by a layer of white (the atmosphere) and covered with a hard shell (the sky)."* He also believed *"that the Milky Way is a mass of dense but faintly luminous stars and agreed with earlier philosophers that the moon was a dark body by nature whose light was only that, which it reflected from the sun."* Let us mention that in Shiaraktsi's era the Greek science was in decline, and the Arabic science (particularly famous for its highly developed astronomy in 9th-12th centuries) had not been formed yet, so to maintain and in some cases even to develop this astronomical valuable knowledge was quite a difficult task. Shirakatsi's "Cosmography" has both scientific and literary significance. Shirakatsi used artistic descriptions and illustrations. At the current stage of the development of Astronomy and Astrophysics, Shirakatsi's ideasstill remain relevant (Harutyunian & Mickaelian, 2014). Benik Tumanyan summarized Shirakatsi's cosmological views in his book *"From the History of Armenian Astronomy"* (Tumanyan, 1985) that leads to the following two main conclusions:

1. His geocentric system about the Sun's position and fiery arch are unique and different from those known at that time.
2. Shirakatsi diligently studied the works of his time and before him, he expressed critical approach, combined it with his own research and gave probable and natural conclusions.

Shirakatsi's works have historical, astronomical, geographical, religious, literary and terminological significance.

When discussing and comparing Universe structure in various regional traditions, myths, folk songs and phraseological units, one comes across "seven worlds", "seven earths" and "seven layers" concepts. By drawing parallels between scientific and mythological Earth and Heaven, one finds similar number of layers on both of the ancient and modern thinking. It is necessary to consider the details about the tripartite structure of the Universe and their interconnection with axis, which is either a column or a Cosmic Tree (Arm. *Kenatz Tsar*). We study the perception of the Universe structure and celestial objects and its significant impact on culture and worldview of the people of the Armenian Highland.

2. STRUCTURED UNIVERSE

Celestial bodies and phenomena had greatly influenced the ancient humans' consciousness; it has found its reflection in the language, folklore, religious beliefs and different arts. From the early beginning of humankind's existence, the civilizations founded their religious beliefs, which heavily included the World Creation. Throughout the history World creation myths can be found almost in all cultures. Numerous civilizations based their new religious beliefs on the past cultures; thus showing the significant evolution of Cosmic views throughout the centuries. By studying ancient Armenian's pantheistic and cosmological perceptions, world views, notions and beliefs we see that according to different parts of the country, the Universe was created differently and had diverse structures from one region to another.

Shirakatsi shared the ideology of Armenian Apostolic church that the creation of the Universe and everything that existed in the Universe and on the Earth was created by God. "*God has no beginning, he is immortal and perfect, he is the beginning and the creator of all visible and conceivable essences*" (Hakobyan, 1956). According to Sassun's people imagination the World is on the back of the fish, and the fish is on the surface of the sea and the earthquakes are caused by fish movements (Petoyan, 2014). Fish-like stone monuments are discovered in Geghama mountains and elsewhere in the Armenian Highland. These monuments are covered with numerous images of animals (ox or bull's head, a pair of storks or other birds), plants (Cosmic Tree) and water (drop of water from top to bottom) carved on the belly of the fish. These carvings symbolize the ancient Armenians or even pre-Armenians cosmological worldview (Petri, 1964). The further answers concerning the cosmological worldviews are reflected on the folk songs; there the Cosmos is on the back of Leiatan-Leko fish and on the horns of an ox. These perceptions are identified with the Muslim's views of Cosmos, since in surah 68:1 it is written that Allah created the Fish Nun, he lifted up the water vapour, and the Heaven was split off from it, and the Earth was spread out upon the back of the fish. The fish became agitated, and as a result, the Earth was shaken up. It was steadied by means of the mountains, he continued, which were proudly located over the Earth like towers.

According to the people of Javakhk (Northern Armenia) the Earth is a circular valley, like a cart wheel and its opposite sides are alike. Both the Earth's apparent surface and the opposite side are covered with mountains, valleys, fields, seas and rivers, however, in the opposite side seas are on the mountains and the mountains are under the sea. The flat land is bread-like, somewhere convex; somewhere concave and is placed on the horns of a huge ox. And when the naughty flies sit on the huge ox eyes and begin to bite, suck his blood, calf pain starts to shake and shake his head, from which the Earth was rocked by earthquakes and come forth (Harutiunian, 2000). Here the bread can be associated not only with the Earth, but also with the Sun. In Armenia the bread is used to be cooked in *tonir* which is a type of underground oven, having round shape and inside is the fire. Hence, since the ancient times *tonir* was worshipped as a symbol of the Sun in the ground.

2.1. *Tripartite Universe*

The early Armenians' vision of the Universe structure is preserved in the ancient "Vahagn's Birth" folk-song, recorded by the 5th century historian Moses of Khoren. Vahagn or Vahagn Vishapakagh (Vahagn the Dragon Reaper) was a god of fire and war worshiped anciently and historically in Armenia. Sometime during Ancient history, he formed a "triad" with Aramazd and Anahit. Vahagn is identified with the Greek deity Heracles. All the gods, according to the Euhemerist belief, had been living men; Vahagn likewise, was introduced within the ranks of the Armenian kings, as a son of the Orontid Dynasty (or Yervanduni dynasty, 6th-4th century B.C.). According to "Vahagn's Birth" song the Universe has tripartite structure: Sky-Earth-Sea.

"In travail were heaven and earth,
In travail, too, the purple sea!
The travail held in the sea the small red reed."
(Khorenatsi, 5th c.)

Moreover, the Sea is perceived as an image of underworld with chaotic waters that surround and limit the Universe from all sides (Ter-Poghosyan, 1911). The chaotic waters or the Space Ocean images are also reflected in Armenian folk traditions recorded in the 19th century. In Alashkert Province (Western Armenia) people believed that between the heavens and the Earth there was a deep ocean "Black water", which was set for Earth punishment, and once it could spout out from its shores and destroy the Earth as it was in the time of flood (Figure 1).

Figure 1. An illustration of tripartite structure of the Universe: Sky-Earth-Sea (Black Waters)

In such cases the Earth and the heavens are torn down and Black water rinse the entire world (Njdehyan, 1902). And if heavens is torn down, the world would be destroyed (Petoyan, 2014). Underworld's apocalyptic chaotic Black waters' visions are also reflected in Armenian heroic poem Daredevils of Sassun (Arm. *Sasna Tzrer*). Interestingly, Mesopotamian and Egyptian sources also illustrate the land surrounding chaotic waters and the depth of the sea is often associated with the graves. All Egyptian Universe creation myths have common element that world was derived from chaotic waters, called Nu.

The vision of tripartite structured Universe (or World) is also preserved in the late oral folk traditions. Again according to people from Alashkert, the world consists of three main parts or floors and the difference of the floors is hidden in the belts (Njdehyan, 1902). The first floor is the underground world, the second floor is surface world, and the third one is the floor of heaven. All of the floors are inhabited, and all floors are populated with people, the only difference is in their belts. Inhabitants of underground world tie the belt on their legs, the second floor inhabitants tie the belt on their stomachs, and people from the third floor tie it on their breasts (Njdehyan, 1902). The ideas of tripartite structured Universe confirmed among Armenians by the adoption of Christianity as they related it with Holy Trinity.

According to the Armenian mythology, the Universe has triptych vertical structure. The sky is on the upper floor, the Earth or the ground is on the middle floor and on the lower floor is the abyss, which is often symbolized with the bottomless sea. The space trinity is opposing to each other, but it is actually a whole. Moreover, the bottomless sea or ocean that has the meaning of lower floor, with its contents, surrounds the Universe from all sides. The Earth is situated in the world's great ocean and is surrounded by chaotic waters, which is separated from the Leiatan fish with encircled body or back. The Sky too is often imagined as a Sea (purple sea), which is a part of the chaotic water that limits the Universe from all sides, and in that sense it is equivalent to the lower world. In another myth, Heaven and Earth are described as husband and wife and between them is the purple Sea. As S. Harutyunian mentioned in his book "In the World of Armenian Ancient Myths" (Harutyunian, 2000), the Heaven is also a city, which is sometimes called Priceless City. It is surrounded by high stoned walls and copper gates. This Priceless City is the temple of the light, or opal-built, highly arched shining palace. Here everything is wonderful; there is no darkness, no cold. This is the home of immortality. The outside of the city is the paradise, here come honest people, who angels and sit on the sunny chairs (Abeghyan, 1971). Three-levelled cosmology also exists at ancient Israelites ideas, they thought that the Universe was made-up of a flat-disc that moves on the water: above is Heaven and underworld is below. Mesopotamian cultures believed that the underworld was morally neutral (Del Olmo Lete, 2008). We can assume that Armenians most possibly expressed the ideas of three-levelled Universe by the influence of Hellenistic philosophers. According to ancient Armenians vision, the three zones of the Universe; the Heaven, the Earth and the underworld are connected with the axis. This axis is either a column or a tree (Eliade, 1998). The center of the Universe is located at the crown of the tree or the column, which is the closest to the sky.

2.1.1. *Cosmic Tree or Kenatz Tzar*

Trees are noteworthy in many of the world's mythologies and religions, and have been associated with profound and sacred meanings throughout the ages. The early cosmologies recount that Cosmic Tree is grown in the centre of the Universe and includes all the spheres of human and Earth existence. Cosmic Tree is a universal image and is also known as World Tree, Life Tree, Sacred Tree, Tree of Knowledge and are portrayed in various religions and philosophies (Mettinger, 2007). These terms cross cultures and are widespread motif in various mythologies, folklores and religions. It alludes to the interconnection of all life on our planet and can be considered as a metaphor for common descent in the evolutionary sense (Giovino, 2007). In most of the cultures, Cosmic Tree has a holy meaning and symbolizes the link between Earth, Heaven and Underworld. Nearly in every culture, branches of the tree reach the sky, the tree itself is on the Earth, and roots are deep into the Earth, so it draws links between three worlds: Heaven, Earth, and Underworld. Cosmic Tree is also metaphor for awareness and enlightenment. It is an evergreen tree, therefore, we can say that it has the same symbol as the colour green: it symbolises fertility, vitality, abundance, growth, rebirth and renewal.

In numerous myths, tree is a fundamental part of the Universe structure. Gods and Divine beings make a trip from world to world by climbing the tree. Particular world trees include *világfa* in Hungarian mythology, *Ağaç Ana* in Turkish mythology, *Modun* in Mongolian mythology, *Yggdrasil* in Norse mythology, the *Oak* in Slavic and Finnish mythology, *Kien-Mu* or *Jian-Mu* in Chinese mythology, and *Ashvattha* in Hindu mythology. The world tree is widespread in Lithuanian folk painting, and is frequently found carved into household furniture such as cupboards, towel holders, and laundry beaters (Straižys, 1990).

According to ancient Armenian's mythological notions, the Universe has three zones, Sky, Ground, and the Underworld, and these three zones are interconnected with an axis. This axis is either a column or the Cosmic Tree. In Armenian culture, the basic meanings of the Kenatz Tzar (Cosmic Tree) are more vivid in Jan gyulums (folk songs), plays, epic, and so on, which was subsequently mixed with religious and spiritual views. Kenatz Tzar is known as a religious symbol and was drawn on walls of fortresses and carved on the armor of warriors. The branches of the tree were equally divided on the right and left sides of the stem, with each branch having one leaf, and one leaf on the apex of the tree (see Figure 2).

Figure 2. Pomegranate Cosmic Tree with animals

Human like figures, possibly servants, stood on each side of the tree with one of their hands up as if they are taking care of the tree. According to the ancient Armenians' perceptions, Kenatz Tzar could be considered either pomegranate tree, apricot tree or apple tree. In Grigor Narekatsi's "Vardavar" and "Revelation" (Arm. Haytnutyun) plays have the same expression of the idea of the tree of life, as it is presented in the miniatures, especially in Jesse's (in the Bible the father of David, represented as the first in the genealogy of Jesus Christ) tree images. Cosmic Tree is also expressed in other plays of Narekatsi, such as "Tsovin and Navin" and "Karos Khachin". Cosmic Tree images are also found in Nerses Shnorhali's and Grigor Tatevatsi's works. From the Vedic literature, Ghevond Alishan mentioned this tree as the most common symbol of God, he pointed out that this tree is in upside-down position and life originates from here as it is Brahman itself. Later on in Armenian conundrums, we find the concept of "Cosmic tree in Hindu city" (Harutyunian, 1965).

Since the middle of the 1st millennium B.C., Solar calendar was used, which contained 365 days (12 months by 30 days and an additional month of 5 days) (Broutian, 1997). In Solar calendar, Kenatz Tzar was among worshiped Gods. Kenatz Tzar in Solar calendar is symbolized as solar year; it is especially reflected in the following conundrum:

"There is a tree, it has 12 branches,
Each branch has 30 apples,
One half of each apple is black, the next half is white" (Harutyunian, 1965).

In one of the medieval prayer, the tree is described as branchless, leafless and an eagle footless and headless is sitting on its top. This praying contains missing parts from both the free and the fowl; it shows the imperfect model of the Universe, semi chaotic condition. They believed that because of these semi chaotic conditions they have illnesses and bad fortunes.

There are some healing prayers that picture Kenatz Tzar with different stuffs and animals. They believed that the man's health condition is related to the gradual layout of stuffs and animals on the tree. They thought that man's microcosm is in disorder when he gets sick, thus he has a direct relationship with Cosmic Tree, especially with tree particularly stuffs or animals (Harutyunian, 2000). This relationship is largely accomplished through the sacrifice an animal or fowl on the tree.

2.1.2. *Four partite Universe*

The ancient Greeks believed that there were four elements from which everything was created: Earth, Water, Air, and Fire. This theory was suggested around 450 B.C., and it was later supported and developed by Aristotle. Shirakatsi in the 7th century followed the ancient Greek scholars' theory and confirmed the four elements, finding that in the Universe everything consists of these elements, he added: "*The Spirit is a divine breath it is intangible, although created.... But the Heavens and the Earth, and anything else that exists among them are made up of four elements, which are equally mixing to each other. Fire features warmness and dryness, Air – temperature and humidity, Water – humidity and coldness, Earth – coldness and dryness. And they contacted to each other for transfering their features*" (Hakobyan, 1956). Shirakatsi's considerations are similar to the modern states of matter: Earth (solid), Water (liquid), Air (gas) and Fire (plasma). Aristotle added the fifth element, *aether*, as the quintessence, reasoning that whereas fire, earth, air, and water were earthly and corruptible, since no changes had been perceived in the heavenly regions, the stars cannot be made out of any of the four elements but must be made of a different, unchangeable, heavenly substance (Lloyd, 1968). Shirakatsi separates the upper and inner skies. To his mind, the top of the sky, which Aristotel called "aether", and the Chaldeans, "mirrored fire", has no subject and no body: it is a simple fire, which did not derive of anything, and nothing had derived from it. This "air" covers all the elements. Shirakatsi also expressed his perception of the Universe layers with these four elements. At the center of our world is the Earth, as the planet we live on. Running over and around the Earth is Water, in the form of lakes, rivers, and oceans. Water fertilizes and the Earth, giving it life. Over the spheres of Earth and Water is that of Air, the atmosphere in which we live. And finally, Fire lights up the Sun, Moon, and stars.

2.1.3. *Seven Layered Earth and Heaven*

Seven is one of the most significant number across religions, spiritual traditions and cultures. The Bambaras, the Buddhist and the Talmud divided the Sky into seven degrees of perfection. The number seven is also widely spread in Islam: the Koran has seven esoteric senses, according to Muslim mysticisms (sometimes they talk about seven senses); there are seven consonants, called *sawakit* and there are seven towers in Mecca; during the pilgrimage to Mecca, they have to take seven turns of the Kaba and seven courses between the mounts Cafa and Marmia. In Armenian culture, the dead people's soul remains seven days beside the tomb. At the beginning of the Bible, the first use of number seven is associated with completeness; it is related to the creation week in Genesis 1. God spends six days creating the Heavens and the Earth, and then rests on the seventh day.

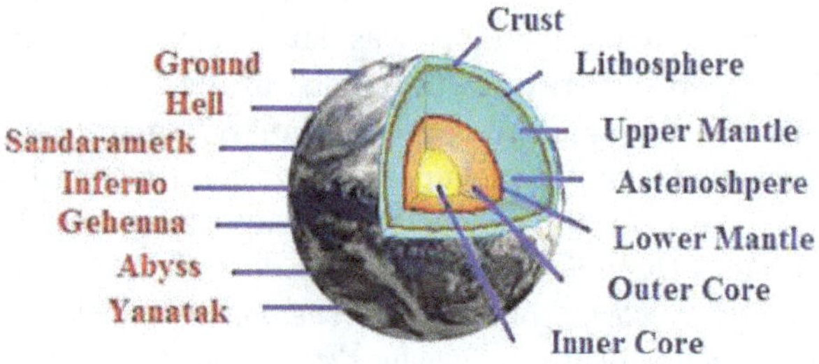

Figure 3. The left of the figure is the mythical Earth and the right is the scientific Earth.

By comparing Universe structure in various regional traditions, myths, folk songs and phraseological units, we very often come across to the concepts of "Seven Heavens" (Seven heavens is a part of religious cosmology found in many major religions such as Islam, Judaism, Hinduism and Catholicism (Davidson, 1967)) and "Seven Earths". Armenians in their turn divided the Earth and the Heavens into seven layers (Figure 3 and 4).

As Njdehyan mentioned the Armenian sky, since the ancient times, was divided into seven layers, the first layer is the Earth's Atmosphere, the Clouds are in the second layer, the third one is the Moon, the forth is the Sun, the fifth are the stars, the sixth is the room for angels and saints, and the seventh layer is for the God. Earth layers gradually fall from the top to down, and the layers bear the following names: Ground, Hell, Sandarametk, Inferno, Gehenna, Abyss and Yanatak. Four of these seven names (Hell, Sandarametk, Inferno, Gehenna) are coming from Iranian, Greek and Jewish traditions, it is the same concept with different names. Ancient Persians also imagined Earth's seven layers and called it *ziggurat,* which is equivalent to Cosmic Mountain (Eliade, 1957). In some way, in modern science too, both the Earth and the Heaven have seven layers.

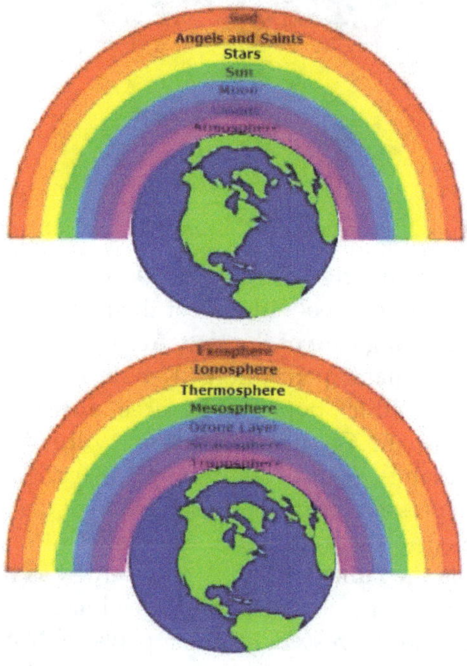

Figure 4: The left of the figure is the mythical Heaven and the right is the scientific one.

The Seven Heavens refers to the layers of our atmosphere (Troposphere, Stratosphere, Ozone Layer, Mesosphere, Thermosphere, Ionosphere and Exosphere). The Seven Earths refer to the layers of the Earth (Crust, Lithosphere, Upper Mantle, Astenoshpere ("Sic"), Lower Mantle, Outer Core, and Inner Core). The seven layers of the Earth may also refer to seven continents (North America, South America, Australia, Asia, Africa, Antarctica and Europe).

ACKNOWLEDGEMENTS

We would like to express our gratitude to the anonymous reviewers for their constructive comments. This work was partially supported by the project "Armenian Cultural Astronomy and World Heritage" funded by National Academy of Sciences of the RA, V. Ambartsumian Byurakan Astrophysical Observatory (BAO) and Armenian Astronomical Society (ArAS). We gratefully acknowledge the support and generosity of the Swiss Agency for Development and Cooperation SDC, without which the present study could not have been completed. The authors would especially like to thank the scientific committee of SEAC 2015 *"Astronomy in*

Past and Present Cultures" conference and also we thank the personnel of the Editorial Board for their support and technical cooperation.

REFERENCES

Abeghyan, M. (1971) *The psalm*. ArmSSR Acad. Sci. M. Abeghyan Institute of Literature. Yerevan. ArmSSR AS publ.

Anania Shirakatsi (1962) *Cosmography, trans. from Ancient Armenian, preface and comments by K. S. Ter-Davtyan & S. S. Arevshatyan*, Yerevan.

Broutian, G. H. (1997) *The Armenian Calendar*. Ejmiatzin, Echmiatzin Publishing House.

Eliade, M. (1998) *Symbolism, the Sacred and the Arts* New York, The Crossroad Publ. Company.

Eliade, M. (1987) *The Sacred and The Profane: The Nature of Religion* Orlando, FL, Harcourt Brace Jovanovich.

Danielyan, E. (1987) Mythological origins of natural philosophy. *Science and Techniques*, Yerevan, 1987, N. 5, p. 5-7.

Davidson, G. (1967) *Dictionary of Angels: Including the Fallen Angels*. New York, The Free Press.

Del Olmo Lete, G. (2008), *Mythologie et Religion des Sémites Occidentaux*, Volume II, Leuven - Paris, Peeters Publishers.

Giovino, M. (2007) *The Assyrian Sacred Tree: A History of Interpretations*. Fribourg, Academic Press.

Hakobyan, T. (1956) *Nature of Armenian Highland according to Anania Shirakatsi's "Ashkharhacuyc"*: Scientific works of Yerevan State University, No. 58, geographical scientific series.

Harutyunian, S. (1965) *Armenian National Riddles* ArmSSR Acad. Sci. Publication, Ejimatsin, Ejimatsin Publishing House.

Harutiunian, S. (2000) *Armenian Mythology* Beirut, Hamazgayin Vahe Setyan Publishing House.

Harutyunian, H. A., Mickaelian, A. M. (2014) Anania Shirakatsi's Cosmographical Works and the Methodology of his Natural Science. In H. A. Harutyunian, A. M. Mickaelian & E. S. Parsamian (eds.) *Astronomical Heritage in the National Culture, Proceedings of the Archaeoastronomical Meeting dedicated to Anania Shirakatsi's 1400th anniversary, held 25-26 Sep 2012 in Byurakan, Armenia. Yerevan*. NAS RA "Gitutyun" Publ. House, pp. 87-97.

Hewsen, Robert H. (2001) *Armenia: A Historical Atlas*. Chicago, University of Chicago Press. [1]

Lloyd, G. E. R. (ed.) (1968) *Aristotle: The Growth and Structure of his Thought* Cambridge, Cambridge Univ. Pr. pp. 133–139.

Mettinger, Tryggve N. D. (2007) The Eden Narrative: A literary and religio-historical study of Genesis 2-3. Winona Lake, JN, American Oriental Society.

Njdehyan, G. (1902) *Paravashunch* Tbilisi, "National Journal" IX book.

Petoyan, V. (2014) *Ethnography of Sassun* Yerevan, Lusakn Publishing House.

Petri, W. (1964) Anania Shirakatsi - an Armenian Cosmographer of the 7th century. *Journal of the German Eastern Society*, Vol. 114, Munich.

Straižys, V. (1990) *Gods and Goddesses of the Ancient Balts* (in Lith.), Vilnius, Lietuvos Dangus (Sky of Lithuania).

Ter-Poghosyan, G. (1911) *Period of Worship of Ancestors Soul* Vienna, Collection of literary monument Publishing House.

Tumanian, B. E. (1985) *History of Armenian astronomy*. Yerevan, YSU Publishing House.

20

ASTRONOMY IN THE MEDIEVAL *LIBER FLORIDUS*

S. Draxler and M.E. Lippitsch

Institute of Physics, Karl-Franzens-Universität Graz, Austria

Corresponding author: Sonja Draxler (sonja.draxler@uni-graz.at)

ABSTRACT

The *Liber floridus* is a medieval encyclopaedia, authored by a certain Lambert, Canon of Saint-Omer. The work was finished in 1121 and deals with a multitude of different topics, from religion and history to geography and astronomy. The Gent manuscript, which is the autograph, contains 287 folios of parchment. Detailed descriptions were given by Delisle (1906) and Derolez (1998).

The topics of Lambert's astronomy are divers. The movement of the sun and moon, the calendar, and the constellations are described. Several planetary diagrams are given, followed by astrological drawings as well as various geographical, meteorological, and astronomical remarks. These astronomical texts and figures so far have not been the object of detailed studies. In this paper an overview on the astronomical topics of the manuscript is given, several pages are described in more detail, and possible sources are discussed.

KEYWORDS: Liber Floridus, medieval astronomy, planetary astronomy

1. INTRODUCTION

Ego Lambertus ... libellum istum ... contexui, ut ... fideles apiculae ... saporis coelestis dulcedinem haurirent ...Floridum intitulavi, quia ... rerum mirandarum narratione praepollet (Lambertus a. S. Audomaro, 1121, f3v). "I, Lambert, have compiled this book, in order that faithful honeybees may suck the sweetness of the celestial juice. I called it 'flowering', because it is exquisite in the narration of admirable things." With these words the author of the book, Lambert of St. Omer, in the *prologus* of his *opus magnum,* the *Liber floridus*, strives to whet the potential reader's appetite.

It is really a grand endeavour the reader is invited to by the author, or, more correctly, the compiler of the *Liber floridus*. Already the sheer dimensions are awe-inspiring. The Gent manuscript that has been shown to be the autograph, contains 287 folios of parchment, 31 cm high and 21 cm wide, completely covered on recto as well as verso, by text and illustrations, 37 m^2 of information. And this is not the whole story, because the manuscript is not complete: The last part known from other manuscripts, *Apocalypsis depictus*, a story of the Apocalyps in comic book format, is missing.

The book was compiled around 1121 by Lambert, Canon of the Church of Our Lady in Saint-Omer, a small town in the north of France near the Belgian border. Very little is known about this man, in fact only what he tells us about himself in the book. In the *prologus* he introduces himself by providing his name, Lambert. He identifies himself as the author of the book and refers to his father Onulph, former canon of the same church. Furthermore, there is a portrait on f13r showing him writing the book. However, any information about his life is missing; even the dates of birth and death are unknown.

The table of content of the Gent manuscript lists 161 sections with cosmographical, biblical and historical topics. Lambert wanted, as he tells us, to select the best parts of his predecessor's works, for example from Isidore of Seville or the Venerable Bede, to preserve their work for future generations.

About 30 % of the content is devoted to religious topics. 25 % deal with historical matter. Geography makes up about 12 %, natural history 10 % and astronomy 8 %. Interestingly, only 4 % are related to *computus* and calendar. This proves that, contrary to a widespread believe, fixing the Easter date was not the dominant motive for astronomical studies in medieval times.

In the following centuries the *Liber Floridus* was copied several times. Table 1 shows a list with libraries containing copies of the book. The oldest one, written in the same century as the original, is housed in Wolfenbüttel, Germany; two copies are from the 13th century, five from the 15th century, and one from 1512. However, none of these copies is identical; sometimes parts of the text have been omitted, sometimes the order of the pages was changed. In addition, three partial copies exist in Belgium and England.

Table I. Libraries owning copies of the Liber Floridus

Libraries	Produced in
Wolfenbüttel, Herzog August Bibliothek	12th cent.
Paris, Bibliothèque Nationale, Ms. Lat. 8865,	13th cent.
Leiden, University Library, Ms. Voss. Lat.	13th cent.
Paris, Bibliothèque Nationale, Ms. Lat. 9675	1429
The Hague, Royal Library, Ms. 27A23	1460
Chantilly, Musée Condé, Ms. 724	15th cent.
Dounai, Municipal Library, Ms. 796	15th cent.
Genua, Bibliotheca Durazzo-Giustiniani, Ms. AIX9	15th cent.
The Hague, Royal Library, Ms. 128C4	1512

Source: http://www.liberfloridus.be/kopieen_eng.html (Retrieved Jan 8, 2016)

The astronomical content of the *Liber Floridus* is voluminous and covers all important parts of knowledge of that time: There are notes on the division of time, calculations of the Easter date and various calendars, as well as a description of stars and constellations. There are several explanations and planetary diagrams showing the movement of sun, moon and planets, sometimes mixed with astrological content, especially when describing characteristics of the planets. Other planetary diagrams give astronomical content in combination with geographical, meteorological or religious information. Furthermore, there is a copy of Cicero's *Somnium Scipionis* in the edition of Macrobius (Macrobius, 1994).

2. DESCRIPTION OF SELECTED PAGES

In this section some interesting pages with astronomical content will be described in detail. All the folio numbers refer to the Gent manuscript. The original Latin text is set in italic letters; English translations are given in quotation marks.

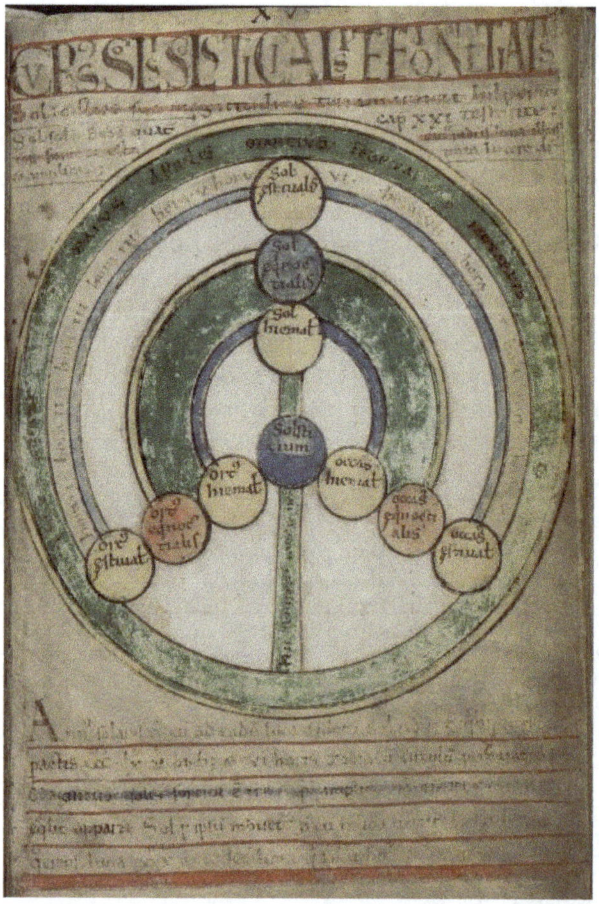

Figure 1. Cursus solaris solsticialis et equinoctialis

2.1. *f 25r: Cursus solis solsticialis et equinoctialis*

One of the first astronomical diagrams is found on f25r and named *Cursus solis solsticialis et equinoctialis* (Figure 1). It consists of a diagram with short texts above and below it. The text above the diagram says:

Sol octies sua magnitudine terram vincit Hilperico capitulo XXI testante, cuius radiis luna illustrate lucere dicitur. Sol secundam Bedam quater terrae fortior est et amplior. "The sun beats the earth in magnitude eight times, as attested by Helperic in chapter 21; by his rays illuminated, the moon is said to shine. According to Beda, the sun is four times stronger and wider than the earth."

Lambert here cites Helperic of Auxerre and his statement in the *Liber de Computo* (Helperic, 1844) correctly, with only a small error in the number of the chapter: the text (*Sol autem, quia octies sua magnitudine terram vincit*) is in chapter 20 instead of 21.

The diagram shows the path of the sun at different times in the year. In the outmost green circle there are the months of the year, followed by hours, starting with *hora prima* on the left hand side to *hora decima* on the right. At summer solstice in June the sun rises early in the morning, called *ortus aestivalis*. In December at winter solstice the sun rises latest. The highest point of its path the sun reaches always in hour VI, named *sol aestivalis, sol aequinoctialis* and *sol hiemalis*. The diagram shows that the path of the sun is shorter in winter than in summer.

Below the diagram there is a description of the sidereal year with 365 days and six hours:

Annus solaris est cum ad eadem loca syderum sol redit ex quibus egressus est, peractis $CCC^{tis}LX^{ta}V^{que}$ *diebus et VI horis zodiacum circulum peragratis.* A solar year is, when the sun returns to the same place of stars from where it had started, after 365 days and six hours and after the zodiac is walked through".

2.2. *f 25v: Cursus lunaris et anni descriptio*

The next page, f25v, bears the title *Cursus lunaris et anni descriptio* (Figure 2). The outer circles show the path of the moon during one year, the phases of the moon, and the c. 30 days of a lunar month.

The inner circles provide geographical and meteorological information. In the red circle there are the terms *Oriens, Auster brumalis, Occidens solsticialis* and *Solsticium*, which stand for descriptions of the cardinal directions. The next, uncoloured circle gives the names of winds as described by Seneca (1998) and Isidore of Seville (2008), with *Subsolanus, Favonius, Nothus* and *Septentrio.* The following circle shows pictures of the winds alternating with the description *Dies, Solis occidens, Nox, Solis ortus.* The next circle gives the names of the four qualities *calidus*

Figure 2. Cursus lunaris et anni descriptio

warm, *sicca* dry, *frigidus* cold and *humidus* wet, followed alternatingly by the names of three of the elements, *ignis* fire, *aer* air and *aqua* water, but with *ventus* wind instead of *terra* earth. There follow the names of the seasons *Estas, Autumnus, Hiemps* and *Ver*. In the center there is the term *Annus*.

The text above the diagram is a mixture of information from Helperic of Auxerre and Venerable Bede (Beda, 1844). It says, for example, *Lunam terra maiorem Beda affirmat*. "The moon is larger than the earth, Bede confirms". In fact, we can find this text in Venerable Bede's *De natura rerum*, chapter 19, where he writes *lunam vero terra esse maiorem*. This is interesting as Isidore of Seville about 100 years earlier wrote in his Etymologia, XX: 48 *Sicut autem sol fortior est terrae, ita terra fortior [est] lunae per aliquam quantitatem*, "As the sun is stronger than the earth, so the earth is stronger than the moon in a certain quantity."

2.3. *f 26v - 46r: Various calendars*

The following pages, from 26v to 46r, contain different calendars. One of them, for example, counts all days from year 1 till year 1291. In addition it gives information on the Easter date and the phases of the moon as well as historical records. The description of historical events ends in 1119, probably the date of production of this part of the book.

2.4. *f 89r - 91v: De ordine et positione signorum*

These folios give short descriptions of more than 40 constellations, with nice illustrations. Lambert is rather consistent in his text and drawings. His numbers are realistic, as can be seen from comparisons between his data and modern ones. For example he reports 22 in *Ursa maior*, and there are 22 brighter than magnitude 5, indeed. Nevertheless, in the drawings the stars' position within a constellation is governed more by the intent to achieve a good agreement with the mythological figure. However, with this procedure Lambert is in good company, because it is true for most medieval writers, including even Apian in the 16th century. A change comes about only with Bayer's *Uranometria* in 1603. Anyway, we can admire the pretty style of Lambert's drawings.

2.5. *f 92r: Calculations on the spheres*

This folio deals with cosmic distances. There is the size of the sun (diameter 140.000 stadia), the earth (half the sun), or the circumference of the sun's orbital sphere (30,170.000 stadia). Lambert's bias to mathematics becomes apparent, for example, in his explanation of the relation circumference/diameter in a circle (triplicationem cum parte VII^{ma} triplicated, with a 7th part), what we now know call the constant π. It is obvious also from the following calculation:

Constat universe terre ambitum habere stadiorum CC^{ta}L milia. Horum XX^{a}II^{a}, id est $\overline{XI}$CC LIIII remanentibus XII, si auferantur, remanebunt $\overline{CCXXXX}$DXXXXIII. „It is sure that the circumference of the earth is 250.000 stadia. The 22th part thereof is 11.454, remaining 12. If removed 11.454, remain 240.543." Repeating Lamberts calculation shows that he is approximately correct except of a permutation of the last two digits. The correct value should be 240.534. But anyway, the purpose of this exercise and the relevance of the result remain unclear.

2.6. *f 93v: De anno mundano Macrobius a Romulo post multa secula annis finitis $\overline{XV}$*

This folio (Figure 3) deals with the "mundane year", a concept already vaguely discribed by Platon (Timaios, 39d), again by Cicero in his *Somnium*

Figure 3. De anno mundano

Scipionis (Dream of Scipio, 6,24) as *annus vertens* and eventually adopted by Macrobius for the time needed for all heavenly bodies to revert to a certain initial position. Cicero (2009) believes this position to be present at the death of Romulus. Since Scipio asserts *Cuius quidem anni nondum vicesimam partem scito esse conversam*, "of this year not yet the twentieth part has elapsed, you should know", the duration of the *annus vertens* has to be more than 11.460 normal years. This number is found accepting the difference of dates between Romulus and Scipio to be 573 years, following Macrobius (1994, 2.11.16). It may be by accident, but an interesting one, that 11.460 is nearly identical with 11.454 given above on f 92r. Lambert, however, gives *quindecim milia ... a discessu Romuli,* "fifteen thousand years from the passing of Romulus", again using Macrobius (1994, 2.11.11) as his source. So he does, continuing with *Ille ergo vere annus vertens vocandus est,* "this therefore in verity has to be called a turnaround year", because it is not only the recurrence of the sun, but of every luminary in the sky. And he repeats Macrobius' unchallengeable argument, *Sicut asserunt physici,* "so assure the physicists". Maybe he regarded a physicist himself?

The text is written in a circular diagram, showing the spheres of the universe. In the centre we read *terrae globus,* in the volume of the first sphere (uncoloured) *radiis in luna* (with the complementary word *solis* over the next sphere. Following is the lunar sphere (green), carrying the moon in five (!) phases, namely (counter-clockwise from bottom right) new light with the text *prima,* waxing crescent with *VII prima,* full moon with *luna XIIII anno I,* third quarter with *XXIII,* and waxing crescent (?) with XXX. There is a fairly unreadable text between new light and waxing crescent. Then follows the sphere of Mercury (blue) with *Mercurius annis XX* on top. The next is the sphere of Venus (red). At the lower right, already in the sphere of the sun, there is a star and the words *Venus Vesper.*

The following, very broad, uncoloured sphere is that of the sun. It is filled with the text on the mundane year discussed above. On top, there is a 16-fold rosette with the inscription *sol ortus* inside and *sol XVIIII* above, the latter referring to the Metonic cycle. On bottom there is another rosette, with sol occasus inscribed. From this rosette to the upper left and upper right each the word *solis* is written, extending over the spheres of Venus and Mercury. Then follows (in green) the sphere of Mars, with *Mars annis XV* (?), that of Jupiter (uncoloured) with *Jupiter annis XII* on top and *Jupiter* on bottom, and finally the sphere of Saturnus (red) with *Saturnus annis XXX cursum explet* on top and the interesting statement *A terra usque ad Saturnum sunt stadia DCLXXXVIII, miliaria* $\overline{LXXX}$ *DCCCCXXXVII* , "from earth to Saturn there are 688 (should read 688.000) stadia, 80.937 miles".

This gives 8.5 stadia per mile. The conversion factor between stadia and miles, the use of these two units and the origin of the given planetary distances will occupy our interest again when dealing with f 94v. The least common multiple of the years given for the turnaround of each planet in the diagram, should give the duration of the mundane year. And, luckily, the result is 16.416, fitting rather well to the "fifteen thousand years" given by Lambert.

The page closes with the somewhat out-of-place sentence *Exemplar lunam lumen habere a sole,* "Example: that the moon has her light from the sun".

2.7. *f 94r De Astrologia secundum Bedam*

f94r is entitled *De Astrologia secundum Bedam.* The introductory text in the upper left corner of the drawing says *Stelle a sole illuminantur et sunt immobiles et cum celo fixe perpetuo motu feruntur. Stelle autem magnitudinem terre uincunt Tullio testante, que globose et rotunde circulos suos conficiunt celeritate mirabili.* "The stars are illuminated by the sun and are immobile and fixed with the sky they are carried in perpetual motion. The stars, however, surpass the size of the earth, as Tullius [Cicero] testifies, and globosely and round they perform their circles with admirable swiftness."

The dominant subject of the drawing is a representation of the night sky with a number of constellations. On this "star map" are superimposed three excentric circles inscribed Sol, Venus and Mars, representing the orbits of these luminaries. The text below the diagram reads: *Duo sunt extremi uertices mundi, quos appellamus pilos Sepemtrionis et Austri, quorum alter a nobis semper videtur, alter uero numquam. In eo qui a nobis cernitur, tria sunt signa constituta, duo scilicet Arcturi, hoc est Maior Ursa et Minor Ursa, et Serpens et cetere stele et signa XII, ut superius notatum est.* "Two are the extreme vertices of the world, that we call Pillars [*recte* Poles] of North and South, of those the first is always visible for us, the second, however, never. In the one we can discerne three signs are arranged, namely two of Arcturus, that is the taler and the smaller she-bear, and the dragon and other stars and 12 signs, as was annotated above." This is the adaptation of a text (uncertainly) ascribed to the 2nd century Roman author Hyginus (Hasper, 1861). The text is found in a number of medieval manuscripts.

In the upper right corner there is a text that could be an introduction to the following f94v. It reads *E quibus unum globum possidet Saturnus, alterum vero Iovis tertiumque Mars, quartum quoque Sol, quintum Venus, sextum uero Mercurious septimumque Luna radiis solis accensa. Supra lunam eterna sunt omnia, et infra nihil nisi mortale.* "From these Saturn possesses one globe, the second, however, Jupiter and the third Mars, also the forth the Sun, fifth Venus, but the sixth Mercury and the seventh the Moon, kindled by the rays of the Sun."

2.8. *f 94v Circuli VII^{em} cursusque VII^{em} planetarum*

This page is the most complex regarding astronomy. It can, therefore, be presented here only without going into depth: This will be done in another publication. Here the mere description must be sufficient.

The page starts with a description of the heavenly spheres, which together form the cosmos. The text above the diagram again is taken from Macrobius (1994 Vol. II.17.2 f).

Novem ordinibus uel globis conexa sunt omnia, quorum unus est celestis extimus, qui reliquos omnes complectitur, summus ipse Deus arcens et continens ceteros, in quo sunt infixi illi qui uuoluuntur stellarum cursus sempiterni. Cui subiecti sunt septem qui uersantur retro contrario motu atque celum. Summus ille celi stellifer cursus, cuius conuersio est concitatior, acuto et excitato mouetur sono, grauissimo autem lunaris infimus.

"In nine ranks or globes all are connected, from which one is the outermost heaven, that comprises all the remaining, the highest God fencing and containing the others, in which are fixed those performing the eternal movements of the stars. Beneath of it are seven, turning backward against the motion and the sky. That highest course of the sky is the star carrier, whose revolution is more rapid, moved with a sharp and thrilling sound, but with the gravest the lowermost of the moon."

Then there is a planetary diagram with seven concentric circles for the planetary spheres including sun and moon. The planets are ordered following the Chaldean or Plinian order (Eastwood, 1997): Moon, Mercury, Venus, Sun, Mars, Jupiter and Saturn. In the centre there is the earth. Here it is shown as a TO-type world map first designed by Isidore of Seville (Etymologiae XIV, 2). In this map Asia takes up one half of the whole, while Europe and Africa share the other half. The combination of a TO-map with a planetary diagram is rare.

Each planet is characterized by a symbol and a description giving the orbital period, and in addition two numbers for the distances between the planetary spheres in stadia and miles. The text says:

Saturnus XXX annis cursum explet. A Terra ad Saturnum stadia DCLXXXVIII, miliaria uero $\overline{LXXXD}$ *XXXVIIς*. "Saturn in 30 years passes the course. From Earth to Saturn are 688 stadia, but 80,937.5 miles."

Iovis XII annis cursum explet. Ab eo usque ad Saturnum stadia LXII ς, miliaria uero $\overline{VII}$ *DCCCXIIς*. "Jupiter in 12 years passes the course. From him to Saturn are 62.5 stadia, but 7,812.5 miles."

Mars XV annis cursum explet, a quo usque ad Iouem stadia LXIIς, miliaria uero $\overline{VII}$ *DCCCXIIς*. "Mars in 15

Figure 4. *Circuli VII^{em} cursusque VII^{em} planetarum*

years passes his course, from there to Jupiter are 62.5 stadia, but 7,812.5 miles."

Sol XVIIII annis cursum explet, a quo ad Martem stadia CXXV, miliaria uero $\overline{XV}$ *DCXXV*. "The sun in 19 years passes its course; from there to Mars are 125 stadia, but 15,625 miles."

Lucifer VIII annis cursum explet, a quo ad Solem stadia CLXXXVIII, miliaria uero $\overline{XXIII}$ *CCCCXXXVIIς*. "Lucifer in 8 years passes the course; from it to the Sun are 188 stadia, but 23,437.5 miles."

Mercurius XX annis cursum explet. Inde stadia LXIIς, miliaria $\overline{VII}$ *DCCCXX*. "Mercury in 20 years passes the course. From here are 62.5 stadia, 7.820 miles.

Luna per annum cursum explet. Ad Mercurium stadia LXII ς, miliaria $\overline{VII}$ *DCCCXIIς*. The Moon passes the course in one year. To Mercury are 62.5 stadia, 7.812.5 miles.

A Terra ad Lunam stadia CXXV, miliaria $\overline{XV}$ *D et XXIIς*. From the Earth to the Moon are 125 stadia, 15,822.5 miles.

The information regarding the orbital periods is taken from Isidore of Seville; he describes the time a planet needs to reach the same position on heaven regarding longitude and latitude.

Table 2 summarizes all the numbers which can be found in the diagram. If we assume that Lambert neglected the sign for 1000 in the stadia, all the values are multiples of 62.500 stadia or 7.812,5 miles. This gives a factor of 8 between stadia and miles, a correct conversion factor between these ancient Roman measures of length. It is not clear why for the distance between Earth and Saturn Lambert uses a conversion factor of 8.5 here and on f 93v.

The numbers given for the distances do not have a real astronomical background but follow a scale first known from Alexandros Lychnos (Alexander of Ephesos) in the 1st century BC. He wrote a hexameter poem describing the harmony of the spheres (Irby-Massie et al., 2013). Alexandros' scale relates planetary distances with musical intervals. The basic distance was 63.000 (or 62.500) stadia, a number which fits well with Eratosthenes' value for a quarter of the earth's circumference. This unit is set equal to the interval of a half tone in music. The distance between earth and moon corresponds to two times the basic unit, the distance between moon and mercury is one time the basic unit, and so on. Thus the distances between the planetary spheres form a musical gamut; details cannot be given here for lack of space. Combining astronomy and music has a long tradition in Greek sciences dating back to Pythagoras, of whom is said that he believed in the planets and stars moving according to mathematical terms corresponding to musical notes (Riedweg, 2005). Later this was called the harmony of the spheres.

The text below the diagram gives properties supposed to be characteristic for the various planets: *Inter celum terramque hec VII sidera pendent: Saturnus candidus natura gelidum. Iouis clarus et temperatus. Mars feruidus et sanguineus. Sol ardens clarus. Lucifer refulgens est iocundus. Mercurius radians resplendet. Luna blanda terra est maior. Sol uero quater terra amplior est Beda testante et octies maior Macrobio et Hilperico affirmantibus.* "Between heaven and earth these seven planets are pending: Saturn, white and of cold nature. Jupiter bright and temperate. Mars flamy and sanguine. The sun burning bright. Lucifer reflective is pleasant. Mercury shines radiant. The friendly moon is greater than the earth. But the sun is four times larger than the earth according to Beda and eight times larger as affirmed by Macrobius and Helperic."

Table 2: Planetary distances given in folio 94v

planet	years	distance (stadia)	distance (miles)	factor stadia/mile
Saturn	30			
↑		62.500	7.812,50	8,0
Jupiter	12			
↑		62.500	7.812,50	8,0
Mars	15			
↑		125.000	15.625,0	8,0
Sun	19			
↑		187.500	23.437,50	8,0
Venus	8			
↑		62.500	7.820,00	7,9923
Mercury	20			
↑		62.500	7.812,50	8,0
Moon	1			
↑		125.000	15.822,5	7,9
Earth				
↓		688.000	80.937,5	8,5
Saturn				

2.9. *Further astronomical content*

Only a selection of interesting astronomic material could be given here. There is more in the book, e. g. on f88v (distribution of the year over the zodiacal signs, seasonal day-lengths), 91v (time spent by sun and moon in each zodiacal sign), 95r-95v (constellatons), 135v (time reckoning), 222v-224v (Scipio's dream, by Cicero), and 225v-228r (planets, eclipses etc.). Each of these folios would deserve becoming subject of a special investigation.

3. CONCLUSION

The astronomical content in the *Liber Floridus* is distributed over the whole book, shows no systematic order, is redundant, often mixes astronomy with geography or meteorology, sometimes astrology and religion. Most of the content is not genuine but taken from the literature.

For us the book may look a little bit confusing. It seems, however, that in medieval times it was rather popular, as it was copied several times. Anyway nearly every part of the text is a challenge for the student of the history of astronomy.

ACKNOWLEDGEMENTS

We thank the Universiteitsbibliotheek Gent for generously making available the original manuscript MS 92 of the *Liber Floridus* in high resolution in the internet.

REFERENCES

Apian, P. (1540) *Astronomicum Caesareum*. Peter Apian Ingolstadt

Bayer, J. (1603) *Uranometria*. Christoph Mang Augsburg

Beda Venerabilis, *De Natura Rerum Liber*. Chap. 19. Migne, J. P. (1844) Patrologia Latina. Vol. 90: Col 0187-0278A

Cicero, Marcus Tullius, *De re publica liber sextus*. BiblioBazaar Charleston South Carolina (2009)

Delisle, L. (1906) Notices sur les manuscrits du Liber Floridus de Lambert, chanoine de Saint-Omer. In: *Notices et extraits des manuscrits de la Bibliothéque National et autres bibliothèques* **38**, 577-791

Derolez, A. (1968) *Liber floridus; Lamberti S. Audomari canonici Liber floridus : codex autographus Bibliothecae Universitatis Gandavensis*. Story-Scientia Gent

Derolez, A. (1998) *The autograph manuscript of the Liber Floridus. A key to the encyclopedia of Lambert of Saint-Omer*. Corpus Christianorum, Autographa Medii Aevi 4. Brepols Publishers Turnhout

Eastwood, B.S. (1997) Astronomy in Christian Latin Europe c.500–c.1150. *Journal for the History of Astronomy* **28**, 235-258

Helperic of Auxerre. *Liber de Computo*. Migne, J. P. (1844) Patrologia Latina Vol. 137, Col: 0017-048D. http://www.documentacatholicaomnia.eu/04z/z_0980-0980__Helpericus_Monachus_S_Galli_Liber_ De_Computo__MLT.pdf.html (Retrieved January 7, 2016)

Hasper, L.W. (1861) *Hyginus, Philosophus*. Dyk'sche Buchhandlung Leipzig

Irby-Massie, G. L., Keyser, P.T. (2013) *Greek Science of the Hellenistic Era: A Sourcebook*. Routledge London

Isidore of Seville, *Etymologiae libri*. Translated an annotated by Lenelotte Möller. Marixverlag Wiesbaden (2008)

Lambertus a S. Audomaro (1121) *Liber Floridus*. Universiteitsbibliotheek Gent, MS 92; http://adore.ugent.be/OpenURL/app?id=archive.ugent.be:018970A2-B1E8-11DF-A2E0-A70579F64438&type=carousel&scrollto=0 (Retrieved Jan 8, 2016)

Macrobius, Ambrosius Theodosius, *Commentarii in somnium Scipionis*. James Willis (ed.) Teubner, Stuttgart/Leipzig (1994)

Platon, *Timaios*. Platon Werke Band 7, 4. Aufl. Wissenschaftliche Buchgesellschaft Darmstadt (2005)

Riedweg, C. (2005) *Pythagoras: His Life, Teaching and Influence*. Cornell University Press

Seneca, Lucius Annaeus, *Naturales quaestiones* /Naturwissenschaftliche Untersuchungen: Lat. /Dt. (Reclams Universal-Bibliothek). Otto Schönberger, Eva Schönberger (eds.) (1998)

21

SURVEY, ARCHAEOASTRONOMY AND COMMUNICATION: THE MAUSOLEUM OF THEODORIC IN RAVENNA (ITALY)

M. Incerti*, G. Lavoratti and S. Iurilli

Architecture Department of Ferrara (Italy)

Corresponding author: manuela.incerti@unife.it

ABSTRACT

The Mausoleum of Theodoric (520 ca.), one of Ravenna's most globally renouned monuments and UNESCO heritage site, has been object of archeoastronomical research in the past years. This paper examines the orientation of the building as well as other elements of the architecture, especially the small openings of the building, to investigate their possible archaeoastronomical significance. The architectural survey was carried out with a laser scanner device (Faro-Focus 3D, software Scene) while the textures were extracted using digital photomodelling software. Starting from the archaeoastronomical analysis, a 3D model was developed to check astronomical phenomena and allow the communication of scientific content through several forms of media. This second part of the contribution is part of the experiments conducted by the research group (Incerti and Iurilli 2014), (Incerti and Iurilli 2015), focusing on new means of multimedia communication, both interactive and not. The research looks at virtual models as a means of *edutainment,* aiming to maximise the fruition of artifacts and cultural sites.

KEYWORDS: Theodoric, architectural survey, archaeoastronomy, edutainment, digital museum

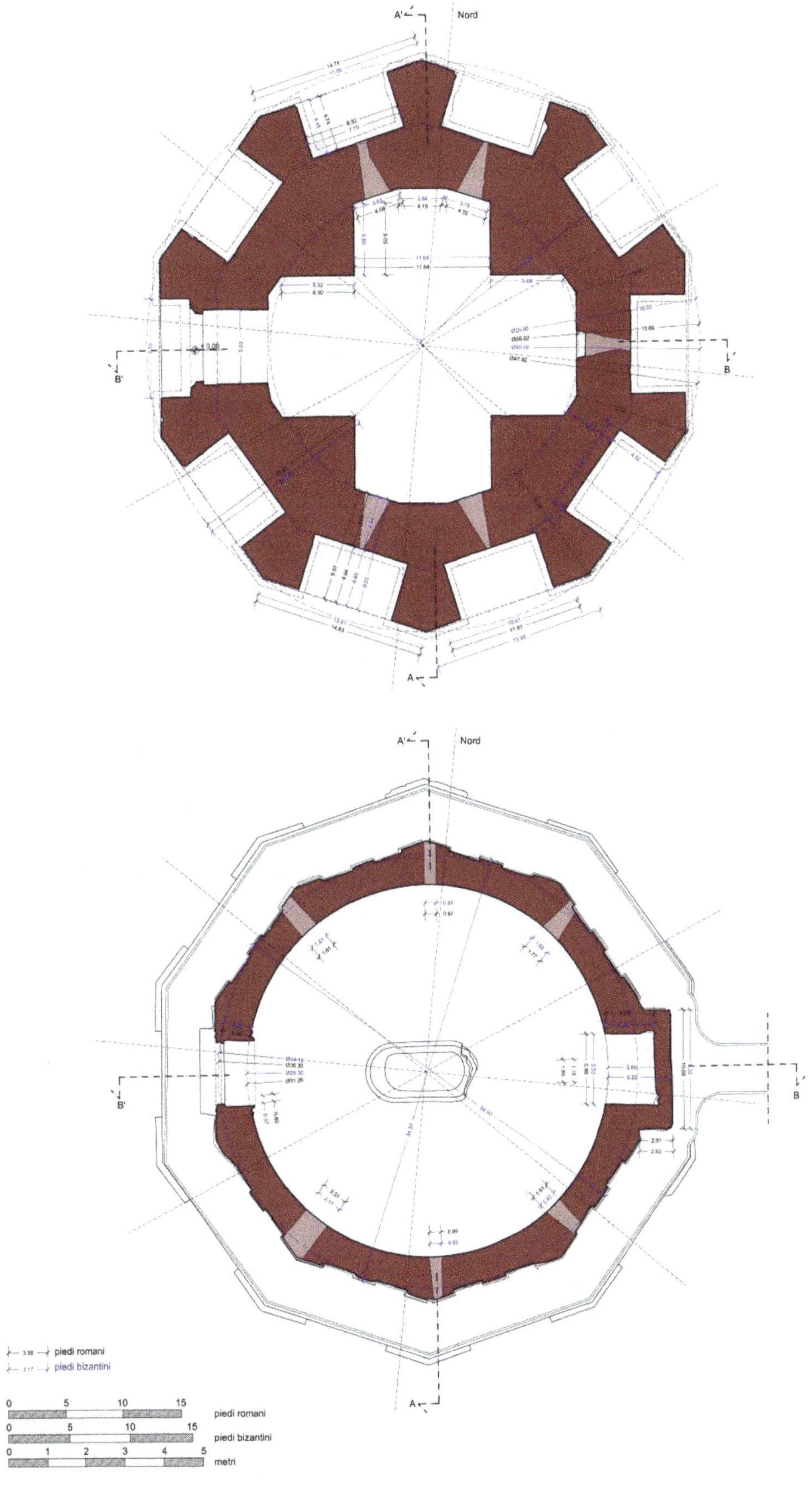

Figure 1. Mausoleum of Theodoric, Ravenna. Plans of the ground floor and first floor. Graphic elaborations of 3d survey.

1. INTRODUCTION

The most ancient information regarding the Mausoleum of Theodoric is included in a fragment of text written by an anonymous reporter known as Valesiano (*Anonymus Valesii),* from the name of the French academic Henri Valois, who edited the piece in 1636. The second part of the text, regarding the period in which Odoacre and Theodoric ruled (years between 474 and 526), narrates the foundation of the building: *"Se autem vivo fecit sibi monimentum ex lapide quadrato, mirae magnitudinis opus, et saxum ingens quod superponeret inquisivit".*[1] Meaning *"When still living, he erected a funeral monument for himself in squared blocks, a work of marvellous grandeur, and requested that an enormous rock be found to cover it".*[2] According to this document, the construction of the building was ordered by Theodoric himself before his death (30th August 526). It was then built in carefully squared blocks, a rarity for this time and practice, and covered with a "*saxum ingentem*", evidently purposefully demanded (*inquisivit*) by the King.

The commissioner of the work, King Theodoric was born around 454, son of Theodemir from the Amali family. At the young age of 12 he was sent to Constantinople as a hostage, and remained at the court of Leo I the Thracian until 472. Scholars do not agree on the terms and type of education he received in the East, in any case it is undeniable that during his kingdom he showed great attention to architecture. This is testified by the restoration of the ancient buildings of Rome, new buildings in Verona, in Pavia and, above all, in Ravenna.

2. THE ARCHITECTURE: HISTORICAL NOTES

The mausoleum is on two levels: the ground floor has decagonal plan in the external profile and a Greek cross interior, while the upper floor has a decagonal exterior and a circular interior space.

Like all monuments of Ravenna, now a UNESCO heritage site, the building has been the object of specialized studies (the most important of which date back to the last century,[3] and survey campaigns: the oldest, also published by Heidenreich (108-114), are those of Giuliano da Sangallo (1445-1516),[4] Baldassarre Peruzzi (1481-1536),[5] Antonio da Sangallo il Giovane (1483-1546)[6] and Giovanni Battista da Sangallo il Gobbo (1496-1552).[7]

In its long history, the small, central plan building was supported by other bodies, including a lighthouse, and later annexed to the church of a monastery (Santa Maria della Rotonda) where it was used as a "choir" (Guberti, 1952, 8-16). In 1748, a large part of the first level of the Mausoleum was still interred in an area often subject to flooding (Guberti, 1952, 17). In 1774 two flights of symmetrical stairs were built on the west side to allow access to the upper floor. Further campaigns of restoration and archaeological excavation followed in 1808, 1810, 1825, 1839, 1844 (Guberti, 1952, 19). In 1918-19 the eighteenth century stairs were removed and in 1927 the surrounding area was released as well as the first access ramp built. Further interventions were necessary in 1945 because of the damage caused by the war[8], as in 1977[9] and finally in 1998, the year in which the restoration of the stone facing stone of Aurisina[10] took place.

In the present study, some of the issues discussed relate to elements of the architecture that are possibly connected to orientation and to their archaeoastronomical meanings. It is therefore absolutely necessary to verify the authenticity and dating of the elements involved in the analysis in order to avoid misinterpretation of the data.

2.1. *The question of the "unfinished"*

Some small arches appear on the external face of the second floor, which may hint at the past presence of a loggia, perhaps lost or never finished. In this regard, the question of "unfinished" and the possible different dating of the two levels introduced by some authors do not appear to interfere with our observations. All reconstructions hypothesised for the second floor, amongst which one must remember the extremely accurate and sophisticated one of De Angelis of Ossat (very accurate graphics) (De Angelis of Ossat 1962), never involve the openings but only address the presence and shape of the portico, lower compared to the system of windows.

---

[1] *(Muratori, 1738), text available at link in bibliography ("Anonymus Valesianus: Chronica Theodericiana," n.d.).*

[2] *Translation in (Bovini 1977, 10-11).*

[3] *Amongst the most important and extensive works we point out (Haupt, 1913; Guberti, 1952; Heidenreich and Heinz, 1971) whose studies were actually carried out in 1938, and (Bovini, 1977). See also (Gotsmich, 1958; Johnson, 1988).*

[4] *(Sangallo, 1910), f. 38-37, see also the Geymuller codex, Uffizi, of which Heidenreich publishes an image (Heidenreich and Heinz, 1971, fig. 65, p. 63).*

[5] *Uffizi, Dis. 441 A (Heidenreich and Heinz, 1971).*

[6] *Antonio da Sangallo, Uffizi. Dis. 1406 A, 888 Av, 887 Ar, (Heidenreich and Heinz, 1971). On the same subject see also Uffizi. Dis. 875, 878, 896 (De Angelis d'Ossat, 1962, 97, tav. V)*

[7] *Uffizi Dis. 1563 Av, 1563 A, 1394 A, (Heidenreich and Heinz, 1971).*

[8] *The substitutions are highlighted by the different colours of the ashlars in the photographs of the time, see also (Conti and Berti, 1997).*

[9] *(Bovini, 1977, I-XV). Regarding restorations see (Piazza, 2013, 84-86) from the same volume see pages 24-25, 35-27, 56-57, 72 and especially the profile by Paola Novara pages 111-116.*

[10] *The recently identified construction material is the stone of Aurisina (Piazza et al., 1998; Bevilacqua, Fabbri, Grillini, and Iannucci, 2003).*

2.2. *The flooring*

The current flooring of both rooms, as previously mentioned, is certainly not original: in 1557 Leandro Alberti mentions traces of a mosaic floor, evidently placed in the top cell, since the bottom was buried underground. [11]

Regarding the progress of this element historians have reported a major failure of the ground on the Eastern side, which led to a drop of 14 cm in the ground floor and a 6 cm drop of the upper floor. The difference in height between the two levels led scholars to believe that an initial failure occurred during the construction of the ground floor. For this reason the upper floor was probably put in place "leveling" the already installed plan, which however, later another slight lowering. Some drawings stored in the Superintendency of Ravenna [12] show an accurate survey of the pavement slabs dated 1903, on which many measurements are noted as well as the measurement of elevation at a point indicated by the letter A. The existing pavements were put in place during the works of the biennium 1975-77 (Novara, 2013, p. 116).

2.3. *The small apse*

On the eastern side of the top space there is a small apse, whose function has been questioned by many historians: its height cannot accommodate an altar or an officiant or even the great porphyry sarcophagus (today placed at the center of the space). On the keystone of the arch is a large Latin cross, the only sculptural element of the interior, which highlights its relevance in the project. The small space, whose floor was slightly lower than that of the rest of the room is, according to scholars, contemporary to the building (Messina, 1980; De Angelis d'Ossat, 1962, 128-129).

2.4. *The sarcophagus*

According to tradition, the remains of King Theodoric were conserved in the great sarcophagus of red porphyry measuring 305x190x101 cm. The tub is characterized by four rings on the top edge and two lion heads at the bottom center of the side faces. The sarcophagus' troubled history has been well documented, its movements traced by Ambrogi (Ambrogi, 1995, 109-111), recalling its relocation to the site in 1913. There is no certainty regarding the original orientation of the object, which is however considered by scholars to be consistent with building, originally arranged in the East-West direction (the current one).

2.5. *The small windows*

The wall of the ground floor has a thickness of about 140 cm and is pierced by 6 splayed narrow slits, with an approximately horizontal intrados, arranged on three sides (each of which is divided into three segments). Their sizes vary in width from 11 to 25 cm, a minimum height of 40 cm, and 70 cm, and are arranged as follows:

(1a, 2a) Northern wall: 2 small windows placed in the first and third segments (intrados lowered outward approximately 6 cm)

(3b, 4b) Eastern wall: 2 small windows vertically aligned with the axis of the building.

(5c, 6c) Southern wall: 2 small windows placed in the first and third segments (intrados lowered outward approximately 6 cm).

The decagonal part of the upper level ends with a series of cylindrical section bands, the highest of which is decorated by the famous "tenaglia" (pincer) shaped design. The central and receding one, about 77 cm thick, is smooth and perforated by 11 windows. Arranged roughly in the directions North-South, East-West with the two diagonals at 45 ° (directions of the compass rose), the small openings have dimensions that vary from 40 cm in height for the windows on the North-South axis, to 62 cm on the diagonal axes. Starting from the entrance, we have:

(7-8-9-10) 4 small windows with rounded arches, combined two by two (West)

(11) 1 small rectangular window with overhead arch between two small straight sections (North-West)

(12) 1 small window with rounded arch (North)

(13) 1 small rectangular window with overhead arch between two small straight sections (North-East)

(14) 1 small Greek cross window (East)

(15) 1 small rectangular window with overhead arch between two small straight sections (South-East)

(16) 1 small window with rounded arch (South)

(17) 1 small rectangular window (South-West).

The opening are almost unanimously considered contemporary to the founding of the building, excluding the rectangular South-Western one which was clearly enlarged at a later date (Guberti, 1952, 59; De Angelis d'Ossat, 1962, 94)[13]. In various pictures preceding the last restoration one can notice the presence of a twelfth opening positioned on the back wall of the small apse, probably added during the period

[11] *Sources in (Fagiolo, 1972, 148-149).*

[12] *Dating from 1903 numbered 2080 and 2081.*

[13] *According to (Heidenreich and Heinz 1971, 58-60) the small windows on the diagonals are not original but date back to the Middle Ages.*

when the building became a church choir (Messina, 1980, 33; Guberti, 1952, fig. 26 pp. 40 and 59).

Keeping the axes of the openings described above into consideration, the internal lighting system of the two rooms, formed by 17 openings (11 + 6) can be traced back to 12 different directions: 5 in the lower deck and 8 in the upper one. Of these, only one - the Eastern one - follows the trend of and lower system.

Below the windows is a protruding band (of about 8 cm) on which inscriptions laid out on 3 different levels have been recently found (Novara, 2013, p. 116, see also the volume at p. 85), (Piazza, 2009), investigated and restored in 2012 (results not published yet).

2.6. *The cupola*

Even the great monolith that covers the building was subject to a great number of specialized studies that have investigated physical, technological, figurative, historical and design aspects (Dyggve 1957; Bean 1972; Tabarroni 1973; Bianco Fiorin 1993).

The inner diameter is about 925 cm and the height on the springing is of about 190 cm. A large crack, which popular tradition blames on a bolt of lightning, marks the southern side where the lighthouse was built adhering to the building.

12 protruding elements are present on the outer edge of the roofing. Their triangular holes convey the image of a "royal crown". Historians have often questioned the real function of these elements and their figurative origin, (Bean 1972). The assumption is that they were used for the passage of cables and ropes necessary for its positioning, as hypothesised by Antonio da Sangallo with a drawing already published (Heidenreich and Heinz, 1971, fig. 65, p. 63), may be considered unfounded because of the enormous weight of the monolith and the common laying techniques of the time (Tabarroni 1973). What all scholars emphasize is the lack of regularity in the arrangement of the dodecagon traced by the protruding elements, for it is not aligned with any of the geometries of the building. The monolith is in fact slightly rotated in relation to the main axis of the building, which led to the unanimous conclusion of a faulty, unfixable installation due to the creation of the dangerous lesions on the southern side.

The names of the apostles and evangelists are inscribed on the vertical face of the elements in the sequence (from the door, clockwise): Lucas, Marcus, Mathias (?), Matteus, Felippus, Johannes Jacobus, Andreas Paulus, Petrus, Simeon, Thomas.[14] The elements are closed with a gable roof, almost simulating a small sarcophagus, except for one: that of San Pietro has a flat roof. This led to believe the existence of a terminal made of a different –and perhaps more valuable– material (gone missing), highlighting the figure of Peter, founder of the Church (Tabarroni, 1973, 141)

3. THE ARCHITECTURAL SURVEY

The architectural survey was carried out with a Faro focus3d scanner [15]; 30 stations were performed covering the interior and exterior. The individual clouds have been registered with the aid of spherical targets. At a later date two different photographic campaigns, needed for the reconstruction of the three-dimensional, textured model, were made: the first relating to the exterior, the second to the interior. The exterior shots were performed by a compact Lumix DMC-TZ7. Interior shots, due to matters of critical illumination, were produced with a digital SLR camera on a tripod. In this stage 406 photographs were elaborated and 188 targets used as follows:
Exterior: 123 photos, 38 targets
Exterior walkway details: 96 photos, 146 targets
Lower interior: 76 photos, 0 targets
Top interior: 111 photos, 4 targets.

4. SURVEY DRAWINGS: METHODS AND PROCEDURE

The pointcloud available, which is particularly dense and with limited bands of occlusion, was sectioned by horizontal and vertical planes in order to extract the two-dimensional canonical drawings (plans, elevations and sections) used to effectively describe dimensions and geometries. A thin slice (thickness of 1 cm) was extracted from the cloud for each cut-plane to represent the elements dissected, re-drawn in field vector by interpolation of the points and exported for the realization of raster images definitive 1:50 scale. The choice of using a "slice" of points so reduced yet still sufficiently detailed was possible thanks to the particular density of the pointcloud, which provided a high degree of detail, also on particularly elaborate portions such as shell decorations on the interior brackets. At the same time, the reduced thickness of the slice has guaranteed a higher precision on the section itself, likening it more to a geometric plan. The elements in the projection are obtained by re-drawing, in vector terms, screenshots specially extracted from the pointcloud and orthophotos produced by software for digital photomodelling. Particular attention was paid to the architectural

[14] *Regarding possible reasons for this sequence see (De Angelis d'Ossat, 1962, 129-131, fig. 9; Heidenreich & Heinz, 1971, p. 14; Fagiolo, 1972, 108, 119, 142, 149, 152-4; Tabarroni, 1973, 133-141).*

[15] *Survey M. Incerti and P. Lusuardi, dated 30/04/2015. Data handling software Scene 5.3, elaboration of data M. Incerti. We would like to thank the Compagnia delle Misure for the use of the tool as well as the Soprintendenza per i Beni Architettonici e Paesaggistici di Ravenna for the permissions accorded.*

details, for which ad hoc screenshots with appropriate definition were taken, suitable to describe even the most complex of moldings.

On the rare occlusion bands,[16] the texture has been reproduced (where possible). This has been done by integrating the gaps with specially shot photograms in order to make the final elaborate complete in all its parts. The areas treated this way have nonetheless been graphically differentiated through the use of superimposed layers and textures to emphasise the diffrent material reliability compared to the rest of the survey.

The final elaborates are therefore the result of the overlapping of parts in section and projection obtained through the procedures described above. This work format, now well established in the science world, ensures greater metric control of the architecture (through comparison and contamination of drawings obtained through different processes, distinct survey campaigns and conducted with different instruments). It also allows for detailed graphics containing material and chromatic information that a traditional survey would not have been able to render.

Plans, elevations and sections thus obtained, describing the fundamental geometry of the architectural artefact with a metric reliability that allows for the analysis reported in the next section. It also reproduces the exact wall texture, the state of conservation, the grain, color and deterioration of different materials, allowing the expert operator to perform numerous other integrated readings.

5. ARCHEAOASTRONOMIC ANALYSIS

5.1. *The orientation*

The survey of the orientation was structurally created by Giuliano Romano (Romano, 1995) who first published the Azimut value of 84,5°.[17] The building is rotated, compared to the equinoctial direction, of a considerable amount that should not be overlooked during the alignment operations. The azimut from the apse (North)[18] is aligned with the rising of the sun on the 26-27 March and the 13th of September at the time of construction (pre 526). The first date is very close to 25th of March, celebration of the Annunciation of Mary, a celebration which was already documented in the *Martirologio Gerominiano* dating back to the IV/V century.[19]

Following the correction of the slight rotation, the survey methodology involved specially processed survey drawings. By overlaying graphics to the 4 main astronomical directions (solstices and equinoxes), the small windows placed on the 45° direction were not found to be perfectly aligned in relation to the center. Despite this, it is clear that these windows allow the entry of light during the two solstitial dates. [20]

5.2. *The windows*

Only 3 of the 17 windows, those on the North side, two downstairs and one on the top floor, do not receive significant sunlight. All others are involved in important moments of the astronomical year. The behavior of sunspots on horizontal and vertical surfaces were analyzed through plans and sections, height and azimuth angles were traced, through specific software, to the ephemeris. Among the episodes we note:

- *window (14).* Cross-shaped window, the rising sun penetrates through it on the day of the Annunciation, shining on the wall placed in front, on the axis of the cell. On the day of the summer solstice, about an hour after sunrise, the light spot passes over the stone sarcophagus.
- *window (3b, 4b).* Behaves like the previous for the rising sun of the 25th March. It should be noted that the phenomenon is no longer visible due to the construction of the new upper floor access system.
- *windows (7-8-9-10).* The sun penetrates them at sunset on the days of the equinoxes, illuminating the previously mentioned emerging scripted band.
- *window (16).* At noon of the winter solstice, the light spot approximately reaches the intersection between the wall and floor. Light only enters this window around noon in the winter solstice period because of its shape.

The windows at a 45° direction still welcome the rays of the sun at dawn and dusk during solstice

[16] As previously mentioned, the pointcloud and orthophotos obtained through digital photographs are very detailed and devoid of distracting occlusion areas. It was nevertheless necessary to intervene with the described procedures to limit the parts of the architecture where undercuts and non-removable elements (balustrades, fixed decor elements etc.) would not allow a complete description of the product.

[17] At the time it was not considered necessary to repeat the measurements because of the tools used and their precision. The investigations have therefore been carried out at a second date through the analysis of the behaviour of light in the interior in relation to the theoretical hypotheses elaborated with the aid of graphic tools.

[18] The horizon was untarnished because the building was situated near the sea shore.

[19] The festivity is documented of the 25th of March (ante diem octavum Kalendas Aprilis) in (De Rossi and Duchesne, 1894). We thank E. Spinazè for the notice. On this date the Hilaria celebrations took place in Rome in honour of the goddess Cibele, mother of the gods (Vaccaj 1927, 52-65).

[20] The building was the subject of an archaeoastronomical study in 2000, today still in form of a manuscript, by the scholar M. Kerner, 2000 cited in a report of the Soprintendenza per i Beni Architettonici e Paesaggistici di Ravenna (Piazza, 2009). An image of the work is published, but not consultable, showing lunar and solar symbols, perhaps referring to lunar standstills as well as solstices and equinoxes.

dates. Phenomena of this kind may have been used for purposes of a calendar to mark the advent of a particular date of the year, or for the computation of time: in other words, to indicate a precise moment of the day.

5.3. *The dome*

The results of the survey allowed us to verify that the arrangement of the protruding elements follows the cardinal directions with rather accurate approximation. The item marked with the name of Paul, founder of the Church (the only one with a flat roof), is aligned to South. Aligned to the East is Jacobus, West Lucas, North Matteus. This condition is a possible explanation for what scholars consider an "executional mistake".

6. DIGITAL MODELS FOR RESEARCH VERIFICATION AND DISSEMINATION

The investigation described above, and the survey supporting it, generated a large amount of data. Their analysis and reading produced significant results, providing new information on the archaeoastronomic characteristics of the Mausoleum.

To be disseminated effectively, such results require a visual support that allows the storage of different types of data (morphological, dimensional and chromatic features, archaeoastronomical analysis, external metadata, video and processed media, etc...), and, on the other hand, facilitate an immediate understanding of complex phenomena by the addressee of the information.

A detailed 3d model of the entire building has been created starting from the digital survey, in order to allow a direct visualization of the luminous phenomena affecting the inner spaces on some particular dates of the year.

The model is a low-polygon mesh made of quadrilaterals (quad mesh), textured with a UV mapping method, whereas the high resolution textures are derived from the orthophotos coming from the SFM[21] survey. The model was ideated to be optimized for future use in real time applications.

The 3d maquette, oriented and placed in a Cartesian space reference, was put in relation with a light of the directional type, which simulates the parallel rays from a source similar to the sun, and is therefore the best suited to virtually reproduce the tour of sun around the mausoleum. The light is put on an animated path that reproduces the movement of the sun on the ecliptic. Every moment of animation (keyframe)[22] was created by parametrically inserting the values (position and rotation) derived from the calculation of ephemeris in "key" times. In particular, the exact time of sunrise has been inserted as the beginning of the path, and the sunset time as the end. This time range was then further divided into half hour fractions to cover the whole course of the day.

After setting the animation, the positional coordinates of the sun during the intermediate hours come automatically from the interpolation of the data provided, and can be read simply by moving a cursor on the timeline: the construction phase of the model thus becomes a moment of assessment and comparison of the calculations previously made (ephemeris).

The procedure was repeated for four significant dates (solstices and equinoxes). The model thus obtained was used as a kind of virtual laboratory for the observation of the effects of light within the burial cells, in an ideal condition as free of obstacles and external elements that, at present, occlude the passage of the sun rays.

The observations made experimentally on the model confirmed the results previously recorded from two-dimensional drawings with extreme precision. In addition, the simulation also provides an "emotional" component, increasing the impact of the observation of archaeoastronomical events. The high level of detail of the model and the photo textures, the use of volumetric lights and the general photorealism of the scene allow the user to observe the phenomenon of light in an accelerated form as it may observed inside the mausoleum.

The system is very powerful and effective from a communication point of view, capable of involving the user in an engaging learning experience.

Some short animated sequences −related to a specific stage− have currently been made; however in the future the program will include the creation of a complete interactive application[23]. It will display the behavior of light inside the mausoleum by interactively moving a cursor on a timeline, or simply by entering parameters such as date and time.

---

[21] *Structure from motion (SFM) is a range imaging technique; it refers to the process of estimating three-dimensional structures from two-dimensional image sequences, which may be coupled with local motion signals. We used this technique to assemble a first version of the 3d model of the mausoleum, from which we extracted detailed ortophotos and basic morphological data.*

[22] *A key frame in animation and filmmaking is a still-image that defines the starting and ending points of any smooth transition. It is commonly called "frame" because its position in time is measured in frames on a strip of film.*

[23] *Similar experiences carried out by our research group are described in: (Incerti, Foschi, Iurilli, and Velo, 2014; Incerti and Iurilli, 2014; Incerti and Iurilli, 2015).*

Figure 2. From left, counterclockwise: daylight simulation within the upper cell of the Mausoleum (vernal equinox, 21th of March, rising and sunset; summer solstice, 21th of June, sunrise) and full textured 3d model of the building.

7. CONCLUSION

To conclude, the archaeoastronomical investigation has certainly given significant results which have extended our knowledge of this extraordinary building into topics which had been previously unexplored. The database of measurements produced by the advanced survey offers additional insight, for example on the geometrical and numerical aspects in relation to units of measurement, and on the figure of Severino Boezio, imprisoned and murdered by Theodoric. Lastly, we tested the important contribution of digital models, both during analysis and in the transmission of complex, layered content such as historical and astronomical knowledge.[24]

Figure 3. The Mausoleum of Theodoric (Photo by M. Incerti)

[24] *Paragraphs attributed as following: M. Incerti (1-3, 5-7), S. Iurilli (6), and G. Lavoratti (4).*

REFERENCES

Anonymus Valesianus: Chronica Theodericiana (n.d.) Retrieved November 1, 2015, from http://www.thelatinlibrary.com/valesianus2.html

Bevilacqua, F., Fabbri, R., Grillini, G.C. and Iannucci, A. (2003) Il mausoleo di Teodorico: la pietra di Aurisina. Tecniche e strumenti di lavorazione. F. Lenzi (Ed.), *L' Archeologia dell'Adriatico dalla preistoria al medioevo. Atti del Convegno internazionale: Ravenna 7-9 giugno 2001*, 572–580. Firenze, All'Insegna del Giglio.

Bianco Fiorin, M. (1993) Il monolite del mausoleo di Teodorico. In: *Teoderico il Grande e i Goti d'Italia. Atti del XIII Congresso internazionale di studi sull'alto Medioevo, Milano, 2-6 novembre 1992, vol II pp. 601-610 Spoleto, Centro italiano di studi sull'alto Medioevo.*

Bovini, G. (1977) *Il Mausoleo di Teodorico.* Ravenna, Longo.

Conti, G. and Berti, M. (1997) Lacune e integrazioni murarie nel mausoleo di Teodorico a Ravenna. In *Lacune in architettura: aspetti teorici ed operativi. Atti del convegno di studi, Bressanone, 1-4 luglio 1997,* 361–378. Marghera, Edizioni Arcadia Ricerche.

De Angelis d'Ossat, G. (1962) *Studi ravennati: problemi di architettura paleocristiana.* Ravenna, Edizioni Dante.

De Rossi, G. B. and Duchesne, L. (Eds.) (1894) Martyrologium Hieronymianum. In *Acta Sanctorum Novembris, tomo II, pars prior* , 140. Bruxellis, Société Belge de Librairie.

Dyggve, E. (1957) Mausoleo di Teodorico: le origini della cupola. *Corso di Cult. sull'Arte Ravenn. e Biz. Univ. degli Stud. di Bol. Ist. di Antich. Ravennati e Biz. Ravenna.*

Fagiolo, M. (1972) Theodericus, Christus, Sol. Nuove ipotesi sul mausoleo. *Arheoloski Vestnik*, Vol. 23, 83–117.

Gotsmich, A. (1958) Il Mausoleo di Teodorico. *Felix Ravenna Università Degli Studi Di Bologna, Dipartimento Di Archeologia, Centro Di Studi per Le Antichità Ravennati E Bizantine Giuseppe Bovini.*, Vol. 3., Ser. 25, 56–71.

Guberti, V. (1952) *Il Mausoleo di Teodorico detto anche "La Rotonda"*. Ravenna, Arti grafiche.

Haupt, A. (1913) *Das Grabmal Theoderichs des Grossen zu Ravenna.* Leipzig, Seemann.

Heidenreich, R. and Heinz, J. (1971) *Das Grabmal Theoderichs zu Ravenna.* Wiesbaden, F. Steiner.

Incerti, M., Foschi, P., Iurilli, S. and Velo, U. (2014) Rilievo, conoscenza e comunicazione: il complesso di san Michele in Bosco e il Chiostro ottagonale dei Carracci. Survey, insight communication: the complex of San Michele in Bosco and the Octagonal Cloister by Carracci. C. Vernizzi and P. Giandebiaggi (Eds.), *Italian survey and international experience. 36° Convegno Internazionale dei docenti della Rappresentazione - Undicesimo congresso UID*. Roma, Gangemi.

Incerti, M. and Iurilli, S. (2014) From survey data to virtual environment. Two case studies. *SCIRES-IT*, Vol. 4 (2), 87–108. http://doi.org/10.2423/i22394303v4n2p87.

Incerti, M. and Iurilli, S. (2015) Schifanoia Palace in Ferrara a virtual storytelling for its knowledge and dissemination. J. C. Torres, H. Graf, F. Remondino, J. Barceló, P. Brunet, H. Susan and L. Duranti (Eds.), *2nd International Congress on Digital Heritage 2015. 28 sept- 2 oct. 2015, Granada* (pp. 41–49). IEEE Conference Publications.

Johnson, M. J. (1988) Toward a History of Theoderic's Building Program. *Dumbarton Oaks Papers*, Vol. 42, 73–96. http://doi.org/10.2307/1291590.

Messina, A. (1980) La cosiddetta "scarsella" del Mausoleo di Teodorico. *Felix Ravenna Università Degli Studi Di Bologna, Dipartimento Di Archeologia, Centro Di Studi per Le Antichità Ravennati E Bizantine Giuseppe Bovini.*, Vol. 1979, 117 , 29–38.

Muratori, L. A. (1738) Anonymus Valesianus. *Rerum Italicarum scriptores, XXIV, IV* (p. 640, col. X). Mediolani, Ex Typographia Societatis Palatinae. Retrieved from http://gutenberg.beic.it/view/action/nmets.do?DOCCHOICE =2307661.xmlanddvs=1446379827278~115andlocale=itandsearch_terms=DTL10andadjacency=and VIEWER_URL=/view/action/nmets.do?andDELIVERY_RULE_ID=7anddivType=

Novara, P. (2013) Mausoleo di Teodorico. A. Ranaldi and P. Novara (Eds.), *Restauri dei monumenti paleocristiani e bizantini di Ravenna patrimonio dell'umanità, Parte I. Architetture*, 114–116. Ravenna, Comune di Ravenna, Archidiocesi di Ravenna-Cervia, MIBAC.

Piazza, V. (2009) *Mauseleo di Teodorico.* Scheda Soprintendenza per i Beni Architettonici e Paesaggistici per le province di Ravenna, Ferrara.

Piazza, V. (2013) Mausoleo di Teodorico. A. Ranaldi and P. Novara (Eds.), *Restauri dei Monumenti paleocristiani e bizantini di Ravenna patrimonio dell'umanità*, pp. 84–86. Ravenna, Comune di Ravenna, Archidiocesi di Ravenna-Cervia, MIBAC.

Piazza, V., Bevilacqua, F., Grillini, G. C. and Pinna, D. (1998) Ravenna - Mausoleo di Teodorico. Esiti delle ultime indagini e progetto. G. Driussi and G. Biscontin (Eds.), *Progettare i restauri: orientamenti e metodi-indagini e materiali. Atti del Convegno di studi, Bressanone 30 giugno-3 luglio 1998*, 251–260. Venezia, Arcadia ricerche.

Romano, G. (1995) *Orientamenti ad sidera. Astronomia, riti e calendari per la fondazione di templi e città. Un esempio a Ravenna.* Ravenna, Edizioni Essegi.

Sangallo, G. (1910) *Il libro di Giuliano da Sangallo, codice vaticano Barberiniano latino 4424 riprodotto in fototipia con introduzione e note di Cristiano Huelsen.* Lipsia, Harrassowitz.

Vaccaj, G. (1927) *Le feste di Roma Antica* (II). Torino, Fratelli Bocca Editori.

22

ARMENIAN ARCHAEOASTRONOMY AND ASTRONOMY IN CULTURE

Areg Mickaelian*[1] and Sona Farmanyan[2]

[1]Byurakan Astrophysical Observatory (BAO), National Academy of Sciences, Byurakan 0213, Aragatzotn Province, Armenia
[2]M. Abeghyan Institute of Literature, National Academy of Sciences, 15 Grigor Lusavorich Str., Yerevan 0015, Armenia

Corresponding author: Areg Mickaelian (aregmick@yahoo.com)

ABSTRACT

A review is given on archaeoastronomy in Armenia and astronomical knowledge reflected in the Armenian culture. Astronomy in Armenia was popular since ancient times and Armenia is rich in its astronomical heritage, such as the names of the constellations, ancient observatories, Armenian rock art (numerous petroglyphs of astronomical content), ancient and medieval Armenian calendars, astronomical terms and names used in Armenian language since II-I millennia B.C., records of astronomical events by ancient Armenians (e.g. Halley's comet in 87 B.C., supernovae explosion in 1054), the astronomical heritage of the Armenian medieval great thinker Anania Shirakatsi's (612-685), medieval sky maps and astronomical devices by Ghukas (Luca) Vanandetsi (XVII-XVIII centuries) and Mkhitar Sebastatsi (1676-1749), etc. For systemization and further regular studies, we have created a webpage devoted to Armenian archaeoastronomical matters at Armenian Astronomical Society (ArAS) website. Issues on astronomy in culture include astronomy in ancient Armenian cultures, ethnoastronomy, astronomy in Armenian religion and mythology, astronomy and astrology, astronomy in folklore and poetry, astronomy in arts, astrolinguistics and astroheraldry. A similar webpage for Astronomy in Armenian Culture is being created at ArAS website and a permanent section *"Archaeoastronomy and Astronomy in Culture"* has been created in ArAS Electronic Newsletter. Several meetings on this topic have been organized in Armenia during 2007-2014, including the archaeoastronomical meetings in 2012 and 2014, and a number of books have been published. Several institutions are related to these studies coordinated by Byurakan Astrophysical Observatory (BAO) and researchers from the fields of astronomy, history, archaeology, literature, linguistics, etc. are involved.

KEYWORDS: archaeoastronomy, historical astronomy, cultural astronomy, landscape archaeology, astroarchaeology.

1. INTRODUCTION

Astronomy is one of the ancient sciences in the world, as well as in the Armenian Highland. On the other hand, Armenia is rather rich in archaeoastronomy and the Armenian culture is rich in evidences of astronomical knowledge, both in ancient times and its modern reflections. Being among the most advanced interdisciplinary and multidisciplinary sciences, archaeoastronomy and astronomy in culture (AAC) involve astronomers, historians, archaeologists, ethnographers, philologists, linguists, artists and other representatives of science and culture. That is why it has such popularity.

Armenian AAC include many creations related to astronomical knowledge, including calendars, rock art, mythology, etc. On the other hand, from the professional point of view, this subject is rather poorly developed in Armenia. There are only individual studies on various related issues (especially many studies related to our medieval great thinker Anania Shirakatsi, see below) but not coordinated actions to manage this important field of investigation. This paper is aimed at presenting a general overview to mention and summarize some recent activities. A review is given on archaeoastronomy in Armenia and astronomical knowledge reflected in Armenian culture.

Several Armenian institutions are related to these studies coordinated by Byurakan Astrophysical Observatory (BAO). To systemization and further regular studies, Armenian Astronomical Society (ArAS) has opened a webpage dedicated to AAC (www.aras.am/Archaeoastronomy/astronomyancientarmenia.html). Armenian astronomers are members of SEAC, IAU working groups on *Astronomy and World Heritage, Archaeoastronomy and Astronomy in Culture* and other international organizations.

2. ARMENIAN ARCHAEOASTRONOMY

Armenia is one of the most ancient countries in the world. On the other hand, the Middle East (where the Armenian Highland is located) is regarded as the cradle of the civilization.

Astronomy in Armenia was popular since ancient times and Armenia is rich in its astronomical heritage (Herouni, 2006; Mickaelian, 2008; 2014; Parsamian, 1999, 2014; Tumanian, 1985). Armenian archaeoastronomy includes:

- names of the constellations,
- ancient observatories,
- Armenian rock art with astronomical content,
- ancient Armenian calendar and other (medieval) calendars,
- astronomical terms and names,
- records of astronomical events by ancient Armenians,
- Armenian medieval great thinker Anania Shirakatsi's astronomical heritage,
- medieval sky maps,
- medieval astronomical devices, etc.

It has been suggested that the **first division of the sky into constellations** was made a few thousand years ago in the Armenian Highland and nearby areas (Olcott, 1911; Flammarion, 1880). This is based on the names of Zodiacal and some other constellations. The main arguments are the period of visibility, precession calculations, animals involved (that lived in these areas), archaeological studies, cultural traces in the history, folklore, etc. In addition (surprisingly), Zodiacal constellation signs are similar to ancient Armenian symbolism. Many other papers indicate that the first division of the sky into constellations in any case was done in the area of Mesopotamia (e.g. Belmonte, 2015).

Figure 1. Zorats Karer. Stones Nos. 60, 62 and 63 and a schematic explanation how our ancestors could make observations of the celestial bodies through the holes. Photo by one of the authors

Two of the ancient observatories discovered in Armenia, namely **Zorats Karer** (or Karahunge) and **Metzamor** are especially well known. Karahunge is the Armenian twin of the Stonehenge and is considered to be even older (Parsamian, 1985a; 1985b;

Bochkarev & Bochkarev, 2005; Herouni, 2006; Martirosyan; 2015). It may be considered as one of the widespread megalithic structures in Europe and Asia (Scuderi et al., 2014). **Figure 1** shows how our ancestors could observe the celestial objects through the holes made in the stones. Beside these two observatories, Sevsar, Astghaberd, Portablur and Agarak have also been revealed.

The **Armenian rock art** (numerous petroglyphs of astronomical content) is especially unique and famous (Tokhatyan, 2014; 2015; Ter-Gulanyan, 2014). Most of such carvings are found in historical Armenia and neighbouring countries. We give in **Figure 2** examples of the Armenian rock art and extracted examples of figures from different stones. Original stone with dots and other elements should be related to the calendar. A number of celestial objects (Sun, Moon, planets, comets and stars) are often present.

Ancient and medieval **Armenian calendars** are well studied and described (Tumanian & Mnatsakanian, 1965; Badalian, 1970; Broutian, 1997; 2015; Tumanian, 1972). At least 3 of them are well-known: *Hayots Bun Tvakan* (Armenian Main Calendar, begins on 11 August 2492 B. C.), *Hayots Mets Tvakan* (Armenian Great Calendar, begins on 11 July 552 A. D.) and *Hayots Poqr Tvakan* (begins on 11 August 1084 A. D., Armenian Small Calendar). Ancient Armenian names of the months, days and even of each hour are survived and presently used in the religious (Apostolic) calendar.

Astronomical terms and names used in Armenian language since II-I millennia B. C. (Armenian names of the planets, constellations and some stars) may be divided into two types: 1) those originated in Armenia and being unique, and 2) those taken from ancient Greek and transformed (translated or transliterated) into Armenian (Harutyunian, 2015; Yeghikian, 2015).

Records of astronomical events by ancient Armenians have always been important for the history, particularly for the establishment of exact or possibly true dating. Their role in historical records is well established (e.g. Polcaro et al., 2008), as they help in revealing many unknown to historians facts and dates. Armenians have observed Halley's comet in 87 B. C. and this event was carved on the coin by Armenian king Tigranes II the Great (Gurzadyan & Vardanyan, 2004). According to some authors, the Supernovae explosion in 1054 on the location of the present Crab nebulae was observed in Armenia in May 1054 compared to the Chinese observations in July of the same year (Pskovskij, 1982).

Armenian medieval great thinker **Anania Shirakatsi's** (612-685) astronomical heritage is huge (Anania Shirakatsi, 1962; Conybeare, 1897; Semyonov, 1953; Nazaryan, 2013; 2014; Petri, 1964; Arevshatyan, 2014; Danielyan, 2014; Eynatyan, 2014; Harutyunian & Mickaelian, 2014; Mirumyan, 2015). Among his 29 works, 13 have relation to astronomy (calendars, tables, papers about Earth, Moon and sky). His views were rather advanced for his time. A webpage dedicated to Anania Shirakatsi is available at ArAS website (www.aras.am/FamousAstronomers/shirakatsi.html).

Figure 2. Examples of the Armenian rock art

Other medieval Armenian astronomy-related historians and scientists were Moses Khorenatsi (410-490), Davit Anhaght (VI c.), Hovhannes Yerznkatsi (1230-1293), et al.

Similar to Johann Bayer, Jan Hevelius, Nicolas Louis de La Caille and some others, Armenian scientists have left **medieval sky maps** with the names of constellations and bright stars. Some lists have been survived since Moses Khorenatsi and Anania Shirakatsi, but the most important full sky map was produced by Mkhitar Sebastatsi (1676-1749) in 1749, *"Astghalits Erkinq"* (**Figure 3**). 63 constellations are drawn with their Armenian names coming from earlier centuries (the mythological ones) or translated from European maps (those introduced during 17th-18th centuries).

Several astronomical instruments were found from the Middle Ages. They are described by B. E. Tumanian (1958; 1985). Particularly interesting is the astronomical-geodetical device astrolabe built by Ghukas (Luca) Vanandetsi (XVII-XVIII centuries).

Figure 3. Armenian medieval sky map "Astghalits Erkinq" by Mkhitar Sebastatsi (1749)

3. ARMENIAN CULTURAL ASTRONOMY

Cultural astronomy is sometimes regarded as the cultural interpretation of archaeological evidence relating to astronomy (Iwaniszewski, 2015a). On the other hand, the astronomical knowledge has been reflected in various cultural creations, from ancient calendars (discussed in Section 2) to modern arts and handicrafts (Stavinschi, M. 2010a; 2010b). Cultural astronomy is represented as a set of interdisciplinary fields studying the astronomical systems of current or ancient societies and cultures. In recent years, considerable attention has been paid to this sphere, particularly international organizations were established, conferences are being held and journals are being published. Armenia is rather rich in various evidences of cultural astronomy.

Here we list the main topics related to Astronomy in Culture (most of them also given in Ruggles, 2015). Issues on astronomy in culture have been described in a number of papers, including by the authors of this paper (Farmanyan & Mickaelian, 2015a; 2015b; Harutiunian, 2000):

- astronomy in ancient Armenia and related cultures (heavily related to Armenian archaeoastronomy),
- ethnoastronomy, which is the study of the knowledge, interpretations, and practices of contemporary cultures regarding celestial objects or phenomena and is tightly connected with the national culture,
- astronomy, religion and mythology (Farmanyan, 2015; Farmanyan & Mickaelian, 2014; Vardumyan G., 2015),
- astronomy and astrology (even though astrology is regarded as false science, it has had a great influence in the history of astronomy and is part of astronomy in culture),
- astronomy in folklore and poetry (Harutyunian, 2014; Arakelyan, 2015),
- astronomy in arts. Astronomy in music (Vardumyan A., 2015),
- astronomy in fashion (clothes, carpets, etc.),
- astronomical terms and astrolinguistics (Harutyunian, 2015; Yeghikian, 2015). As mentioned, many constellations and stars, the 5 planets visible by naked eye, and many other astronomical terms in Armenian are known since the I-II millennia B. C.,
- astroheraldry, i. e. astronomical signs on state flags, coats of arms, coins and banknotes, stamps, etc. (Mickaelian & Farmanyan, 2015). Astronomical signs used in other fields may be combined as astrosymbolism. It is strongly reflected in various Armenian writings: petroglyphs, hieroglyphs, syllabic, and modern alphabets (**Figure 4**).

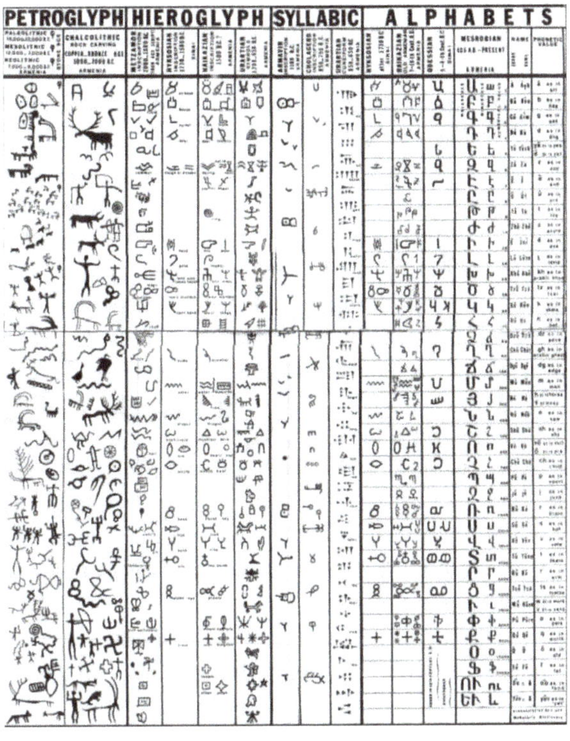

Figure 4. The evolution of Armenian writing from petroglyphs to hieroglyphs, syllabic, and modern alphabets

A number of recent studies in the Armenian cultural astronomy were carried out by Farmanyan & Mickaelian (2014), Farmanyan (2015) and others.

A similar to Archaeoastronomy webpage for Astronomy in Armenian Culture will be created at ArAS website and a permanent section *"Archaeoastronomy and Astronomy in Culture"* has been introduced in ArAS Electronic Newsletter "ArASNews" (www.aras.am/ArasNews/arasnews.html).

4. SUMMARY

We reviewed the Armenian astronomical heritage reflected in archaeoastronomical sites and documents, as well as in ancient, medieval and modern culture.

During the recent years (2007-2014), we have organized a number of meetings, where archaeoastronomy was involved, including those completely devoted to this subject. A number of books have been published. The last meeting, held in 2014 was the first meeting on *Relation of Astronomy to other Sciences and Culture and Society* and for the first time it involved the subject Astronomy in Culture. The list of these **meetings related to AAC** is as follows:

- Joint European and National Astronomy Meeting (JENAM-2007), Special Session #6: *"Archaeoastronomy"*, 21 Aug 2007, Yerevan.
- ArAS VIII Annual Meeting *"Astronomy and Society"*, Session *"Archaeoastronomy"*, 7 July 2009, Byurakan.
- Archaeoastronomy seminar dedicated to Anania Shirakatsi's 1400th anniversary, 13-14 July 2011, Byurakan.
- Archaeoastronomical meeting *"Astronomical Heritage in the National Culture"* dedicated to Anania Shirakatsi's 1400th anniversary combined with ArAS XI Annual Meeting, 25-26 Sep 2012, Byurakan (Harutyunian et al., 2014).
- Meeting *"Relation of Astronomy to other Sciences, Culture and Society"* (RASCS) combined with ArAS XIII Annual Meeting, Sessions"Archaeoastronomy" and "Astronomy in Culture", 7-9 Oct 2014, Yerevan (Harutyunian et al., 2015).

Abstracts of JENAM-2007, the meeting *"Astronomical Heritage in the National Culture"* (2012) and RASCS (2014) are available in Astrophysical Data System (ADS). Proceedings books have been published for the latter two.

A number of institutions in Armenia are involved in AAC. So far, BAO coordinates AAC related actions. UNESCO (UNESCO-IAU Astronomy and World Heritage Initiative (AWHI), e.g. Ruggles, 2013) has officially recognized BAO as the coordinator in Armenia in the field of *Astronomy and World Heritage*, a sub-project of UNESCO World Heritage project. BAO representatives also are involved in IAU Working Groups *Astronomy and World Heritage* (WGAWH) and *Archaeoastronomy and Astronomy in Culture* (WGAAC) and the European Society for Astronomy in Culture (SEAC, Societé Européenne pour l'Astronomie dans la Culture). In Armenia, there are several institutions connected to AAC and involved in related inter- and multi- disciplinary studies:

- NAS RA V. Ambartsumian Byurakan Astrophysical Observatory (BAO) (www.bao.am),
- NAS RA Institute of History (www.academhistory.am/en),
- NAS RA Institute of Archaeology and Ethnography (www.archaeology.sci.am),
- NAS RA M. Abeghyan Institute of Literature (language.sci.am/en),
- NAS RA H. Acharian Institute of Language (litinst.sci.am/en),
- Matenadaran, Mesrop Mashtots Institute of Ancient Manuscripts (www.matenadaran.am).

One can regard **Astrotourism** (which is part of the Scientific Tourism) as part of Astronomy in Culture. Since 2009 (International Year of Astronomy, IYA-2009, declared by UNESCO), we also develop these activities in Armenia that includes observations of starry skies, acquaintance of tourists to native astronomies with the help of archaeo- and ethnoastronomy (Iwaniszewski, 2015b), visits to modern observatories, space museums and planetaria, as well as popular astronomical lectures.

Astronomical journalism is part of the scientific journalism, as most of topics of the latter relate to cosmic subjects: astronomy/astrophysics, space flights, extraterrestrial intelligence, cosmic catastrophes, and even UFOs and astrology (Stavinschi & Mosoia, 2010). In 2010, we have created scientific journalism in Armenia by opening a group of interested journalists and researchers and circulating related regular press-releases to mass-media. Numerous articles, online info materials, radio and TV programs and interviews followed, so that such items increased by a factor of 10-20.

After the foundation of **IAU South West Asian Regional Office of Astronomy for Development** (SWA ROAD) in Armenia, AAC became one of the main directions of its activities. We consider AAC as the most important and appropriate area of astronomy (and perhaps science in general) that can be close to public and play a significant role in the development. SWA ROAD has won a project funded by IAU OAD and a number of events are planned for 2016, including conferences, study of the field, construction of database and webpage, etc. Having SWA ROAD, now the Armenian AAC project also includes regional studies, so that this large area, where very high concentration of world heritage sites and matters are observed, will undergo combined and comparative investigations.

ACKNOWLEDGEMENTS

We thank the organizers of SEAC-2015, particularly *Prof.* Vito Francesco Polcaro for the kind invitation to the meeting and his hospitality. This work was partially supported by the Swiss Agency for Development and Cooperation SDC. SF would especially like to thank the personnel of Matenadaran, the museum of ancient manuscripts in Yerevan, Armenia.

REFERENCES

Arakelyan, K. (2015) The Metaphorical Applications of Heavenly Bodies and Phenomena in Western Armenian Poetry at the Beginning of the XX Century. *Proceedings of the Meeting "Relation of Astronomy to other Sciences, Culture and Society", 7-10 Oct 2014, Byurakan, Armenia.* Ed. H. A. Harutyunian, A. M. Mickaelian & S. V. Farmanyan. NAS RA "Gitutyun" Publ. House, p. 232-237.

Anania Shirakatsi (1962) *Cosmography, trans. from Ancient Armenian, preface and comments by K. S. Ter-Davtyan & S. S. Arevshatyan,* Yerevan.

Arevshatyan, S. S. (2014) Anania Shirakatsi's Natural Philosophical Views. *Astronomical Heritage in the National Culture, Proceedings of the Archaeoastronomical Meeting dedicated to Anania Shirakatsi's 1400th anniversary, held 25-26 Sep 2012 in Byurakan, Armenia. Yerevan.* Ed. H. A. Harutyunian, A. M. Mickaelian & E. S. Parsamian. NAS RA "Gitutyun" Publ. House, p. 68-75.

Badalian, H. S. (1970) *History of Calendar.* Publishing House of the Armenian Academy of Sciences, Yerevan.

Belmonte, J. A. (2015) Voyages of the Zodiac, an Impenitent Traveller across Lands and Ages. *SEAC 2011 Stars and Stones: Voyages in Archaeoastronomy and Cultural Astronomy, Proceedings of a conference held 19-22 September 2011.* Archaeopress / British Archaeological Reports, p. 28-34.

Bochkarev, N. G., Bochkarev, Yu. N. (2005) Armenian Archaeoastronomical Monuments Carahunge (Zorakarer) and Metsamor: Review and Personal Impressions. *Proceedings of SEAC X Annual Conference: Cosmic Catastrophes, held in Tartu, Estonia, 2002. Eds. Mare Koiva, Izold Pustylnik & Liisa Vesik,* Tartu, p. 27-54.

Broutian, G. H. (1997) *The Armenian Calendar.* Echmiatzin Publishing House, Ejmiatzin, 560 p.

Broutian, G. H. (2015) The Ancient Armenian Calendars' connection with the celestial bodies. *Proceedings of the Meeting "Relation of Astronomy to other Sciences, Culture and Society", 7-10 Oct 2014, Byurakan, Armenia.* Ed. H. A. Harutyunian, A. M. Mickaelian & S. V. Farmanyan. NAS RA "Gitutyun" Publ. House, p. 322-333.

Conybeare, F. (1897) Ananias of Shiraq (A.D. 600-650). *Byzantinische Zeitschiff.* Bd. VI.

Danielyan, E. (2014) Anania Shirakatsi's Cosmographical and Natural Philosophical Views. *Astronomical Heritage in the National Culture, Proceedings of the Archaeoastronomical Meeting dedicated to Anania Shirakatsi's 1400th anniversary, held 25-26 Sep 2012 in Byurakan, Armenia. Yerevan.* Ed. H. A. Harutyunian, A. M. Mickaelian & E. S. Parsamian. NAS RA "Gitutyun" Publ. House, p. 76-86.

Eynatyan, J. (2014) Anania Shirakatsi's Overheads System. *Astronomical Heritage in the National Culture, Proceedings of the Archaeoastronomical Meeting dedicated to Anania Shirakatsi's 1400th anniversary, held 25-26 Sep 2012 in Byurakan, Armenia. Yerevan.* Ed. H. A. Harutyunian, A. M. Mickaelian & E. S. Parsamian. NAS RA "Gitutyun" Publ. House, p. 98-106.

Farmanyan, S. V. (2015) Ancient and Medieval Earth in Armenia. *Proceedings of the Meeting "Relation of Astronomy to other Sciences, Culture and Society", 7-10 Oct 2014, Byurakan, Armenia.* Ed. H. A. Harutyunian, A. M. Mickaelian & S. V. Farmanyan. NAS RA "Gitutyun" Publ. House, p. 246-253.

Farmanyan, S. V., Mickaelian, A. M. (2014) Sun and Sun Worship in Different Cultures. *Astronomical Heritage in the National Culture, Proceedings of the Archaeoastronomical Meeting dedicated to Anania Shirakatsi's 1400th anniversary, held 25-26 Sep 2012 in Byurakan, Armenia. Yerevan.* Ed. H. A. Harutyunian, A. M. Mickaelian & E. S. Parsamian. NAS RA "Gitutyun" Publ. House, p. 36-43.

Farmanyan, S. V., Mickaelian, A. M. (2015a) Cultures across the Universe and the Roots of Armenian Astronomy. *Publ. Astron. Obs. Belgrade,* No. 94.

Farmanyan, S. V., Mickaelian A. M. (2015b) Armenian Cultural Astronomy. *Proceedings of the Meeting "Relation of Astronomy to other Sciences, Culture and Society", 7-10 Oct 2014, Byurakan, Armenia.* Ed. H. A. Harutyunian, A. M. Mickaelian & S. V. Farmanyan. NAS RA "Gitutyun" Publ. House, p. 223-231.

Flammarion, C. (1880) Astronomie populaire.

Gurzadyan, V. G., Vardanyan, R. (2004) Halley's Comet of 87 BC on the coins of Armenian king Tigranes? *Astronomy & Geophysics,* Vol. 45, No. 4, p. 4.06.

Harutiunian, S. (2000) *Armenian Mythology.* Hamazgayin Vahe Setyan Publ. House, Beirut, 527 p.

Harutyunian, H. A. (2014) Armenian Vahagn God as birth of four cosmic elements. *Astronomical Heritage in the National Culture, Proceedings of the Archaeoastronomical Meeting dedicated to Anania Shirakatsi's 1400th anniversary, held 25-26 Sep 2012 in Byurakan, Armenia. Yerevan.* Ed. H. A. Harutyunian, A. M. Mickaelian & E. S. Parsamian. NAS RA "Gitutyun" Publ. House, p. 23-28.

Harutyunian, H. A. (2015) Astronomical Terms and Names as an Indicator of the Science Level. *Proceedings of the Meeting "Relation of Astronomy to other Sciences, Culture and Society", 7-10 Oct 2014, Byurakan, Armenia.* Ed. H. A. Harutyunian, A. M. Mickaelian & S. V. Farmanyan. NAS RA "Gitutyun" Publ. House, p. 272-280.

Harutyunian, H. A., Mickaelian, A. M. (2014) Anania Shirakatsi's Cosmographical Works and the Methodology of his Natural Science. *Astronomical Heritage in the National Culture, Proceedings of the Archaeoastronomical Meeting dedicated to Anania Shirakatsi's 1400th anniversary, held 25-26 Sep 2012 in Byurakan, Armenia. Yerevan.* Ed. H. A. Harutyunian, A. M. Mickaelian & E. S. Parsamian. NAS RA "Gitutyun" Publ. House, p. 87-97.

Harutyunian, H. A., Mickaelian, A. M., Parsamian, E. S. (Eds.) (2014) Astronomical Heritage in the National Culture. *Proceedings of the Archaeoastronomical Meeting dedicated to Anania Shirakatsi's 1400th anniversary and XI Annual Meeting of the Armenian Astronomical Society, held 25-26 Sep 2012 in Byurakan, Armenia.* Yerevan, "Gitutyun" Publ. House of NAS RA, 220 p.

Harutyunian, H. A., Mickaelian, A. M., Farmanyan, S. V. (Eds.) (2015) Relation of Astronomy to other Sciences, Culture and Society, *Proceedings of XIII Annual Meeting of the Armenian Astronomical Society, held 7-10 Oct 2014 in Byurakan, Armenia.* Yerevan, "Gitutyun" Publ. House of NAS RA, 460 p.

Herouni P. M. (2006) *Armenians and Ancient Armenia: Archaeoastronomy, linguistics, ancient history.* "Giteliq" Publ. House, Yerevan, 320 p.

Iwaniszewski, S. (2015a) Cultural Interpretation of Archaeological Evidence Relating to Astronomy. *Handbook of Archaeoastronomy and Ethnoastronomy,* ISBN 978-1-4614-6140-1. Springer Science + Business Media, New York, 2015, p. 315.

Iwaniszewski, S. (2015b) Astrotourism and Archaeoastronomy. *Handbook of Archaeoastronomy and Ethnoastronomy,* ISBN 978-1-4614-6140-1. Springer Science + Business Media, New York, p. 287.

Martirosyan, H. (2015) Sisian, Zorats Karer, Tsits Karer, Karahunge. *Proceedings of the Meeting "Relation of Astronomy to other Sciences, Culture and Society", 7-10 Oct 2014, Byurakan, Armenia.* Ed. H. A. Harutyunian, A. M. Mickaelian & S. V. Farmanyan. NAS RA "Gitutyun" Publ. House, p. 299-321.

Mickaelian, A. M. (2008) Preservation and maintenance of the astronomical sites in Armenia. *Proceedings of the International Conference "Starlight: A Common Heritage", held in La Palma, Canary Islands, 19-20 April 2007.* Eds. Cipriano Marín & Jafar Jafari, IAC, Spain, p. 403-411.

Mickaelian, A. M. (2014) Armenian Astronomical Heritage. *Astronomical Heritage in the National Culture, Proceedings of the Archaeoastronomical Meeting dedicated to Anania Shirakatsi's 1400th anniversary, held 25-26 Sep 2012 in Byurakan, Armenia. Yerevan.* Ed. H. A. Harutyunian, A. M. Mickaelian & E. S. Parsamian. NAS RA "Gitutyun" Publ. House, p. 44-55.

Mickaelian, A. M., Farmanyan, S. V. (2015) Astroheraldry. *Proceedings of the Meeting "Relation of Astronomy to other Sciences, Culture and Society", 7-10 Oct 2014, Byurakan, Armenia.* Ed. H. A. Harutyunian, A. M. Mickaelian & S. V. Farmanyan. NAS RA "Gitutyun" Publ. House, p. 287-296.

Mirumyan, K. (2015) Anania Shirakatsi's Cosmology in the Context of the Era's Civilization. *Proceedings of the Meeting "Relation of Astronomy to other Sciences, Culture and Society", 7-10 Oct 2014, Byurakan, Armenia.* Ed. H. A. Harutyunian, A. M. Mickaelian & S. V. Farmanyan. NAS RA "Gitutyun" Publ. House, p. 362-373.

Nazaryan, L. S. (2013) *Anania Shirakatsi.* Anania Shirakatsi College, Yerevan, 242 p.

Nazaryan, L. S. (2014) Anania Shirakatsi's Life and Activities. *Astronomical Heritage in the National Culture, Proceedings of the Archaeoastronomical Meeting dedicated to Anania Shirakatsi's 1400th anniversary, held 25-26 Sep 2012 in Byurakan, Armenia. Yerevan.* Ed. H. A. Harutyunian, A. M. Mickaelian & E. S. Parsamian. NAS RA "Gitutyun" Publ. House, p. 58-67.

Olcott, W. T. (1911) *Star Lore of All Ages.* New York, G. P. Putnam's Sons.

Parsamian, E. S. (1985a) On Astronomical Meaning of the Small Hill of Metsamor. *Communications of BAO,* Vol. 57, p. 92-100.

Parsamian, E. S. (1985b) On Possible Astronomical Significance of Megalithic Rings of Angelacot. *Communications of BAO,* Vol. 57, p. 101-103.

Parsamian, E. S. (1999) On Ancient Astronomy in Armenia. Proceedings of the International Conference Oxford VI and SEAC 1999, Ed. J. A. Belmonte, La Laguna, p. 77-81.

Parsamian, E. S. (2014) Archaeoastronomy in Armenia. *Astronomical Heritage in the National Culture, Proceedings of the Archaeoastronomical Meeting dedicated to Anania Shirakatsi's 1400th anniversary, held 25-26 Sep 2012 in Byurakan, Armenia. Yerevan.* Ed. H. A. Harutyunian, A. M. Mickaelian & E. S. Parsamian. NAS RA "Gitutyun" Publ. House, p. 12-22.

Petri, W. (1964) Anania Shirakatsi - an Armenian Cosmographer of the 7th century. *Journal of the German Eastern Society,* Vol. 114, Munich.

Polcaro, V. F., Valsecchi, G. B., Verderame, L. (2008) The Gaugamela Battle Eclipse: An Archaeoastronomical Analysis. *Mediterranean Archaeology and Archaeometry,* Vol. 8, Issue 2, p. 55-64.

Pskovskij, Yu. P. (1982) Supernova having created the Crab nebula, *Zemlya Vselennaya,* No. 3, p. 36-40.

Ruggles, C. L. N. (2013) UNESCO's Astronomy and World Heritage Initiative: Progress to Date and Future Priorities. *American Astronomical Society, AAS Meeting #221,* id.91.01.

Ruggles, C. L. N. (2015) *Handbook of Archaeoastronomy and Ethnoastronomy.* ISBN 978-1-4614-6140-1. Springer Science + Business Media, New York.

Scuderi, A., Polcaro, V. F., Mercadante, F., Lo Cascio, P., Maurici, F. (2014) Astronomically oriented megaliths of the Monte Jato area (Sicily). *Mediterranean Archaeology and Archaeometry,* Vol. 14, No. 1, p. 155-165.

Semyonov, L. (1953) Anania Shirakatsi as an astronomer. *Ejmiatzin,* No. 7-8.

Stavinshki, M. (2010a) Astronomy in Culture. *Innovation in Astronomy Education,* Cambridge University Press, p. 75-84.

Stavinschi, M. (2010b) Astronomy in Culture. *Proceedings of the 9th International Conference of the Hellenic Astronomical Society,* held 20-24 September 2009 in Athens, Greece. Ed. Kanaris Tsinganos, Despina Hatzidimitriou & Titos Matsakos. San Francisco: Astronomical Society of the Pacific, p. 465.

Stavinschi, M.; Mosoia, C. (2010) Science Outreach: scientists or journalists? *European Planetary Science Congress 2010,* held 20-24 September 2010 in Rome, Italy. p. 143.

Ter-Gulanyan, A. (2014) Ancient Astronomical Hieroglyphs of the Armenian Highland and their Echo in Architectural Structures. *Astronomical Heritage in the National Culture, Proceedings of the Archaeoastronomical Meeting dedicated to Anania Shirakatsi's 1400th anniversary, held 25-26 Sep 2012 in Byurakan, Armenia. Yerevan.* Ed. H. A. Harutyunian, A. M. Mickaelian & E. S. Parsamian. NAS RA "Gitutyun" Publ. House, p. 29-35.

Tokhatyan, K. (2014) Origins Rock Art and Calendar in Armenia and Anania Shirakatsi. *Astronomical Heritage in the National Culture, Proceedings of the Archaeoastronomical Meeting dedicated to Anania Shirakatsi's 1400th anniversary, held 25-26 Sep 2012 in Byurakan, Armenia. Yerevan.* Ed. H. A. Harutyunian, A. M. Mickaelian & E. S. Parsamian. NAS RA "Gitutyun" Publ. House, p. 107-133.

Tokhatyan, K. (2015) Primitive Art and Petroglyphs of Armenia. *Proceedings of the Meeting "Relation of Astronomy to other Sciences, Culture and Society", 7-10 Oct 2014, Byurakan, Armenia.* Ed. H. A. Harutyunian, A. M. Mickaelian & S. V. Farmanyan. NAS RA "Gitutyun" Publ. House, p. 334-344.

Tumanian, B. E. (1958) Armenian newly revealed astronomical device. *Haypethrat,* Yerevan, 31 p.

Tumanian, B. E. (1972) *History of chronology.* Academy of Sciences of Armenia, Yerevan.

Tumanian, B. E. (1985) *History of Armenian astronomy.* YSU Publishing House, Yerevan, 286 p.

Tumanian, B. E., Mnatsakanian, H. H. (1965) Bronze Age belt-calendar. *Mitq.* Yerevan, 56 p.

Vardumyan, A. (2015) The Relations between Astronomy and Music in Medieval Armenia. *Proceedings of the Meeting "Relation of Astronomy to other Sciences, Culture and Society", 7-10 Oct 2014, Byurakan, Armenia.* Ed. H. A. Harutyunian, A. M. Mickaelian & S. V. Farmanyan. NAS RA "Gitutyun" Publ. House, p. 254-262.

Vardumyan, G. (2015) The Universe and Armenian Mythology. *Proceedings of the Meeting "Relation of Astronomy to other Sciences, Culture and Society", 7-10 Oct 2014, Byurakan, Armenia.* Ed. H. A. Harutyunian, A. M. Mickaelian & S. V. Farmanyan. NAS RA "Gitutyun" Publ. House, p. 345-355.

Yeghikian, A. G. (2015) Astronomical Terms in Armenian. *Proceedings of the Meeting "Relation of Astronomy to other Sciences, Culture and Society", 7-10 Oct 2014, Byurakan, Armenia.* Ed. H. A. Harutyunian, A. M. Mickaelian & S. V. Farmanyan. NAS RA "Gitutyun" Publ. House, p. 281-286.

23

ASTRONOMY AND SOLAR SYMBOLOGY IN THE BASILICA OF SAN MINIATO AL MONTE AND IN THE BAPTISTERY OF SAN GIOVANNI IN FLORENCE

Simone Bartolini[1], Marco Pierozzi[2]

[1]*Italian Military Geographic Institute (Florence)*
[2]*Italian Navy Hydrographic Institute (Genova)*

Corresponding author: Simone Bartolini (simo.bartolini@tiscali.it)

ABSTRACT

The Basilica of San Miniato al Monte is one of the oldest churches in Florence: its construction began in 1018 by the Bishop Hildebrand who wanted to celebrate St. Minias, the patron Saint of the city. The Basilica is one of the finest examples of Romanesque in Tuscany and presents intriguing astronomical and symbolic allusions that, however, largely remain undiscovered. This article presents the study on the relationship between architecture, religious symbols and astronomical elements of the church of San Miniato al Monte.
In order to identify possible correlations between the position of the Sun and the decorative components of the Basilica that are illuminated by sunlight, we adopted the GNSS methodology and a Total Station. We found that the figures of Christ and the Virgin Mary on the apse mosaic are illuminated by the Sun during the Easter period, and the crypt altar, containing the relics of Saint Minias, is brightened on the day of his martyrdom.
The preeminent symbolic/astronomical element is certainly the marble zodiac, similar to that of the Baptistery, which Capricorn-Cancer axis coincides with the axis of the Basilica. At the summer solstice a fascinating light phenomenon occurs: the sunlight perfectly illuminates the Cancer sign, thereby confirming the astronomical function of the marble zodiac. Furthermore the sunbeams coming from the windows of the right wall of the nave design a path of light on the ideal axis of the Basilica.
The re-discovered astronomical role of this zodiac led us to hypothesize that the marble zodiac of the Baptistery of Florence was originally placed adjacent to the basement of the columns in front of the northern door of the Baptistery. In this way, on summer solstice day the light beam coming from the central opening of the dome, would have illuminated the Cancer sign, similarly to what happens for the zodiac of San Miniato.

KEYWORDS: San Miniato al Monte, Baptistery of Florence, zodiac, Cancer sign, Capricorn sign, summer solstice, winter solstice, gates of heaven.

1. INTRODUCTION

The history of the Basilica of San Miniato al Monte in Florence dates back to the time of the martyrdom of Minias (A.D. 250). The hagiography tells us that Minias, an Armenian prince, was tortured and beheaded in the presence of the Roman Emperor Decius: Minias then picked up his head and walked on the Mons Florentinus, the hill where the present church was built, where he was buried.

The figure of Saint Minias is very important to the Florentine people: in fact, he his venerated as the first Christian martyr of Florence and for some centuries he was considered the only patron saint of the city. Later, both Saint Minias and Saint John were deemed to be the patron saints of the Church of Florence, as mentioned in a certificate of King Berengar of April 25, 899 (Manetti, 2009). It is then not a coincidence that the two most important Romanesque churches of Florence are dedicated to Saint Minias (San Miniato al Monte) and to Saint John (the Baptistery). Documents of donations from various emperors, between VIII and X century, including Charlemagne, confirms the existence of a place of worship dedicated to Saint Minias, perhaps a Paleochristian Basilica, in the site of the resting place of the martyr (Loris Pegna, 1974). However, as reported by Moreni, in the following decades *"o per l'incursione dei barbari, o per l'antichità, o per altro accidente era ridotta in pessimo stato"* [1] (Moreni, 1794): Bishop Hildebrand then proposed to build a large Basilica dedicated to St. Minias. After Pope Benedict VIII and Emperor Henry II had approved the project, on April 27, 1018, Hildebrand issued a document, the *charta ordinationis* (Pratesi, 1995), which declared the foundation of the Basilica of San Miniato and the subsequent beginning of the construction works, which took almost two centuries to complete. Francesco Gurrieri (Gurrieri, 1988) identified the four construction phases as follows:

- 1018-1068: the wall of the crypt in the apse area;
- 1070-1093: the walls of the side aisles and choir, the first order of the façade;
- 1128-1150: the Colonnade, the wall of the nave, the second order of the façade;
- 1175-1207: the pediment and the floor of the nave.

As reported by the date engraved on the inlaid marble floor near the main entrance of the Basilica, construction works were finalized in 1207, although in later centuries the church underwent renovation. Because of its harmonic architectural proportions, the Basilica of San Miniato al Monte is considered one of the finest examples of Florentine Romanesque. This church is also characterized by an intriguing religious and solar symbolism and complex astronomical allusions that are related to architectural and decorative details, which make this monument unique in its genre. Although the solar/astronomical symbolism of San Miniato al Monte has fascinated numerous scholars in recent decades (Gettings, 1987; Incerti, 2011; Shrimplin, 2011), many aspects still remain undisclosed. In this article we present the study on the relationship between architecture, religious symbols and astronomical features of the church of San Miniato al Monte.

2. SOLAR SYMBOLISM OF THE BASILICA OF SAN MINIATO

The relationship between architecture and astronomy is almost as old as mankind itself. In his treatise "De Architectura" (15 B.C. circa) Vitruvius highlighted the importance of astronomy in art and architecture. During the medieval period, architects widely employed astronomical orientations and sunlight in the construction of religious buildings in order to link the humanity to the celestial realm (Shrimplin, 2011). The illumination of architectural and decorative details by the sunlight evidenced the sacred space circumscribed by the church, the house of God, which is the point of contact between Earth and Heaven (Hani, 1996).

Because little is known about the actual astronomical features of the Basilica of San Miniato al Monte, we investigated the possible correlation between architecture and religious symbols of San Miniato with the path of the Sun.

2.1. *Astronomical survey*

In order to perform a rigorous investigation to verify solar and astronomical correlation within the Basilica of San Miniato, we carried out precise measurements with Total Station Theodolite on the square in front of the façade. In particular, using static GNSS methodology[2], we determined the coordinates of a point in front of the entrance of the Basilica. We then measured a traverse within the Basilica itself in order to detect the coordinates and heights of architectural and decorative elements of symbolic and interest, such as:

- the single-lancet windows on the right wall of the main nave and the apse windows;
- details of Christ, the Virgin Mary and Saint Minias depicted in the apse mosaic;

[1] "Perhaps because of barbarian incursions, or aging, or any other accident [the church] was almost destroyed".

[2] GNSS is the global navigation satellite system based on the US system GPS, the Russian system GLONASS and the European system GALILEO.

- the Cancer and Capricorn signs on the inlaid marble zodiac;
- the edges of the base plate of the columns of the main nave, from which we determined the azimuth of the Basilica's axis, which is 131.5°.

Furthermore, we determined the position of the altar of the crypt with respect to the central single-lancet-window[3] of the crypt to shed light on possible solar-symbolic connections.

2.2. *The altar of the crypt*

During the construction of the crypt, Bishop Hildebrand decided to place the relics of Saint Minias at the centre of the altar, which became the founding and distinctive element of the entire Basilica for its symbolic importance. It is therefore crucial to verify a possible connection between the position of the altar with the sunlight, like it has been observed in other churches (Spinazzè, 2015).

Using a laser distance-meter, we measured the inclination of the sunbeams that would filter through the central window of the crypt[4] with respect to the altar: we observed that the central part of the altar would be illuminated when the Sun is 17° to 24° high over the horizon (see Figure 1) and it is on the axis of the Basilica. From such data we computed the declination values of the Sun and the days of illumination of the altar (Table I).

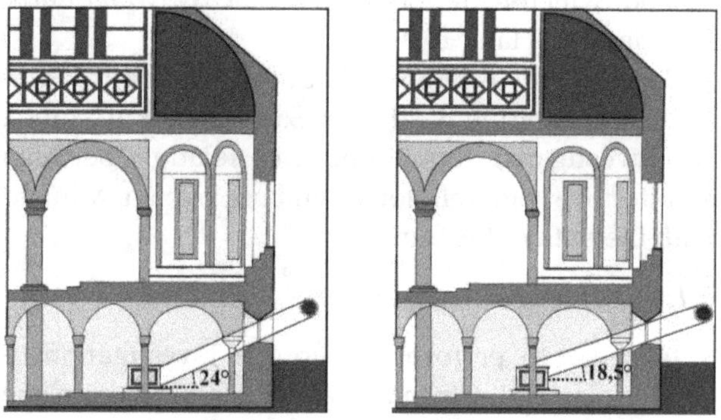

Figure 1. The illumination of the crypt altar containing the relics of Saint Minias.

Table I. The illumination of the crypt altar.

azimuth	height	declination	enlightening days
131.5°	17° to 24°	-9° 02' to -14° 52'	Oct 17 to Nov 3

The period of the illumination of the relics of Saint Minias, at the centre of the crypt altar, comprises the day of martyrdom of Minias, October 25. It should be pointed out, however, that in the XI century, when Hildebrand ordered the construction of the Basilica, the Julian calendar was in use. That this calendar was slightly inaccurate, as it considered the year to be about 12 minutes longer: therefore, the day of October 25 in the first half of the XI century corresponded in actuality to 30-31 October. On these dates, at about 8.50 am, the Sun is about 18°-18.5° high on the Basilica's axis and the sunbeams would brighten the centre of the altar. This event would confirm the hypothesis that the Basilica was conceived with astronomical implications, as the sunlight celebrates Saint Minias illuminating its relics on the day of his martyrdom.

2.3. *The apse and the orientation of the apse windows*

The apse, in Christian architecture, represents the spiritual fulcrum of a church the very place where man meets God. The apse of San Miniato presents many religious symbols and elements of solar and astronomic significance: in fact, the hemispherical semi-dome symbolizes the vault of heaven, and the mosaic, completed in the XIII century, depicts Christ the Pantocrator together with the Virgin Mary and Saint Minias. In addition to the symbolic significance, the apse also presents interesting astronomical implications: in fact, the first window on the left is directed towards the sunrise at the summer solstice (azimuth 56.5°), while the second is oriented towards the sunrise on the day of the equinox (azimuth 90°): the altar and the apse would be flooded[5] with light in these two particular astronomical moments.

2.4. *Solar and lunar illumination of the apse mosaic*

The apse mosaic presents other intriguing connections with the position of the Sun. In particular, the sunbeams filtering through the two windows to the right of the presbytery of the main nave (labelled I and II in Figure 2 a and b) illuminate the figures of Christ and the Virgin Mary in the late afternoon of the months of March and September, especially near the equinoxes (Table II, III and IV).

Because of the inaccuracy of the Julian calendar, in the XIII century the period of illumination of the figures of Christ and the Virgin Mary was delayed of about 6-7 days with respect to the actual Sun position. The illumination of the apse mosaic is particularly evident during the Easter period, when the two sunbeams precisely illuminate the figures of Christ and the Virgin Mary. The figure of Christ the Pantocrator flooded with light reminds us of the resurrection and the victory of light over darkness.

[3] From now on we will refer to "single-lancet-window" as "window".

[4] It should be remembered that while the two lateral windows are closed with glass windows, the central window of the crypt is closed with a slab of alabaster.

[5] The sunlight does not illuminate the altar because the apse windows are closed with alabaster slabs.

Finally, it is worthy of note the wonderful light effect already observed by Gettings (Gettings, 1987). During the period 11-25 April and from 19 August to 2 September (Table V) the sunbeams coming from the first window to the right of the presbytery shines first under the right foot of Christ (see *Figure 2c*) and then moves toward the left foot, where the light disappear. This fascinating light effect may symbolize the greatness of Christ the Pantocrator as the sovereign of all things, including the Sun.

Carrying out additional calculations concerning the position of the Moon in the sky during the year, we observed that, when the Moon is in the phase between the first quarter and the full moon, the moonlight would illuminate the feet of the Virgin Mary through the first window to the right of the presbytery (indicated by I in *Figure 2a*). This would happen when the Moon is 12° to 15° high over the horizon and with azimuth 249°-253°. Such event reminds us of the verse 12.1 in the Book of Revelations: "and there appeared a great wonder in heaven: a woman clothed with the Sun, with the Moon under her feet and on her head a crown with twelve stars". Indeed, for about four months of the year (Table III and IV) the Virgin Mary depicted in the mosaic, who has a crown of stars, represented by the nacres set around her head, is illuminated by the sunlight just before the sunset (see *Figure 2d*). In addition, she would have "the Moon under her feet", which is the lunar light one day a month.

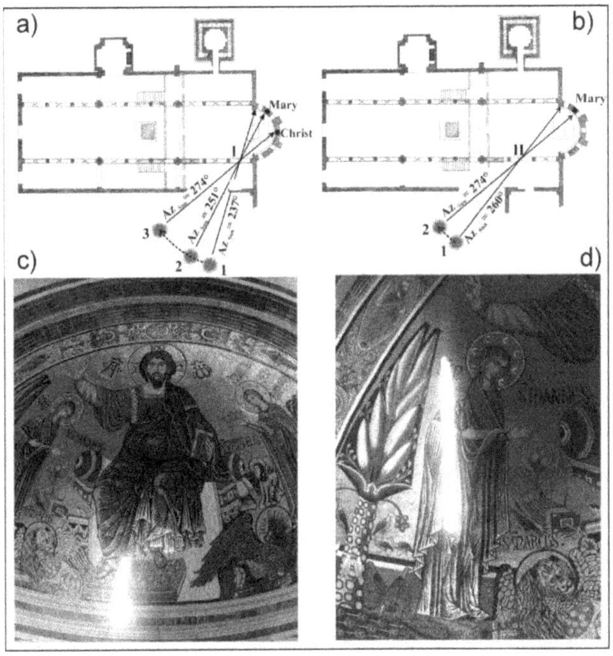

Figure 2. a), b) The sunlight filtering through the windows I and II on the right wall illuminate the apse (see text). c) The sunlight filtering through the window I shines under the foot of Christ the Pantocrator. d) For about four months a year the Virgin Mary is illuminated just before the sunset.

Although this would allow the hypothesis of a cosmic symbology associated to the Virgin Mary, the illumination of Mary's feet by the moonbeams seems more a coincidence that a planned effect, since the day of illumination varies from month to month.

Table II. Illumination of Christ by window I is from Mar. 26 to Apr. 5 - from Sep. 8 to 18

Days of illumination	Solar hour	Azimuth	height	decl.
March 26 and September 18	from 17.12 to 17.30	from 263.1° to 266.2°	from 10° to 6.8°	2°
April 5 and September 8	from 16.55 to 17.12	from 263° to 266°	from 15.9° to 12.9°	6°

Table III. Illumination of Mary by window I is from Feb. 21 to Mar. 15 – from Sep. 30 to Oct. 23

Days of illumination	Solar hour	Azimuth	height	decl.
February 21 and October 23	from 16.46 to 17.08	from 249° to 252.9°	from 5.3° to 1.9°	-11°
March 15 and September 30	from 16.22 to 16.38	from 250.8° to 253.8°	from 15.5° to 13°	-2.5°

Table IV. Illumination of Mary by window II is from Mar. 21 to Apr. 16 - from Aug. 28 to Sep. 23

Days of illumination	Solar hour	Azimuth	Height	decl.
March 21 and September 23	from 17.45 to 17.52	from 267.5° to 269°	from 3° to 1.5°	0°
April 16 and August 28	from 17.12 to 17.42	from 269° to 274.1°	from 15.6° to 10.2°	10°

Table V. The Sun under the foot of Christ by window I from Apr. 11 to 25 - from Aug. 19 to Sep. 2

Days of illumination	Solar hour	Azimuth	Height	decl.
April 11 and September 2	from 17.02 to 17.17	from 265.8° to 268.5°	from 16° to 13.2°	8°
April 25 and August 19	from 16.40 to 16.55	from 266° to 268.6°	from 23.4° to 20.5°	13°

2.5. *The marble zodiac of San Miniato al Monte*

To our knowledge, no historical document somehow reports of a possible astronomical meaning of the marble zodiac of San Miniato al Monte (made in 1207), while it is known that the zodiac of the Baptistery was built with astronomical implications. Even Leonardo Ximenes, in his treatise on astronomical culture in Tuscany (Ximenes, 1757), did not mention the zodiac of San Miniato, while he examined in depth the astronomical significance of zodiac of the Baptistery. Therefore, the zodiac of San Miniato appears to be a mere, though marvellous, decorative element full of cosmic symbology. Only recently, Manetti has hypothesized an astronomical function of this zodiac (Manetti, 2009).

The two zodiacs look remarkably alike: indeed, they might have been designed and manufactured in the same workshop. Both zodiacs are enclosed in a circle which is inscribed in a square and the centre of both is decorated with an inlaid sun. In medieval symbology the circle represented the cosmos and the celestial sphere (the "vault of heaven") whilst the square represented the earth with the implied reference to the four elements (earth, air, water and fire) and the four cardinal points (Incerti, 2011).

The Basilica of San Miniato is oriented to the south-east. We observed that the signs of the two solstices on the marble zodiac lie on the ideal axis of the Basilica: entering the church we can see the Capricorn sign first and then Cancer sign. Therefore, the Gemini sign, and not the Taurus (Gettings, 1987), is oriented to the East (see Figure 3b). This observation led us to hypothesize that the zodiac might have an astronomical significance linked to the solstices.

Although Manetti (Manetti, 2009) suggests that the Capricorn sign would be illuminated early in the morning of the winter solstice, we determined that the sunlight that would filter through the central window of the apse (now closed with a slab of alabaster), would actually light up the bottom of the central door of the Church at 8.25 hours (solar time) of the winter solstice, when the Sun is about 7° high and has an azimuth of 131.5° (see Figure 3a). Therefore, anyone who entered the church prior to the XV Century[6] early in the morning on winter solstice day would have been bathed in sunlight, and walking towards the presbytery would have found the inlaid marble Capricorn sign, which is associated of Christ's birth that in the first centuries of Christianity coincided with the winter solstice.

[6] In the first half of the XV century Michelozzo built the Chapel of the Crucifix, so the sunlight could no longer reach the main door.

We then wondered if there was a more consistent link with the summer solstice. Because this astronomical event is linked with the Cancer sign, we speculated that the inlaid Cancer of the zodiac might be illuminated on the summer solstice day.

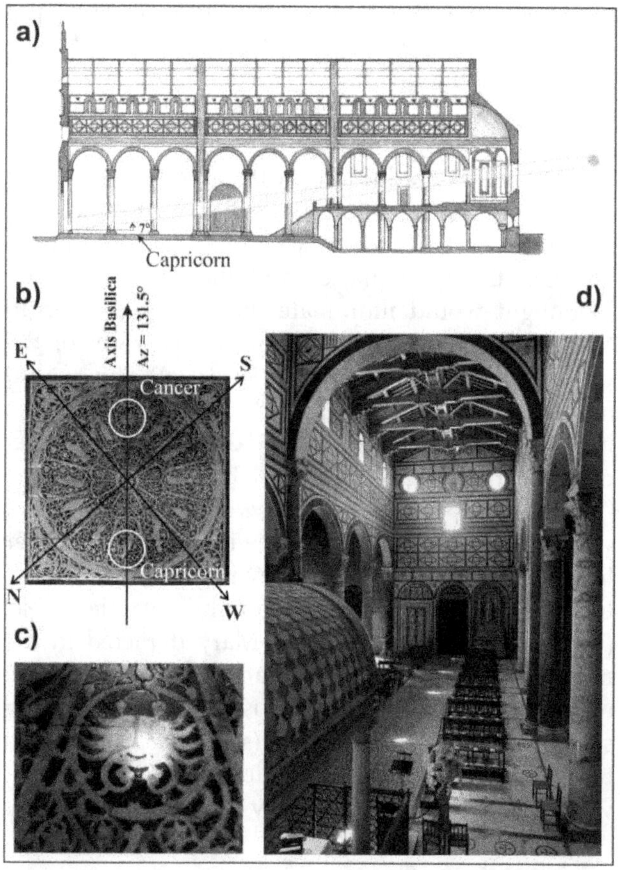

Figure 3. a) The sunbeams filtering through the central window of the apse central window would illuminate the bottom of the central door on winter solstice day. b) The orientation of the marble zodiac. c) The Cancer sign is illuminated by the sunlight at 12.36 (solar time) on summer solstice day. d) The sunlight shines from the windows of the right wall of the main nave (entering the church) to create a path of light.

The rigorous survey and the precise calculation we performed led us to identify the third window of the right wall of the main nave (as we enter the church) as the possible solstice Gnomon from which the sunbeams can illuminate the sign of Cancer. Indeed, on June 21st near to solar noon, at 12:36 of solar time (13:53 civil time), the first sunbeam that filters in the Church through the window mentioned above, illuminates the Cancer sign (Bartolini, 2013). Within 2-3 minutes the light spot, initially only a few centimetres wide, expands progressively and become as big as the inlaid crab shape (see Figure 3c). The precision with which the sunbeam illuminates the Cancer sign does not leave any doubt: the zodiac has been carefully positioned to indicate the summer solstice. This phenomenon is observable for a very short period of time, about four days before and after the summer solstice. Finally, it may be argued that the marble zodiac also assumes a religious function: in fact the illumination of the Cancer sign at summer solstice announces the feast of St. John, the patron Saint of Florence, on June 24. With the rediscovery of the astronomical function of the zodiac of San Miniato, we can add this solstice sundial (the oldest sundial still working!) to the list of the other eight ancient monumental sundials built in Florence.

In order to complete the astronomical study of marble zodiac, we computed the apparent position that the Sun had in early XIII century, when the pavement was completed. In fact, because of the precession of Earth's rotation axis, back in the early centuries of the second millennium AD the Sun appeared higher in the sky: in particular, at noon on the summer solstice the Sun was about 6' of arc higher. Based on the height of the window, we computed that the disc of light illuminating the Cancer sign would be shifted about 2 cm towards the window, an acceptable accuracy for the purpose that could have the marble zodiac.

The illumination of the Cancer sign is a wonderful event that is part of a fascinating light phenomenon occurring at the summer solstice. In fact, at about 12.30 (solar time) the first spot of light appears just under the Chapel of Crucifix; three minutes later another spot of light shines at the centre of the nave floor about six meters away from the preceding one and subsequently a third spot illuminates the Cancer sign. Then, two other spots of light shine every three minutes from one another on the ideal axis of the Basilica, giving the impression that a path of light is taking us towards the main door (see Figure 3d).

3. THE MARBLE ZODIAC OF THE BAPTISTERY OF SAN GIOVANNI

The marble zodiac of San Miniato and the zodiac of the Baptistery of San Giovanni are very similar. Although the one of the Baptistery is more elegant and refined, the two zodiacs might have very well been created in the same workshop. The zodiac of the Baptistery is positioned near the Gates of Paradise, the eastern doors in front of the Duomo, and has a precise cardinal orientation: the sign of Aries is oriented to the East, the Libra to the West, the Capricorn to the North and the Cancer to the South.

The astronomical function of the zodiac of the Baptistery is confirmed by Giovanni Villani, a Florentine diplomat and chronicler of the XIV century who, in a passage of his "Nuova Cronica" ("New Chronicles", one of the first histories of Florence) writes: "And we find, from ancient records, that the figure of the sun carved in mosaic, which says: "En giro torte sol ciclos, et rotor igne," was done by astronomy, and when the sun enters into the sign of Cancer, it shines at mid-day on that place through the opening above, where is the turret"[7].

This clearly suggests that the "sun carved in mosaic" is the centre of the astronomical function of the marble zodiac, because at the summer solstice ("when the sun enters into the sign of Cancer"), the inlaid marble sun "shines at mid-day". In fact, it should be point out that the dome of the Baptistery was originally open, similarly to the Pantheon in Rome, and the sunlight could filter from there.

However, according to Villani, the opening on the dome was closed with a lantern (called "turret" by Rose E. Selfe) built in 1150: therefore, on June 21st ("when the sun enters into the sign of Cancer") of the XIV century the sunlight could no longer filter through the opening to illuminate the inlaid sun on the marble zodiac (Degl'Innocenti, 1994). The zodiac is placed on the floor in proximity of the eastern doors, but according to our calculations, even if the lantern was removed, the sunlight would not illuminate the zodiac because this is not on the meridian of the Baptistery. In another edition of Villani's Cronica of 1729, the phrase reported above is slightly different: "... and when the sun enters into the sign of Cancer, and not in other periods of the year, shines at mid-day the *spera* on that place through the opening above, where is the lantern".

This version provides us additional information, but at the same time adds questions. Certainly, it is more accurate than the previous one because it tells us that the sunlight filtering through the opening on the dome illuminated the floor only in the period of the summer solstice, but it is not clear about the meaning of the word *spera*. This term may refer to an opening, a sphere or a mirror: depending on the meaning we assign to *spera* we have three different interpretations that lead to us very different assumptions.

If we consider *spera* as referred to an opening, according to our calculations when the sun is at noon on the summer solstice the light filtering from the dome would illuminate the north gate of the Baptistery (*Figure 4a*), but not the zodiac.

If the word *spera* refers to a sphere (Degl'Innocenti, 1994), we might assume the presence of a sphere at the centre of the opening on the dome. Accordingly, when the sunlight filtering through the opening casted its shadow on the inlaid sun symbol in the centre of the zodiac it indicated that was the summer solstice day.

7 Villani's Chronicle being selections from the first nine books of the Croniche Fiorentine of Giovanni Villani translated by Rose E. Selfe and edited by Philip H. Wicksteed M.A. London Archibald Constable & CO. LTD. 1906 second edition carefully revised.

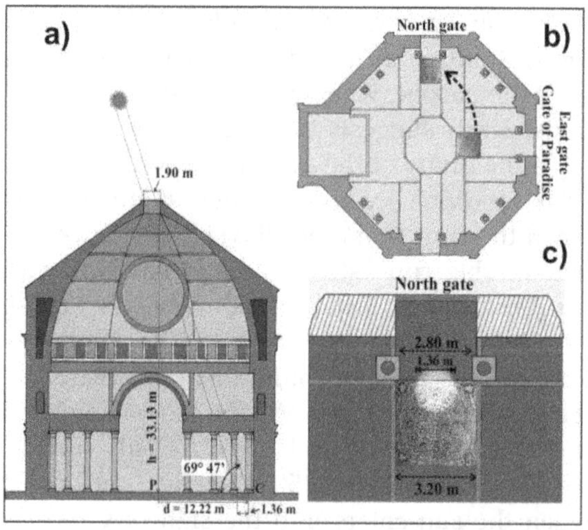

Figure 4. a) Section of the Baptistery of San Giovanni with the solstitial light filtering through the opening. b) Plan of the Baptistery with the hypothesized original position of the marble zodiac. c) In this simulation, the Cancer sign on the marble zodiac is illuminated at noon on the summer solstice in the first half of the XII century.

Filippo Camerota speculates that *spera* could actually stand for mirror. Therefore, he hypothesizes, if a mirror was positioned at the centre of the opening on the dome, the sunlight could have been reflected on the marble carpet despite the occlusion caused by the lantern. In this way, the sunlight would have illuminated the inlaid Sun symbol on the summer solstice day.

It is worthy of note that the floor of the Baptistery underwent renovation in 1351 (Strozzi, 1892) and Camerota himself reports that some works were made in 1200 (Camerota, 2007). At that time, as reported by Villani, the lantern on the dome already existed, and that perhaps the sunlight could no longer filter through the opening. This may suggest that the marble zodiac had been moved from its original position. Based on this information and consistently with our observation on the zodiac of the Basilica of San Miniato al Monte, we propose a new hypothesis about the astronomical role of the marble zodiac of the Baptistery. Although the target of the solstitial sunlight has always been considered the inlaid Sun symbol, Villani, might actually have referred to the Cancer sign. Therefore, considering the Cancer sign and not the Sun symbol on the zodiac as the target of the solstitial light, we can speculate about the original position of the marble zodiac. If in fact the Cancer sign was conceived to be illuminated on the summer solstice day - or to be obscured by a sphere - then the squared zodiac might have been positioned with one side along the base of the columns, rotated of 180° and then with the Cancer sign oriented to the Northern Gate (*Figure 4b*).

In order to verify this hypothesis, we measured the height of the opening on the dome and the horizontal distance from the supposed position of the marble zodiac. The results that we obtained confirm the possibility that the zodiac was "done for astronomy" and, in particular, that it was positioned with the Cancer sign to the north in order to indicate the summer solstice and the upcoming feast of Saint John, the patron Saint of Florence. In *Figure 4. a)* we simulate the position of the light disk deriving from the sunbeams filtering through the opening on the dome at noon on the summer solstice of the year 1100: the sunlight illuminates the cancer sign.

Our hypothesis would be supported by the rediscovered astronomical function of the marble zodiac of the Basilica of San Miniato al Monte, which is clearly positioned with a solstitial orientation, as demonstrated by the illumination of the Cancer sign on the summer solstice day.

4. CONCLUSION

In this article we illustrate that several architectural and decorative elements of the Basilica of San Miniato al Monte are linked with the path of the sun and specific astronomical events.

The unusual orientation of the church, to the south-east instead of east, like many Romanesque churches, might have three reasons associated with three distinct astronomical moments:

- to celebrate Saint Minias, as the sun illuminates its relics on the day of his martyrdom, October 25;
- to enlight the devotees: on the winter solstice day the sunlight would have illuminated the bottom of the central door and whoever entered the church;
- it has also been proposed a connection with the northern lunistice (Incerti, 2011).

Perhaps, the most plausible hypothesis is the first one, because the crypt where the relics of Saint Minias are preserved was the first portion of the church to be built. Indeed, we computed that the sunbeams filtering through the central window of the crypt (now closed with a slab of alabaster) would illuminate the relics of Saint Minias, at the centre of the altar, in a period of the year that comprises the day of martyrdom of Minias, October 25. In addition to that, it is known that the orientation of other Romanesque churches is related with the death day of the Saint to whom the church is dedicated.

The second hypothesis may be acceptable, especially if we consider that there are two paths of light on the axis of the Basilica towards the main entrance: one for the winter solstice and the other for the summer solstice.

Because of the anomalous orientation of the axis of the Basilica, Incerti hypothesizes a connection with the northern lunistice, because the Moon azimuth is 311° (Incerti, 2011). However, this eventuality appears to be less probable, because a lunar orientation of the church would subvert the actual criteria applied in the Middle Ages.

Many symbolic and decorative elements of the apse are connected with the solar illumination. In fact, the sunbeams filtering through the two windows to the right of the presbytery illuminate the figures of Christ and the Virgin Mary depicted in the mosaic in the late afternoon near the equinoxes. Furthermore, during the periods of mid-April and late August the sunlight shines first under the right foot of Christ and then moves toward the left foot. This fascinating light effect may symbolize the greatness of Christ the Pantocrator as the king of all things.

The preeminent symbolic/astronomical element is the marble zodiac. Although the zodiac of San Miniato al Monte was thought to be a simple decorative element, we attested its actual astronomical function. In fact, we observed that on summer solstice day the sunbeams illuminate the inlaid Cancer sign which is the only, among the twelve signs on the marble, to be inserted in a circle, like it was a target for the sunlight. The illumination of the Cancer sign is part of a light phenomenon that occurs at the summer solstice: in fact the sunbeams filtering through the windows on the right wall of the main nave create a path of light on the ideal axis of the Basilica, which takes us towards the main door. These events confirms that the position of the windows have been carefully calculated with respect to the centre of the main nave in relation to the summer solstice. Philosophers like Plato, Porphyry and Macrobius linked these two astronomical events to the descent of souls from the stars of the celestial sphere to Earth and the ascent of souls of the virtuous to the heaven.

Our findings confirm that the Basilica of San Miniato al Monte was projected and built with astronomical significances, with a particular focus on the solstices. Furthermore, the rediscovery of the astronomical function of the zodiac of San Miniato allows us to hypothesize that the zodiac of the Baptistery, was conceived to indicate the summer solstice by way of the illumination of the Cancer sign by the sunlight. Accordingly, we speculate that the zodiac has been moved from its original position, perhaps in 1351, because, in line with our calculations, today the zodiac of the Baptistery would not be illuminated by the solstitial sunlight, even if the lantern on the dome was removed.

ACKNOWLEDGEMENTS

We would like to thank Dr. Federico di Gesualdo for his thoughtful comments on our study and for the essential revision of this manuscript.

REFERENCES

Bartolini, S. (2013) *Sun and Symbols- The zodiacs in the Basilica of San Miniato al Monte and in the Baptistry of San Giovanni in Florence*, Firenze, Polistampa.

Camerota, F. a cura di (2007) *La linea del Sole – Le grandi meridiane fiorentine*, Firenze, IMSS – Edizioni della Meridiana.

Camerota, F. (2011) *Lo specchio solare del Battistero*, in Barbolini, S. and Garofalo G. *Le meridiane storiche fiorentine*, Firenze, Polistampa.

Degl'Innocenti, P. (1994) *Le origini del bel S. Giovanni: da tempio di Marte a Battistero di Firenze*, Firenze, CUSL.

Gettings, F. (1987) *The secret zodiac: the hidden art in medieval astrology*, London, Routledge & Kegan Paul.

Gurrieri, F., Berti, L. and Leonardi, C. (1988) *La Basilica di San Miniato a Firenze*, Firenze, Cassa di Risparmio di Firenze.

Hani, J. (1996) *Il simbolismo del tempio cristiano*, Roma, Edizioni Arkeios.

Incerti, M. (2011) *Lo sguardo di San Miniato al Monte in Firenze,* Atti del XI Convegno-Società Italiana di Archeoastronomia, Bologna 28-30 October 2011.

Lopes Pegna, M. (1974) *Firenze dalle origini al Medioevo*, Firenze, Del Re.

Manetti, R. (2009) *La lingua degli angeli*, Firenze, Polistampa.

Moreni, D. (1794) *Notizie istoriche nei confronti di Firenze* – Lettera III, Firenze.

Pratesi, F. (1995) *La splendida Basilica di San Miniato a Firenze*, Firenze, Octavo.

Shrimplin, V. (2011) *Church of San Miniato al Monte: Astronomical and Astrological Connections,* in The Inspiration of Astronomical Phenomena VI. Astronomical Society of the Pacific Conference Series vol. 441.

Spinazzè, E. (2015) PhD thesis, Dottorato in Storia delle Arti, Ciclo XXVII *La luce nell'architettura sacra del X-XII secolo dalla Romandie alla Toscana*, Università Ca' Foscari Venezia.

Strozzi, L. (1892) *Vita di uomini illustri di casa Strozzi*, Firenze, Salvatore Landi.

Villani, F. (1979) *Croniche di Giovanni, Matteo e Filippo Villani*, Torino, Aquilecchia.

Ximenes, L. (1757) *Del vecchio e nuovo gnomone fiorentino…*, Firenze, Stamperia Imperiale.

24

AN HYPOTHESIS OF AN ASTRONOMICAL SYSTEM OF LIGURES APUAN IN GARFAGNANA

Mauro Peppino Zedda

Isili, Sardinia

Corresponding author: Mauro Peppino Zedda (maurozedda@tiscali.it)

ABSTRACT

In the course of my research on the orientation of Romanesque churches in Tuscany, I came across a very interesting case. The entrance and the axis of two churches in Barga (St. Cristoforo's dome and St. Frediano's church in Sommocolonia - one of Barga's boroughs) look at the same point on the horizon: the Monte Forato mountain, one of the peaks of the Apuan Alps which is characterized by an impressive natural rock arch. Strikingly, Monte Forato's profile reproduces the so called "face" of the "Omo Disteso" (literally "Lying man") , about which there are various legends.

An astronomical sense has been indeed found about the geographic axis of the two churche, which are turned towards the same point. Observing the sunset at winter solstice, from St. Frediano's church in Sommocolonia, we can see that the sunset point coincides with Monte Forato. While from St. Cristoforo's dome, the moonset point coincides with Monte Forato at the minor southern standstill. We may further observe that the moonset point coincides with the Monte Forato, at the major southern standstill, looking from the pre-Romanesque church (of Longobard age) St. Michele a Perpoli (Gallicano).

Is it a fortuitous case? In this paper I will try to demonstrate that this system, of astronomic and geographic coincidence, originates from a form of religious syncretism which envisaged the construction of the churches in the ancient holy sites of the Ligures Apuan. The alignment system is perfectly identical to the one found in the British Islands (Ruggles, 1999) and in Sardinia (Zedda 1992; 2009; 2013).

We are in front of an ancient custom of building alignment clearly widespread among people in prehistoric ages. Once the prehistoric peaple found a natural reference point, well-characterized on the horizon, they chose and they marked the observation point (with a menhir, a stone circle, a dolmen or something else) from where they could observe the Sun and the Moon rising and setting at a chosen view point in their standstill points.

KEYWORDS: Tuscany, Ligures Apuan, Barga, Gallicano, Garfagnana, archaeostronomy.

1. INTRODUCTION

During my second campaign of measurement of the Romanesque churches in Tuscany (October - November 2011) I found that the orientation of the entrance axis of St. Cristoforo in Barga and St. Frediano in Sommocolonia converge towards the Monte Forato, an enchanting peak of the Apuan Alps, that derives its name from the presence of a big natural rock arch (see fig. 1).

Figure 1. Arch of Monte Forato

Moreover the profile of those peaks draw the profile of an "Omo disteso" ("Lying man" so called from Barga's inhabitants) where the hole in the rock corresponds to the "mouth" of the "Omo disteso" (see fig. 3).

Monte Forato's arch is located in direction 242° from Barga's dome and 235,5° from St. Frediano.

The orientation of Barga's dome (from the entrance to the apse) is 63° of azimuth with 7° of altitude while the orientation of St. Frediano is 56° of azimuth with 7° of altitude. Obviously the entrances of these two churches are oriented in the opposite directions, that are 243° and 236°, converging with very good approximation toward the direction of the arch of Monte Forato.

Is this a fortuitous geographical coincidence?

No, I do not think so.

Why did they choose two orientations, 56° and 63°, which in the opposite direction (236° and 243°) look, with very good approximation, at Monte Forato?

Why among so many possibilities both axis of orientation look at the most singular target on the horizon?

The astronomical declination of the orientation of Barga's dome is +24° while St. Frediano is + 29°. So Barga's dome orientation, even if it does not coincide with the peak characterizing the Tuscan Romanesque churches (see Zedda 2012) is, anyway, included in the arch of horizon where the sun rises and that we could consider as characteristic of Romanesque churches. Vice-versa St. Frediano's church shows an anomalous orientation that could be explained through the religious and astronomical syncretism that characterized those sites.

2. DATA ANALYSIS

From the winter solstice in 2011 until the first decade of January 2012 I carried out a further study campaign in Garfagnana and in Serchio's valley.

For explanatory reasons, in the face profile of the "Omo disteso" I distinguished four points:

The upper part of the "forehead", the tip of the "nose", the "mouth" and the tip of the "chin".

St. Frediano's church (lat. 44°05'31.14"; long. 10°29'26.31") is slightly more than 15 Km far from Monte Forato. The face of the "Omo disteso" profile occupies 2°29' of the horizon that could be observed by Sommocolonia.

From St. Frediano (fig. 2) we can see:

The upper part of the "forehead" at an azimuth of 233°33', alt. 1°30', dec. -24°22'

The tip of the "nose" at an azimuth of 235°05', alt.1°50', dec. -23°07'.

The "mouth" at an azimuth 235°32', alt.1°39', dec. -22°58'.

The tip of the "chin" at an azimuth of 236°02, alt. 1°44', dec. -22°35'.

Figure 2. "Omo disteso" from Church of St. Frediano

St. Cristoforo's dome (lat. 44°04'24.00"; long. 10°29'09.01") is 13.6 Km far from Monte Forato.

The visage profile of the "Omo disteso" occupies 2°52' of the horizon visible from Barga's dome.

From Barga's dome we can see (fig. 4):

the upper part of the "forehead" at an azimuth of 239°36', alt. 2°48', dec. -19°24'; Lunar dec. -18°41';

the tip of the "nose" at an azimuth of 241°20', alt. 3°10', dec. -17°58', Lunar dec. -17°16';

the "mouth" at an azimuth of 241°50', alt. 3°, dec. -17°46', Lunar dec. -17°04'.

the tip of the "chin" at an azimut of 242°24', alt. 3°05, dec. -17°20' Lunar dec. -16°37'.

Figure 3. "Omo Disteso" (literally "Lying Man") from Barga's Dome

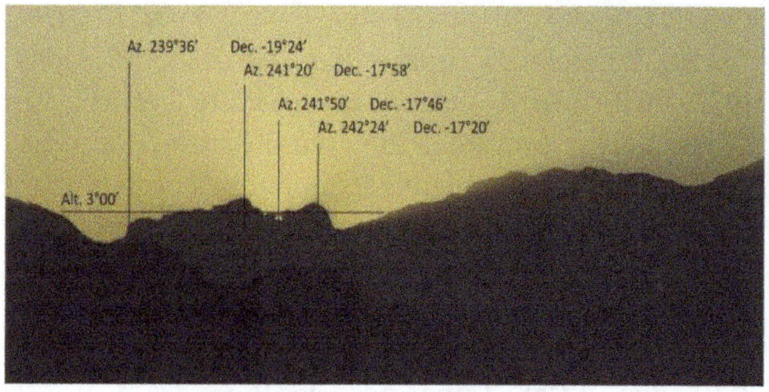

Figure 4. From Barga's Dome

In fig. 5 we can observe the indication of the point of sunset at winter solstice in 2000 b.C. and in fig. 6 the sunset at winter solstice in 2011 from St. Frediano.

Figure 5. From St. Frediano, indication of the point of sunset at winter solstice in 2000 b.C. (Zedda 2012)

From the data analysis we can state that, in the same point where the sun set in 2000 b.C. at winter solstice (observing from St. Frediano's church), in the same period it was possible to observe the moon set, in the same point, at the minor southern standstill, but from the Barga's Dome.

Figure 6. Sunset at winter solstice 2011 from St. Frediano

Today the Lunar declination at the minor southern standstill is -18°17′ while in 2000 b.C. was -18°36′. So we can state that, at that astronomical epiphany, moon set at the forehead of the "Omo disteso".

If, for what concern the sun and the stars, the Earth could be considered as a simple point, this is not correct for the astronomical data concerning the Lunar declination.

In fact, due to the "relatively short" distance between the Earth and the Moon, one observer at the Barga latitude would see the Moon under an angle lightly different from an equatorial observer.

According to the "standard formula" the data of the theoretical astronomical declination are referred to the line passing from the center of the Earth and from the center of the Moon, while observing the Moon from Barga, there is a short difference of about 40′. So it is necessary to correct the standard declination (referred to an observer positioned in the center of the Earth), with the declination referred to a real observer positioned in the surface of the Earth. The difference between the two data is minimal at the equator and maximum at the poles.

I calculated the correction using the software kindly given to me by Clive Ruggles. So, at Barga's latitude, the asterism with declination of -19°24′, sets at 239°36, with an horizon altitude of 2°48′, with the exception of the Moon. The Moon sets at the aforementioned azimuth and altitude when its declination is -18°41′.

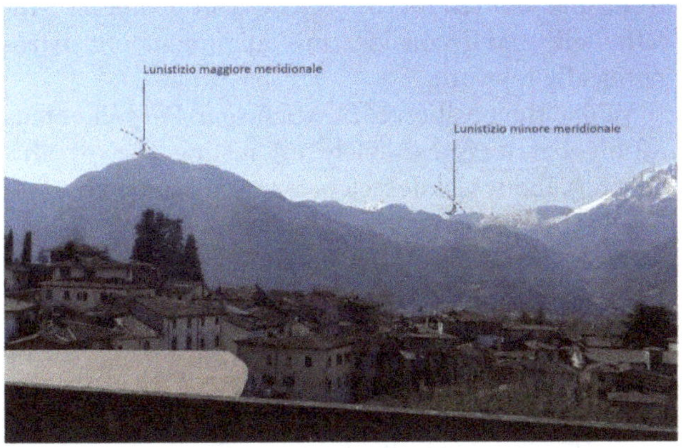

Figure 7. From Barga's Dome, position of moonset at major and minor standstill. (from Zedda 2012)

According to these data, I can formulate the following hypothesis:

i) both churches were built in sites already and previously "sacralized" by the Ligures Apuan, who chose them according to astronomical criteria;

ii) such sites continued to be considered sacred in the Roman period;

iii) still in the Christian period there was the sense of the sacredness of the direction towards Monte Forato and it determined the orientation of both the churches.

It is kwon that both in Barga and Sommocolonia areas there are archeological ruins of the Roman and Ligures Apuan period. Stefano Borsi speculates that in the site where the dome was built there was a Roman temple and that before there was a settlement of the Ligures Apuan (Borsi 2009).

At this point we must ask ourselves: how many Romanesque or pre-Romanesque churches in Garfagnana can see Monte Forato?

Unluckily I did not find any list of the Romanesque churches in Garfagnana neither in the Serchio's valley. Therfore, I visited all the villages searching Romanesque churches.

I realized that Monte Forato, apart from the above mentioned churches, can be seen from the Romanesque and pre-Romanesque churches in: St. Michele in Coreglia Antelminelli, St. Pietro and Paolo in Fiattone (Gallicano) and St. Michele in Perpoli (Gallicano).

Very probably it was possible to observe Monte Forato even from two little Romanesque churches, today destroyed, St. Quirico and St. Margherita, close to Barga's wood.

I tried to find the ruins of St. Quirico, but I didn't succeed, so I only remember a nice day in good company in the chest nut wood of Barga! After the failing of my research of St. Quirico, I gave up the ruins of St. Margherita church as well.

Monte Forato is located at an azimuth of 253° from the church of St. Michele in Coreglia Antelminelli and I consider this alignment not astronomically relevant.

While, the position of St. Michele in Perpoli versus Monte Forato is extremely interesting (fig. 8), this churh is about 13 kilometers far from Monte Forato.

Observing the profile of the "face" of the "Omo disteso", from St. Michele (lat. 44°05'55.20"; long. 10°26'38.80"), it occupies 3°08' of the horizon.

The upper part of the "forehead" at an azimuth of 221°06', alt. 2°20', dec. -31°03', Lunar dec. -30°16'.

The tip of the "nose" at an azimuth of 222°49', alt. 2°44', dec. -29°45', Lunar dec. -28°58'.

The "mouth" at an azimuth of 223°10', alt. 2°32', dec. -29°43', Lunar dec.-28°56'.

The tip of the "chin" at an azimuth of 223°46', alt. 2°38', dec. -29°23', Lunar dec. -28°36'.

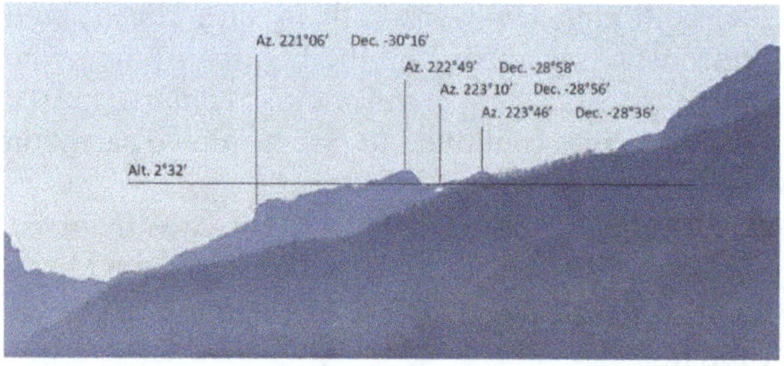

Figure 8. From St. Michele at Perpoli, the declination is "lunar declination"

Nowadays the Moon declination at southern major standstill is a -28°43' while in 2000 b.C. was -29°12'. So we can easily conclude that in such an astronomical phase, if we would have placed ourselves in the area where lately was built St. Michele's church, the moon would have set on the "forehead" of the "Omo disteso".

Even in St. Michele in Perpoli, the moon sets at major southern standstill on the "forehead" of the "Omo disteso". Is it a coincidence? I don't think so. I suppose that the Ligures Apuan chose that the target was the "forehead" of the "Omo disteso".

I found that even the position of the church of St. Pietro and Paolo (lat. 44°05'37.69"; long. 10°26'53.40") in Fiattone, is astronomically relevant.

St. Pietro and Paolo's church is about 12,5 kilometers far from Monte Forato, that is just visible and it appears only in the part from the "forehead" to the "nose", occupying 2° of the horizon circle visible from Fiattone (fig 9).

Figure 9. Frm churh St. Pietro and Paolo at Fiattone

The tip of the "forehead" at an azimuth of 223°48', alt 3°07', dec. -28°51'.

The tip of the "nose" at an azimuth of " 225°45', alt. 3°31', dec. -27°24'.

The tip of the "nose" is the target where Venus set with great precision 2000 b.C. at its southern major standstill.

Nowadays, the declination of Venus ranges between ±26°50'. Four thousand years ago between ±27°18'. So 4000 years ago from the site where the St. Pietro and Paolo was to be built, it was possible to observe the set of "the evening star" at its maximum of southern declination -27°18' on the tip of the "nose" of the "Omo disteso". Is it for chance? It's surely a fact to be considered because it is inserted in a very significant scenario.

It's surely significant that four of the seven Romanesque and pre-Romanesque churches, from where it is possible to observe the Monte Forato, are displaced in an extremely precise system to target the sunset at winter solstice, the moonset at southern minor, medium and major standstill and Venus sets at its maximum southern elongation (see table I).

Table I. Orientation from St. Cristoforo at Barga, St. Frediano at Sommocolonia, St. Pietro and Paolo at Fiattone and St. Michele at Perpoli towards Monte Forato, "Omo disteso" visage

Backsight	Foresight	Azimut	Alt.	Dec.	Lunar dec.	Target
St. Cristoforo Barga	Upper forehead	239°36′	2°48′	-19°24′	-18°41′	Moonset minor standstill southern
"	Tip nose	241°20′	3°10′	-17°58′	-17°16′	
"	Mouth	241°50′	3°00′	-17°46′	-17°04′	
"	Tip chin	242°24′	3°05′	-17°20′	-16°37′	
St. Frediano Sommocolonia	Upper forehead	233°33′	1°29′	-24°22′	-23°39′	Sunset winter solstice and/or Moonset medium standstill southern
"	Tip Nose	235°05′	1°49′	-23°07′	-22°24′	
"	Mouth	235°32′	1°39′	-22°58′	-22°15′	
"	Tip chin	236°02′	1°44′	-22°35′	-21°52′	
St. Michele Perpoli	Upper forehead	221°06′	2°20′	-31°03′	-30°16′	Moonset major standstill southern
"	Tip Nose	222°49′	2°44′	-29°45′	-28°58′	
"	Mouth	223°10′	2°32′	-29°43′	-28°56′	
"	Tip chin	223°46′	2°38′	-29°23′	-28°36′	
St. Pietro Paolo Fiattone	Upper forehead	223°48′	3°07′	-28°51′	-28°05′	
"	Tip Nose	225°45′	3°31′	-27°24′	-26°38′	Venus set major standstill southern

Strabon mentioned the Apuan Alps as "lunae montes" (mountains of the Moon). On the other side of the Apuan Alps (observing from Barga), there is the Roman city of Luni, sacred to the Moon and the wonderful territory of Lunigiana (door of the Moon?).

In my opinion it is possible argue that the astronomical system was built by Ligures Apuan people who lived there, and that, in such sites, in Christian period, some churches were erected inheriting sacred and astronomically oriented locations.

This is an astronomical system based on a point of reference and in different points of observation, typical of the sites studied by A. Thom (1967) and Clive Ruggles (1999) in Great Britain and by me in Sardinia (Zedda 1992, 2004, 2009, 2013, 2014; Zedda and Pili 2001).

From my point of view, the geographical and astronomical reports presented in this paper represents an extraordinary example of astronomical and religious syncretism.

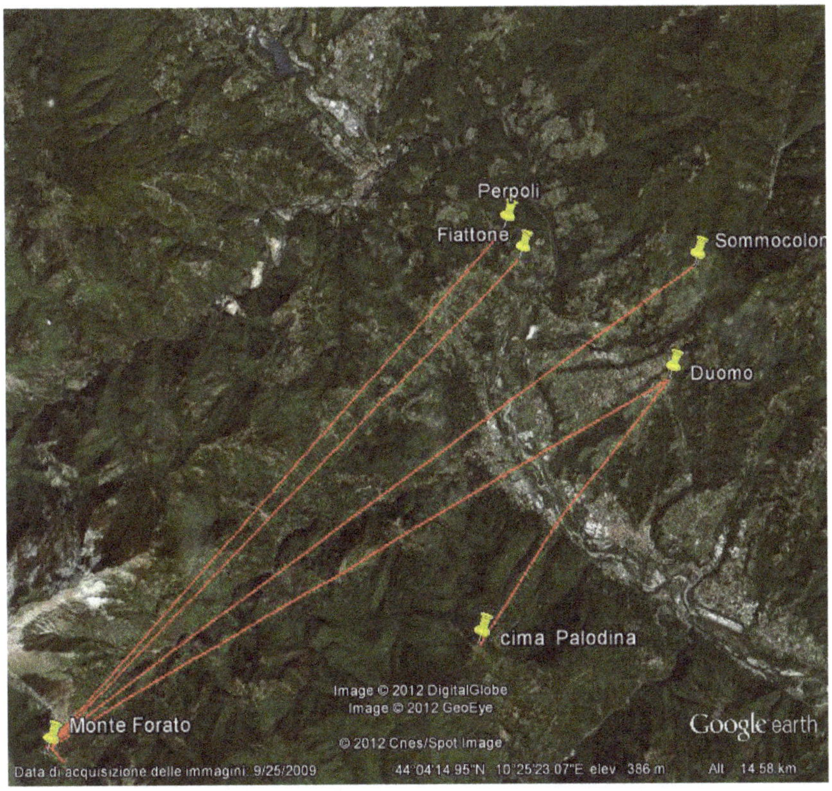

Figure 10. Map of alignments of the astronomical system show in this paper (from Zedda 2012).

ACKNOWLEDGEMENTS

A special thanks to Delfina, Pier Giuliano Cecchi, Paolo Littarru, Roberta Fadda, Donatello Orgiu, Paolo Pili, Carla Sulis, Stefano Carboni and Dante Olianas.

REFERENCES

Borsi S. (2009) *Le origini di Barga e il culto di San Cristoforo*, Melfi.

Borsi S. (2010) Il San Cristoforo di Barga, in *Il Duomo di Barga -Storia, Arte e Spiritualità nei primi tre secoli dopo il Mille*, Pontedera.

Borsi S. (2010a), Risposte del prof. Borsi in Il *Duomo di Barga -Storia, Arte e Spiritualità nei primi tre secoli dopo il Mille*, Pontedera.

Cecchi P. G. (2010) Le tracce Templari in *Il Duomo di Barga -Storia, Arte e Spiritualità nei primi tre secoli dopo il Mille*, Pontedera.

Ruggles C. (1999) *Astronomy in prehistoric Britain and Ireland*, London.

Ruggles C. (2014) *Handbook of Archaeoastronomy and Ethoastronomy*, New York.

Thom A. (1967) *Megalithic sites in Britain*. Oxford.

Zedda M.P. and Pili P. (2001) Archaeoastronomy study on the disposition of Sardinian Nuraghes in the Brabaciera valley" in proceedings of Oxford Vi and Seac 99 *Astronomy and Cultural Diversity*, J. Belmonte and C. Esteban Editors, Santa Cruz di Tenerife.

Zedda M. P. (1992) *I Nuraghi il Sole la Luna*, Cagliari.

Zedda M. P. (2004) *I Nuraghi tra Astronomia e Archeologia*, Cagliari.

Zedda M. P. (2009) *Archeologia del Paesaggio Nuragico*, Cagliari.

Zedda M. P. (2012*) Monte Forato e il Duomo di Barga*, Cagliari.

Zedda M.P. (2013) *Astronomia nella Sardegna Preistorica*, Cagliari.

Zedda M.P. (2014) Sardinian Nuraghes in the Handbook of Archaeoastronomy and Ethoastronomy, editor Clive Ruggles, New York.

25

ASTROLOGY, PROPHECY AND THE BIRTH OF MODERN SCIENCE SEEN THROUGH THE LENSES OF RENAISSANCE ART

Giangiacomo Gandolfi

Zetema/Planetario di Roma - Sovrintendenza BBCC Comune di Roma

Corresponding author: Giangiacomo Gandolfi (giangiacomo.gandolfi@gmail.com)

ABSTRACT

As a follow-up to my previous studies in astrological representations in the XV-XVI century art of Veneto, I will shed light on the *fil rouge* connecting different experiences like the planning of astrological frescoes in Montagnana by Galeottus Martius, the collaboration between Giovan Battista Abioso and Giorgione in the painting of the famous Castelfranco Frieze and the visual prognostication of Giulio Campagnola's Astrologer, namely the artistic representation of highly complex astral theories applied to world history. This commitment to mundane astrology mixed with prophetic and apocalyptic doctrines emerges from an analysis of the whole catalogue of astral artworks in the Italian territory: 10 out of 240 of them reveal this unmistakable character and all are concentrated in the northeastern territory. There are many additional evidences that the recurring theme of astrological divination was in fact a widespread enterprise among humanists, philosophers, theologians and artists of the "Venetian Terraferma" at the time. Its discussion, reverberated across several generations and in nearby territories, reveals an informal "web" of study and research that goes well beyond the lecture halls of the local University, including poets, physicians, painters, heretics, goliardic secret societies and would-be revolutionary astronomers like Copernicus. In many features of the cited instances we find the typical rationalistic approach of the Aristotelian-Averroistic Paduan school inextricably linked to mysticism, neo-platonic magic and theology, a tradition going back at least to Pietro d'Abano and Biagio Pelacani. The same tradition, reinforced by a growing penchant for mathematical modelling and careful measuring, that will lead to the materialism of Pomponazzi, Telesius and Vanini and ultimately to the mature science of Galileo.

Instead of considering this process a slow but glorious liberation from superstition as too many historians still do, I will try to argue that the interest in astrological divination has been a fertile breeding ground for the Scientific Revolution and in particular for the rebirth of Heliocentrism, a controversial thesis in the vein of Dame Frances Yates, akin to that recently advocated by Robert Westman in his debated "The Copernican Question" (2011).

KEYWORDS: Astrology, Copernicus, Italy, Prophecy, Renaissance Art, Scientific Revolution

1. INTRODUCTION

The study of astral representations in the art of Late Middle Age and Renaissance shows how complex was the "science of the stars" at the time. There were many layers in the astrological doctrine accumulated since Antiquity, corresponding to different visions, philosophies and objectives (positional astronomy, genethlialogy, horary, catarchic, talismanic astrology, etc). Very often such complexity is overlooked or discarded by scholars and it should be clear that there was no such thing as an "astrology", but many "astrologies" instead, converging in a multifaceted tradition full of contradictions and misunderstandings, sometimes even among its practitioners.

In this essay I will analyze astral art concentrating on Italy and on a particular strand of thought reflected by a number of sophisticated depictions: historical (or mundane) astrology, that is the study of the influence of the stars on lands, populations and on the destiny of human race (rather than on single individuals). This tradition, connected to ancient astral divination but developed especially by medieval arabic astrologers, was very often associated with joachimite prophecy and apocalyptic thinking.

Tracing this kind of representations along the territory one discovers that they are concentrated around the three major university centres of northern and central Italy, and primarily around Padua and the "venetian terraferma", the same milieu where a strong mixture of prophecy and astral divination was massively produced, very often accompanied - at first sight paradoxically - by a rationalistic attitude steeped in averroism and aristotelianism that bordered sometimes on heresy.

Such a correlation (not necessarily a causation) is explored with a particular emphasis on copernican heliocentrism, probably developed or at least matured in Padua, a case where prophecy and mysticism met innovation without apparent conflicts.

The simultaneous presence of prophetic traditions tinged of astrology and of the mathematical component of the science of the stars in the same place and among the same practitioners should be considered highly meaningful: my final aim is to suggest a cohabitation, a kind of symbiosis, very probably even a fueling role in the birth of modern science.

2. ASTRAL REPRESENTATIONS IN ITALIAN MEDIEVAL AND RENAISSANCE ART

The starting point of my analysis was the construction of a database of astral art on the italian territory between the XIIth and the XVIIth century as complete as possible (Gandolfi, 2016), an upgrade of the previous existing surveys (Mori, 1980; Gotze, 2012)[1].

Figure 1. Atlas of Astral Art in Italy (XII-XVII century)

The resulting atlas - a work in progress consisting at present of about 240 depictions - maps the diffusion of the Science of the Stars in Italy during the Late Middle Age and the Renaissance, allowing to trace territorially different traditions and different sensitivities on the part of courts, patrons, artists and practitioners of the field. Such a diversity is reflected in a variety of representations that will be studied in detail in a future essay, still in preparation. In the meanwhile the particular tradition of historical astrology has been isolated and analyzed in the database, relying on a series of markers that will be discussed below.

2.1. The Historical Astrology subset

The sample so far obtained is composed of only ten representations, i.e. about 4% of the entire database. However the emergence of such sophisticated, abstract and technical celestial doctrines in paintings or frescoes appears so surprising and improbable that even a limited number of cases constitutes an exceptional event. The very presence of the subset in a catalogue strongly dominated by conventional calendrical cycles indicates the relevance of historical astrology in the context of general culture, while the narrow distribution of its representations along the north-eastern Italian coast should be considered particularly worthy of attention.

Figure 2. Atlas of the Historical Astrology Subset

[1] *Gotze's catalogue is very wide but I opted for a more restrict definition of astral art, following the criteria of Giulia Mori.*

2.1.1. Historical Astrology Techniques as Markers

In order to select the works of art meant to convey a message or a metaphor connected to the theme of historical astrology, reasonable markers of this doctrine have to be identified. Usually the artist or the creator of the iconographic program resort to a veiled symbolism or - less often - to explicit references to astrological techniques, with celestial charts, constellations and planetary symbols.

The main techniques used in the field of mundane astrology are: a) the so-called Revolution of the Year, i.e. the astral theme of the time of Equinoxes, often calculated also as the theme of the last sizigy preceding the same phenomena (Bezza, 2008); b) the Magnus Orbis, the arabic Dawr or Mighty Fardar, an abstract cycle of 360 years that associates the zodiacal signs to each planet in all possible permutations (Bezza, 2008); c) the Great Conjunctions Jupiter-Saturn cycle, whose duration of 960 years is divided in subcycles of 240 years in the same astrological triplicity (Buscherini, 2013); d) the Saturnian Revolutions of 30 years, with great historical changes expected each 10 cycles (Buscherini, 2013); e) the Cycle of "Trepidatio" derived from the Alphonsine Tables, an oscillation of the equinoctial points from +/- 9° to 0° with a period of 4077 or 640 or 7000 years, often considered in addition to the Precession of Equinoxes (Ackermann Smoller, 1998); f) the Ptolemaic Astrological Chorography, connecting nations and stars (Feraboli, 1985); g) the Ptolemaic Theory of Luni-Solar Eclipses Influences (Feraboli, 1985); h) the Theory of Cometary Influences (Feraboli, 1985);

2.1.2. Description of the Sample

Each work of art of the selected subset includes an allusion to at least one of the cited techniques, sometimes to a mixture of them, while the dates of the representations range from about 1460 to 1538, a short interval of time during which the debate about prophecy and historical astrology was at its peak. I have studied in detail some of these artworks (located in the Veneto region) in previous essays, but here I offer a brief description of the whole sample:

Figure 3. The Historical Astrology Subset: a) Giulio Campagnola - "The Astrologer"; b) Anonymous - "Three Astrologers"; d) Giorgione - "Frieze of the Liberal Arts"; e) Sebastiano Florigerio - "Saint Joseph and the Child"; f) Giorgione - "The Three Philosophers"; g) Cosmé Tura - "Annunciation"; h) Cosmé Tura - "Madonna of the Zodiac"; i) Ghirlandaio - "Nativity"

1. *Madonna of the Zodiac by Cosmé Tura now in Gallerie dell'Accademia in Venice (ca. 1460), and*
2. *Madonna of the Zodiac by Cosmé Tura now in Galleria Colonna in Rome (ca. 1460);*

These two paintings clearly reveal the commitment of the artist and his entourage to the themes of mundane astrology, in a cultural centre as Ferrara, always characterized by a pervasive passion for the science of the stars (Vasoli, 1977). The Virgin with the Child is encircled in both cases by a thin and translucent zodiac where particularly prominent is the Sign of Virgo, emerging behind the sleeve of the Madonna in the former while central but occulted by her head in the latter. The reference is to conjunctionism, in particular to the ascending first decan of the Virgo, announcing in the starry sky the mistery of Incarnation (Golsenne, 2006).

3. *Annunciation in the enclosures of Ferrara Cathedral Organ by Cosmé Tura, now in Museo del Duomo in Ferrara (ca. 1469);*
   The same artist produced this complex painting for the bishop Lorenzo Roverella, brother of cardinal Bartolomeo Roverella, whose secretary Filippo Bonaccorsi was at the same time fugitive being part of a conspiration against Pope Paul II fueled on the basis of astrological prognostications. Guidoni and Marino (1979) identify a clear allusion to the Horoscope of Religions in the wing of the archangel eclipsing the pagan statue of Jupiter, while Campbell (1997) sees in the same sculpture Mercury instead (but the astrological meaning remains the same), and interprets the whole image as detached and ironic, almost critical towards the arabic doctrine of conjunctionism. Both authors consider the concealed Sphinx in the background as a symbol (appreciated or despised) of the ancient hermetic tradition.
4. *Astrological Fresco in the Rosario Chapel of Montagnana's Cathedral, probably by Angelo Zoppo (ca. 1483);*
   This enigmatic fresco portrays at the same time a solar eclipse in Leo with its paranatellonta and a symbolic apocalyptic scenario where the constellations stand for the protagonists of a mighty culture clash, probably implying the final demise of Ottoman Empire. The sophisticated astrological technique employed for the prophecy is the Magnus Orbis cycle, integrated by a clever use of the Ptolemaic theory of Luni-solar Eclipses influences and by Ptolemaic Astrological Chorography. The author of the program, inspired by the dominican erudite Annius of Viterbo, was the humanist Galeottus Martius, connected to a contemporary paduan goliardic sect, the "Macheronica Secta" involved in quasi-heretical astrological discussions besides crude jokes and the promotion of vernacular poetry (Gandolfi, 2016).
5. *Frescoes in the Sassetti Chapel by Ghirlandaio inside Santa Trinità in Florence (1482-85);*
   This famous cycle has a hidden anti-astrological message, as demonstrated by Peter Porçal (1984). Each wall contains the reference to a different kind of divination, but in polemics with conjunctionism, astral prophecy is blatantly absent, evidencing its incompatibility with Christianity. The Child on the ground in front of the pagan sarcophagus in the Nativity scene symbolizes the complete refusal of ancient astral doctrines and justifies by negation the belonging of the cycle to the sample.
6. *Eastern Wall 'Frieze of the Liberal Arts, by Giorgione in Casa Marta-Pellizzari in Castelfranco Veneto;*
   Indisputably the most complex of all the representations in the sample, the giorgionesque frieze is a real manifesto of Divinatory Arts (D'Amicone, 2010; Gandolfi, 2016), containing horoscopes, technical diagrams, sky charts, astrological talismans, etc. It refers mainly to the apocalyptic climate generated by the last three Great Conjunctions Jupiter-Saturn of the Watery Trigon cycle in 1484, 1503 and 1524 and is the result of the collaboration between Giorgione and the astrologer Giovan Battista Abioso, author of an influent *Dialogus in Astrologiae Defensionem* (1494) that contrasted Pico della Mirandola's *Disputationes Adversus Astrologiam Divinatricem* of the same year.
7. *'The Astrologer', an engraving by Giulio Campagnola created in the Venetian Terraferma and now in many collections, including the British Museum (1509?);*
   Giulio Campagnola was a polymath artist, friend and collaborator of Giorgione, known especially for his fine engravings. This particular print[2] contains a visual prognostication referred to 1509 inscribed on an astrological disk: it relies on the technique of the Revolution of the Year and is generally connected to the calamities generated by the Saturn-Jupiter Conjunction of 1503 (D'Amicone, 1992; Gandolfi, 2014).
8. *'Three Astrologers', a drawing by an anonymous artist (monogram AG) of the Veneto School now preserved in Fondation Custodia in Paris;*

[2] *I have ignored here the many copies of the theme because derivative and devoid of any astrological value.*

This drawing, sometimes attributed to Campagnola, shows three scholars measuring a celestial globe (van der Sman, 2003). Equator and ecliptic are easily discernible, as the symbol of Saturn near the equinoctial point, a configuration referring to the technique of Saturnian Revolutions. An astrological analysis of the artwork is still lacking in literature.

9. *"The Three Philosophers", the masterpiece painting by Giorgione now in the Kunsthistorische Museum in Wien (1509?);*
   One of the uncontested masterpieces of Giorgione, it is the topic of dozens of inconclusive analysis and interpretations in art history. The astrological subtext, however, is indisputable and comes from the parchment in the hands of the old philosopher on the right, containing at least a probable lunar eclipse, the symbols of Jupiter and Saturn and a nocturlabe. Whether a representation of three Magi or of three great astrologers symbolizing the Horoscope of Religions, the painting refers nevertheless to astral divination and to conjunctionism (Settis, 1978; Gentili, 1999).
10. *'Saint Joseph and the Child', a painting by Sebastiano Florigerio in the Cathedral of Cividale del Friuli (1537-38).*
   This almost unknown artwork in Friuli displays an ancient zodiac in the ruined metopes of a pagan temple, standing above Christ and his human father (Tonetto, 2001). The zodiacal sequence begins with Pisces instead of Aries, alluding certainly to Trepidation, moving the vernal point across the former constellation since the divine birth. Even in this case a scholarly analysis of the astrological meaning of the painting is completely lacking.

3. PROPHECY AND ASTRAL DIVINATION IN NORTH-EASTERN ITALY

Between the XIVth and the XVIth century Italy, as many countries in Europe (especially Germany), was obsessed at all levels by astrological divination. This theme was widespread in Politics, Literature, Theology, Natural Science and even among scholars that we usually judge skeptical and highly rational[3].

Many practitioners of the science of the stars were busy with this kind of mundane astrology, trying to confirm through celestial cycles their religious eschatology: Smoller Ackermann (1998) has called them "Apocalyptic Calculators", citing among them Pierre d'Ailly, John of Ashenden, Giovan Battista Abioso, Jean de Bruges, Pierre Turrel.

As shown by Niccoli (1987) a real prophetic craze culminating with the famous deluge expectations for the Jupiter-Saturn Conjunction of 1524 was diffused in Italy at the turn of the century among peasants and citizens, clerics and laity, nobles and humble people, and it was highly permeable to those same astrological suggestions and prognostics. North-eastern regions were particularly affected and especially Veneto: here a strong academic tradition of astronomy and astrology mixed with an increasing interest in Joachimite Prophecies centered in Venice, as demonstrated by the wealth of prophetic pamphlets connecting Last Emperors, Angelic Popes and Sibyls to historical astrology (Reeves, 1969). This production, very common also in Germany, was dramatically boosted by the printing technology introduced by Gutenberg (Green, 2011). As revealing examples we cite the venetian Emperor/Pope Sibylline Prophecy of the second half of XVth century, alluding to a comet and citing Saturn as a malevolent influence (Gandolfi, 2016); the contemporary Gamaleon Prophecy, a clear blending of eschatological visions with conjunctionism (McGinn, 1979) and the seminal italian collection of sybilline prophecies published by Filippo Barbieri in 1482, where the oracle of the Sybilla Chimica included a famous text by the arabic astrologer Abu Ma'shar relating to the first decan of the Virgin (Smoller-Ackermann, 2010).

Numerous manuscripts and books of a more technical nature were produced at the same time in the north-eastern territories by influent astrologers, from the real bestseller *Prognosticatio* of Johannes Lichtenberger published in Venice many times, to the *Prognosticon super Antechristi Adventu* by Johannes de Lubeck, printed in Padua in 1474, to the already cited venetian *Dialogus in Astronomiae Defensione* by Giovanni Battista Abioso and the mysterious *Prognosticon ad Eversione Europae* of 1490-95 by Antonius Arcuatus, a ferrarese student of the science of the stars. All these works developed astrological divination creating theological narratives reflected in the sky and the same tendency extended along the Adriatic coast, including prominent texts as the bolognese *Questio de Fine seu Durabilitate Mundi* (1481) by Johannes Paulus de Fundis and the famous *Pronostica ad Viginti Annos duratura* by Paul of Middelburg, published in 1484 in Urbino, and even across the Apennines, in Genoa, with the *De Futuris Christianorum Triumphis* (1481) by Annius of Viterbo.

This widespread joachimism full of religious overtones and mixed with astrology could be perceived today as a sign of a general sense of paralyzing and impending doom, of a renounce to act on the verge of the final crisis. On the contrary, there is ample

[3] *Pico della Mirandola, for example, refused in the end all astrological techniques but resorted to Kabalah to describe history and the cosmos, calculating in his "Conclusiones Cabalisticae" the exact date of the end of the world.*

evidence that such apocalyptic thinking was not associated to purely passive fatalism: it maintained instead a fundamental ambiguity about the duration of the interregnum between the coming of the Antichrist and the real Judgement Day or it postponed *tout court* the end of the world through further astronomical cycles, provoking an oscillation between the necessity of pro-Church activism in the face of the coming Apocalypse and an exciting sense of deep renewal of society against clerical corruption and machiavellian political manoeuvres, the so-called *Renovatio* (Vasoli, 1981).

4. ARISTOTELIANISM, AVERROISM AND THE BIRTH OF MODERN SCIENCE IN VENETO

Many scholars indicated in the paduan Aristotelian milieu the principal breeding ground for the Scientific Revolution finally started by the new physics of Galileo (Randall, 1940; Wallace 1984) and by the experimental medicine of Girolamo Fabrici, Vesalius and William Harvey (Berti, 1997), focusing on the same territory we have considered so far, mapping the simultaneous diffusion of astral art, mundane astrology and astrologically tinged prophecy. While the late Charles Schmitt scaled down such paradoxical but intriguing idea (Schmitt, 1983), a strong sensation of affinity with the general tenets of later natural science still emanates from some of the scholars and philosophers that gravitated around the paduan university, in particular the so-called "averroists".

As amply remarked by historians of philosophy (Akasoy and Giglioni, 2013), such a label is an invention: almost all Aristotelian scholars were influenced in one way or another by the commentary of Ibn Rushd, but with time the term "averroists" became synonym of "thinkers who pursued Aristotelian philosophy apart from theology and whom we might better describe as secular Aristotelians"(Kristeller, 1980). This very loose, heterogeneous and long lasting school of thought included personalities as different and remarkable as Peter of Abano, Blaise of Parma, Galeottus Martius, Pietro Pomponazzi and Giulio Cesare Vanini, all on the brink of heresy and atheism. Their inclination towards rationalism and determinism came actually from many different influences (often mediated by arabic culture) and combined invariably with a strong astrological vision as a unifying theory of nature and the cosmos, able to explain politics and religion through conjunctionism and especially through the theory of the so-called Horoscope of Religions. The resulting mechanicism, even polluted by extravagant beliefs, has many analogies with science, with an increasing attention for nature and experimentation and with a strong respect for mathematics.

The majority of averroists of this kind, actually, was more involved in logic or in heuristic medicine than in true mathematics applied to physics, but the exception - and the bridge with the academic world of "professional" mathematicians - was again the science of the stars. Next to many qualitative thinkers we find in fact in the same paduan context also astrologers like Abioso with his continuous exaltation of new instruments and accurate measurements in the footsteps of Regiomontanus (Pastore Stocchi, 1987) or scholars like Girolamo Fracastoro and Giovan Battista Amico (Swerdlow, 1996), trying to build an homocentric theory of the cosmos rigorously based on mathematics in order to effectively describe planetary movements. Even humanists without a specific mathematical training like Galeottus Martius invoked the precision of astronomical tables in order to reform astral divination.

But the seduction of mathematics applied to the structure of the cosmos was at the time characteristic in particular of pythagoreanism and neo-platonism, doctrines diffused and appreciated especially in Venice, and not necessarily and always considered opposed to paduan aristotelianism. Mystical numerology was here inextricably mixed with a strong esthetical longing for harmony and symmetry, whose far-reaching consequences for modern science would have been especially evident in the cosmologies of Copernicus and Kepler. And Copernicus, indeed, while maturing his ideas passed his last years in Italy between Padua and Ferrara, which is certainly not a coincidence.

5. THE COPERNICUS CASE

The paradigmatic and seminal role of the copernican revolution in the birth of modern science makes the stay of Copernicus in Padua particularly relevant for our discussion and the (re)birth of Heliocentrism offers a wonderful example of the inextricable mixture of metaphysics and analytic-mathematical approach that characterizes the paduan melting pot with his omnivorous humanism and philosophy.

Only in the XXth century the importance of such a scenario has painstakingly emerged from local researches, that have slowly shown the academic milieu of Copernicus as well as his probable acquaintances (Birkenmajer, 1922 ;Nardi, 1959; Bilinski, 1983; Pepe, 2003). Besides Fracastoro, Pomponazzi, Tiriaca and other professors, he may for example have met (or just read) Celio Calcagnini in Ferrara and he surely knew of Giorgio Valla in Venice.

We are here particularly interested in the evidences of his involvement, however indirect, in the astrological side of the science of the stars and in the mystique of pythagoreanism and neo-platonism that could have supported it. I will list them here: 1) Copernicus' apollonian seal, alluding to a form of solar cult (Mossakowski, 1973); 2) his collaboration with the prognosticator Domenico Maria Novara (Bonoli, 2012); 3) his famous citation of Hermes Trismegistus in *De Revolutionibus* I.10 ; 4) his implicit assent on Rheticus digressions about astrology in *Narratio Prima* (1540); 5) the mention in a letter of 1535 by Bernard Wapowski of a now lost almanac written by himself.

Yet focusing on the Paduan stay we can also add the meaningful paduan (or ferrarese) observations of the Great Conjunction of 1503/4 (discovered in his handwriting in a copy of the Alphonsine Tables by Owen Gingerich in 1973), while a note on a copy of Stoeffler's Almanach by Martinus Biem of Cracow reports a date for the third conjunction of June 1504 "*secundum Copernicum*", meaning that original observations were circulating and that the heliocentric model was at such an advanced stage of elaboration that it allowed the calculation of ephemeris (Birkenmajer, 1922). To all these evidences we should add a much overlooked discovery by Sergio Bettini (1975), able to disclose new perspectives to our knowledge of Copernicus frequentations: the probable portrait of the scientist in the "Marriage of the Virgin", painted in 1505-7 by Campagnola in the paduan Scoletta del Carmine. The link with Copernicus allows to infer the acquaintance with other figures of Campagnola's circle: Luca Gaurico (a prominent astrologer), Giorgione and Abioso[4], connecting further heliocentrism and prognostications.

The first scholar to propose with energy the propulsive role of Hermeticism and Pythagoreanism in the Scientific Revolution and in Copernicanism was Frances Yates, but after many controversies her thesis has been superseded by the awareness of a multiplicity of practices and different approaches in the Natural Magic/Science of the period that taught historians to be more nuanced and cautious in front of generalizations[5]. Nonetheless her intuition of "metaphysical" driving forces behind Science - in opposition to purely internal approaches - remains more vital and valuable than ever. To the point that one of her critics - the historian Robert Westman - has recently proposed another controversial theory about the genesis of *De Revolutionibus*, apparently confirming the scenario discussed so far: Copernicus could have been motivated in his research program by Pico's *Disputationes adversus Astrologiam Divinatricem*. In order to defend the science of the stars as a whole from such an attack, he would have chosen to refound the discipline starting from planetary order (as proposed also by Goldstein, 2002). The new order reflected in fact a satisfying simmetry in the ptolemaic astrological qualities (heat, cold, moisture, dryness) in addition to a well-tuned sequence of velocity-distance relations (Westman, 2011).

6. ASTRAL DIVINATION AND SCIENCE

More than a simple heir of aristotelianism or antiaristotelianism or neo-platonism, Science was the result of the complex dialectic of many components in this Veneto melting-pot, a scenario that irradiated infectious ideas across Europe. Between these ideas heliocentricity had a crucial role as the real building block of the Scientific Revolution, confirming the igniting impact of astral divination both as a main intellectual problem in the field of natural science and as a religious and spiritual concern for individuals and for the Church. Together with other metaphysical stimuli as hermeticism or pythagoreanism, the irresistible fascination for historical astrology was a driving psychological force behind the scene, exactly in the sense advocated by Frances Yates: no technical analysis of mathematics and methods may acknowledge such a key component of the new philosophy of Nature, no internal history of science could calculate its ethereal but essential weight.

It is no exaggeration to state that astral divination was the reason of the centrality of the science of the stars during the late Middle Age and the Renaissance: whether enthusiastically accepted or vehemently denied it was always considered an horizon of meaning for human history itself, encompassing religion, philosophy and science and guaranteed by Nature. That is why, with a few relevant exceptions, all philosophers and astronomers of the time tried hard to revive it one way or another, even when reforming the entire structure of the cosmos (the case of Rheticus, when in the *Narratio Prima* proposes a new technique of mundane astrology based on the motion of the center of Earth eccentric) or when recuperating a pure ptolemaic tradition (the case of Girolamo Cardano, that could not resolve to renounce to arabic conjunctionism in commenting the *Tetrabiblos*). In light of such a centrality in the history of ideas and of territorial coexistence (the paduan stays of giants like Copernicus, Regiomontanus, Vesalius, Harvey and Galileo highlight the absolute relevance of the location), an influence of astral divination on the long and complex process of creation of modern science should have been considered before, but science historians are blinded by a Whig History paradigm and still convinced that astrology is a superstition not worth of a careful analysis.

---

[4] *Abioso was also the author of the preface for the* Epitome *of Regiomontanus, a book Copernicus used intensively.*

[5] *For a fair review of Yates thesis and its opponents in the historiographical context of the Scientific Revolution see Floris Cohen (1994).*

Is it really so paradoxical and unbelievable that astral divination may have contributed to the birth of Modern Science? Considering its many connections in the same historical period to instances of innovation - discoveries or inventions - almost everywhere in Europe, the answer should be surely negative. We give just three impressive examples: the feat of Christopher Columbus, whose manuscript *Libro de las Profecias* demonstrates the importance of Pierre d'Ailly's prophetic-astrological influence as a stimulus for the voyage to the Indies (Moffitt Watts, 1985); the printing of a book about Sibyl's Prophecies among the first experiments of Johannes Gutenberg with a movable type system (Green, 2011); the creation of logarithms by John Napier for calculating the date of apocalypse according to St. John (Fava, 2013).

But going back in time, evidences of a propulsive role of apocalyptic astrology become more compelling: there are many crucial medieval figures that combined a *Weltanschauung* tinged of astral divination with experimentalism and proto-scientific innovations, such as Robert Grosseteste, Albertus Magnus and Roger Bacon in the XIIIth century, all committed to conjunctionism (Thorndike, 1923), and Jean de Roquetaillade and Arnald of Villanova in the XIVth, practicing astral prophecy and alchemy as weapons against Antichrist (De Vun, 2009). On the other hand, even under the pressure of Enlightenment and Positivism, even in our age of unopposed secularism, Science has never ceased to be propelled by metaphysical or religious stimuli (Burrt, 1925), but at this point we have reached the territory of sociology of science and we conclude the discussion.

7. CONCLUSIONS

Starting from an updated catalogue of italian astral art I have identified a subset of sophisticated representations dedicated to historical astrology, concentrated in the north-eastern region, confirming the crucial role of such a doctrine and its influence in different milieus. The analysis of this sample reveals a network of study and research that goes beyond the halls of the local universities, including poets, cardinals, painters, secret societies and astronomers but at the same time it is firmly rooted in the longstanding aristotelian tradition of paduan academism, enriched by neo-platonism. I have suggested on the basis of many correlations that such a large interest in mundane astrology, excited by its confirming role in the domain of religious prophecy and apocalypticism, may be considered a key ingredient in the birth of modern science rising in the same location, as the Copernicus case seems to emphasize.

In this sense the contested Yates thesis has been insightful indeed: it helped appreciate the impact of metaphysics on the apparently most neutral and objective of human endeavours, but at the same time it emancipated as much as possible the consideration for astrology and magic in history. The first sign of attention for this realm in the field of cultural studies was actually Aby Warburg's project of analyzing the psychological meaning of astrology in the history of art and culture. But Warburg, however brilliant and passionate about the theme, despised that ancient astral worship and considered it a relic of the past to be overcome. *"Athens must always be conquered afresh from Alexandria"* he said, implying for Egypt the role of symbol of all superstitions and religious mysteries, in opposition to Greece as the home of rationality. But Alexandria was at the same time the cradle of mysticism and the place where Mouseion and Library stood as temples of knowledge in Antiquity, without contradictions. As convincingly shown by Russo (1996), the first scientific revolution was the hellenistic one, a conquest never deterred by eastern spirituality and metaphysics. On the contrary one could speculate about the necessity of both modes of thought in developing Science, re-phrasing Warburg's aphorism in the light of this dialectic: *"Athens needs always to be stimulated afresh by Alexandria"*.

REFERENCES

L. Ackerman Smoller, "The Alphonsine Tables and the End of the World" in *The Devil, Heresy and Witchcraft in the Middle Ages. Essays in Honor of Jeffrey Burton Russell* cur. A. Ferreiro (ed.), Leiden-Boston, E.J. Brill, 1998

L. Ackermann Smoller, "Teste Albumasare cum Sibylla: astrology and the Sibyls in medieval Europe", *Studies in History and Philosophy of Biological and Biomedical Sciences*, 41, 2010

A. Akasoy, G. Giglioni (eds.), *Renaissance Averroism and its aftermaths*, Dordrecht, Springer, 2013

S. Bettini, "Copernico e la Pittura Veneta", *Notizie da Palazzo Albani*, IV, 2, 1975

E. Berti, "Paduan Aristotelianism and the birth of experimental medicine", *Med Secoli*, 9(1), 1997

G. Bezza, "Tolemeo e Abû Ma'shar: la dottrina delle congiunzioni Saturno-Giove presso i commentatori tolemaici", in *From Masha'allah to Kepler. The Theory and Practice of Astrology in the Middles Ages and the Renaissance*, London, The Warburg Institute, 13-15 Nov. 2008, in press

B. Bilinski, "Il periodo padovano di Niccolò Copernico (1501-1503)" in *Scienza e filosofia all'Università di Padova nel Quattrocento*, A. Poppi (ed.), Trieste, Lint, 1983

L.A. Birkenmajer, "Niccolò Copernico e l'Università di Padova", in *Omaggio dell'accademia polacca di scienze e lettere all'università di Padova nel settimo centenario della sua fondazione*, Cracow, 1922

F. Bònoli (ed.), *I Pronostici di Domenico Maria da Novara*, Olschki, Firenze, 2012

E.A. Burrt, *The Metaphysical Foundations of Modern Physical* Science, London, K. Paul, Trench, Trubner & co., ltd., 1925

S. Buscherini, *L'Astrologia Storica. La Teoria delle Congiunzioni di Giove e Saturno e la Trasmissione dei loro Parametri Astronomici*, Mimesis, Milano, 2013

S. J. Campbell, *Cosmè Tura of Ferrara: style, politics, and the renaissance city, 1450-1495*, New Haven, Yale Un. Press, 1997

S. D'Amicone, "Apocalypsis cum Mensuris: l'Astrologo di Giulio Campagnola", *Venezia Cinquecento*, 2, 1992

S. D'Amicone, *I Cieli di Giorgione. Astrologia e Divinazione nel Fregio delle Arti*, ZeL Edizioni, Treviso, 2010

L. DeVun, *Prophecy, Alchemy, and the End of Time. John of Rupescissa in the Late Middle Ages*, New York, Columbia University Press, 2009

S. Fava, *Environmental Apocalypse in Science and Art: Designing Nightmares*, New York, London, Routledge, 2013

S. Feraboli (ed.), *Tolomeo: Le previsioni Astrologiche (Tetrabiblos)*, Milano, Mondadori, 1985

G. Gandolfi, "Le Metamorfosi dell'Astrologo da Giulio Campagnola a Matthias Gerung", in *Atti del XII Convegno della Società Italiana di Archeoastronomia, Albano Laziale 5-6 ottobre 2012*, La Città del Sole, 2014

G. Gandolfi, "Realismo e Verosimiglianza nel Contenuto Astronomico di Opere d'Arte Rinascimentali. Parte I: il Caso del Fregio Giorgionesco di Casa Marta a Castelfranco (TV)", in *Atti del XIII Convegno della Società Italiana di Archeoastronomia, Sassari 14-16 novembre 2013*, in press, 2016

G. Gandolfi, "L'Eclissi e l'Orbe Magno del Leone: l'Affresco Astrologico del Duomo di Montagnana come Profezia Anti-Turca", in *Atti del XIV Convegno della Società Italiana di Archeoastronomia,Padova 17-18 ottobre 2014*, 2016

G. Gandolfi, "Un nuovo catalogo di Arte Astrale Medievale e Rinascimentale in Italia", *in preparation*, 2016

G. Gandolfi, "The Star of the Sybil: Analysis and History of a Late Medieval Illustrated Prophecy", *in preparation*, 2016

A. Gentili, "Giorgione", *Art e Dossier*, 148, 1999

O. Gingerich, "Copernicus and Tycho", *Scientific American*, 229, 1973

B. Goldstein, "Copernicus and the Origin of his Heliocentric System", *Journal for the History of Astronomy*, 33, 2002

J. Green, *Priniting and Prophecy*, Ann Arbor, University of Michigan Press, 2011

T. Golsenne, "Une astrologie ornamentale: la 'Madone au Zodiaque' de Cosmé Tura" in *L'arte de la Renaissance entre science et magie*, P. Morel (ed.), Somogy, Roma, 2006

E. Gombrich, *Aby Warburg: An Intellectual Biography*, London, the Warburg Institute, 1970

O. Gotze, *Der öffentliche Kosmos*, Munchen, Utz, 2010

E. Guidoni, A. Marino: "Cosmus Pictor. II 'nuovo organo' di Ferrara: armonia, storia, e alchimia della creazione", *Storia dell'arte*, 4, 1969

P.O. Kristeller, "Paduan Averroism and Alexandrism in the light of recent studies" in *Renaissance thought and the arts*, Princeton, Princeton Un. Press, 1980

B. McGinn, *Visions of the End: Apocalyptic Traditions in the Middle Ages*, New York, Columbia University Press, 1979

G. Mori, *Arte e Astrologia*, Firenze, Giunti, 1987

P. Moffitt Watts, "Prophecy and discovery: on the spiritual origins of Christopher Columbus's "Enterprise of the Indies"", *American Historical Review*, 90, 1, 1985

S. Mossakowski, "The Symbolic Meaning of Copernicus' Seal", *Journal of the History of Ideas*, 34, 3, 1973

B. Nardi, "Copernico studente a Padova" in *Mélanges offerts à Etienne Gilson de l'Académie française*, Toronto, 1954

O. Niccoli, *Profeti e popolo nell'Italia del Cinquecento*, Roma-Bari: Laterza, 1987

M. Pastore Stocchi, "Astrologia e Umanesimo: il caso di G.B. Abioso", in *Pagine di storia dell'Umanesimo italiano*, Milano, Franco Angeli, 2014

L. Pepe (ed.), *Copernico e lo Studio di Ferrara. Università, dottori e studenti*, Bologna, CLUEB, 2003

P. Porçal, "La Capella Sassetti in Santa Trinita a Firenze: osservazioni sull'iconografia", *Antichità viva*, 23, 1, 1984

J.H. Randall, "Development of scientific method in the School of Padua", *Journal of the History of Ideas*, 1, 177, 1940

M. Reeves, *The influence of Prophecy in the later Middle Ages: a study in Joachimism*, Oxford, Clarendon Press, 1969

L. Russo, *La Rivoluzione Dimenticata*, Milano, Giangiacomo Feltrinelli Editore, 1996

C.B. Schmitt, "Aristotelianism in the Veneto and the origins of modern science: considerations on the problem of continuity", in *Aristotelismo Veneto e Scienza Moderna vol.1*, ed. By L. Olivieri, Padova, 1983

S. Settis, La "*Tempesta*" *interpretata: Giorgione, i committenti, il soggetto*, Torino, Einaudi, 1978

G.J. van der Sman, *Le siècle de Titien: gravures vénitiennes de la Renaissance*. Zwolle, Waanders, 2003

N. Swerdlow, "Astronomy in the Renaissance" in *Astronomy Before the Telescope*, C. Walker (ed.), New York, St. Martin's Press, 1996

L. Thorndike, *A History of Magic and Experimental Science*, vol. II, New York, Columbia University Press, 1923

T. Tonetto, "Nuove proposte per la pala di San Giuseppe di Sebastiano Florigerio nel duomo di Cividale", *Arte Documento*, 15, 2001

C. Vasoli, "Il Mito dell'Età dell'Oro nel Rinascimento", in *Giorgione e l'Umanesimo Veneziano*, R. Pallucchini (ed.), Firenze, Olschki, 1981

W.A. Wallace, *Galileo and his sources: Heritage of the Collegio Romano in Galileo's science*, Princeton, Princeton UP, 1984

R. Westman, *The Copernican Question: Prognostication, Skepticism, and Celestial Order*, Univ. of California Press, 2011

26

PREHISTORIC ROCK SANCTUARY WITH ARCH NEAR THE VILLAGE OF KOVACHEVITSA, BULGARIA: SPATIAL ORIENTATION AND SOLAR PROJECTIONS

Alexey Stoev[1], Penka Maglova[1], Vassil Markov[2] and Mina Spasova[3]

[1]*Space Research and Technology Institute, Bulgarian Academy of Sciences, Stara Zagora Department*
[2]*South-West University "Neofit Rilski", Blagoevgrad, Bulgaria*
[3]*Institute for the Study of Societies and Knowledge, Bulgarian Academy of Sciences, Sofia, Bulgaria*

Corresponding author: Alexey Stoev (stoev52@abv.bg)

ABSTRACT

Study of spatial - temporal organization of the sanctuary interior of rock - cut monument with rock arch is presented in this work. The sanctuary is located in the Western Rhodopes, village of Kovachevitsa, Garmen Municipality and it is archaeologically dated back to the Eneolithic. The monument is about 20 meters long and it is part of a large rock ridge. The topographical, landscape and microrelief features of this monument are described and the excellent view of the ancient observer to the sky and local horizon is noted. An analysis is made of the structural elements of the sanctuary and possible observational purposes. Astronomical azimuths of the base lines connecting the arch opening and the other elements of the sanctuary are determined. It has been found that sunrise during the summer solstice can be observed from the observational platform through the arch opening. Chronological boundaries of existence of the rock-cut monument are determined. They refer to the era of the late Eneolithic - III - II millenium BC.

KEYWORDS: Rock sanctuary, rock arches, solar projections, extreme rises of the sun.

1. INTRODUCTION

The study of the spatial organization of the rock-cut monuments (rock sanctuaries) and the presence of linear and circular structures gives serious evidence of solar orientations, natural relief and architecture in antiquity associated with the cult of the sun. Therefore, rock-cut monuments are subject of separate and independent archaeoastronomical investigation. The study is based on our knowledge about the place of astronomy in the culture of the ancient archaic societies lived in Bulgarian lands, and the recording, analysis and interpretation of their morphometric modifications (Potemkina et al., 2006).

Figure 1. Rock arch on the territory of the megalithic sanctuary "Tsarevi porti" (Royal gates) near the village of Kovachevitsa, Garmen Municipality

Longtime research has shown that the earliest era in the history of astronomy in Bulgarian lands is associated with observations on the horizon and fixing points of the horizon from one observational point (rises and sets of the sun, moon, stars, planets). Later, special constructions are used for observations of culminations of the sun and the light projections.

On the other hand, in ancient astronomy, structure of morphological elements, height and depth of the horizon, symmetry, fractal, isomorphism and others are parameters that affect us in the perception of relief and generate subjective perceptions of its relationship with the sky, celestial objects and phenomena. The totality of natural objects and natural conditions is a valuable spatial resource used by ancient astronomers and priests (Stoev et al., 2013).

In later times, natural rock alignments were replaced by artificial building constructions, which are now called dolmens. The dolmens are more complex structures - fixed plate (roof) above the ground with supporting elements. It turns out that the morphology of the later dated objects clarifies somewhat morphometric characteristics of the earlier sites. We see some similarities (which illustrate the cultural continuity, tradition) as well as some differences (expression of cultural development and technological evolution). The comparison allows us to draw the following conclusions:

First: The studied rock arches show that besides from the direction of the sun's extreme sunrises and sunsets, the creators of these sacred objects were also attracted from the directions of sunrise and sunset on random days of the year. Probably, they have coincided with important days of the cult calendar of the ancient societies. All these directions were sought in both the design and realization of the monuments from the so called epoch of dolmen building.

Second: The earlier megalithic culture, relying on natural forms and elements additionally incorporated in them, at the end of the Ist millenium BC has been replaced by a new megalithic culture with completely artificially created monuments. Sacred, religious and ideological content of the later structures (reflected in the orientation and ritual practices) has been maintained in the new historical era. The basic principle-orientation in compliance mostly with the Sun God at its extreme rises and sets and occasionally with sunrise and sunset on the equinoxes remained the same in the new era with an addition-orientation towards the highest daily position of the sun during different seasons. This proves in one more way the continuity of evolution in the general and special cultural life of the society of ancient Thrace.

Figure 2. Main structural elements of the rock sanctuary "Tsarevi porti"

Third: The evolution of rock arches is caused not only by improving orientation construction technique, but also by the development of theological doctrines. The most significant religious ideas, practices, customs (expressed in and through orientation) are emphasized. Temple architecture of the sacred monuments is simplified and the astronomically significant directions are further accented with additional rock-cut shapes. Sun worship has undergone major development towards simplification. This process is a demonstration of the parallel processes of territorial, economic and political consolidation among the Late Eneolithic and Thracian tribes. The common observational practice in the investigated natural monuments (rock arches) is associated with probable prognostic features of the rock complexes connected with the daily and annual solar cycle (Maglova et al., 2010).

Rhodope Mountains emerged as a specific cultural area in the Chalcolithic and Bronze Age - contact zone between the Carpathians and the Aegean. From the period of early Eneolith (Karanovo V), in the Rhodopes so far have not found traces of permanent population. Since the second half of the Eneolith begins the expansion of the peoples of the Thracian lowland to peripheral areas. There are large changes in the lands of the central Balkans in the Late Eneolith. It is established tribal unity on huge territory lasted about 400 years - the culture Kodzhadermen-Gumelnitsa-Karanovo VI. Rhodopes fall in the southern periphery of this tribal confederation stretching north across the Danube river.

Figure 3. Groove for outpouring of sacred fluids (libations) on the upper part of the rock arch "Tsarevi porti"

This archaeological culture possessed dual nature of their governance bodies - social and religious. New types of objects - a craft-production centers, rock sanctuaries and probably some gold mines – are organized and acted to the end of that epoch to service the needs of the new social formation.

Figure 4. Main opening of the rock arch "Tsarevi porti" for sun projections viewed from the stone-altar

Such objects are found only in the southeastern part of the mountain, while the southwest part, along the Mesta river, remains less populated and without particularly large settlements. Rather, it remains territory reserved for religious centers associated with major cults of the Great Mother – Goddess and God - Sun (Raduncheva A., 2003).

2. DESCRIPTION OF THE ROCK SANCTUARY "TSAREVI PORTI"

The discovery of the rock arch "Tsarevi porti" in the land of the village of Kovachevitsa, Garmen municipality (Figure 1) was a result of research of the National complex scientific expedition "Thracian sanctuaries from the Western Rhodopes, Pirin and Rila" from 2002 to 2006 (Gotcev and Markov 2005), organized by South-West University "Neofit Rilski" - Blagoevgrad, National Archaeological Institute with Muzeum, Bulgarian Academy of Sciences, Sofia and Plovdiv Regional ethnographic Museum.

Figure 5. Sacrificial altars of the sanctuary "Tsarevi porti"

In 2009, within the frames of the fieldwork during the complex expedition, detailed archaeoastronomical research on spatial-temporal organization of the sanctuary interior of the rock-cut monument is made of specialists from the Space Research and Technology Institute, Bulgarian Academy of Sciences, Stara Zagora Department.

Figure 6. The stone-altar, standing on the line of the main astronomical azimuth of the sanctuary

The part of the Rila-Rhodope massif, where the rock arch is located consists of system of structures composed of igneous rocks (granite, gneiss, schist, marble and others), shaped at the end of the Early Cretaceous. In the early Paleogene the rock massif was subjected to stretching and faulting, which leads to the formation of depressions filled with sedimentary and volcanogenic rocks. It is characterized by a complex tectonic structure.

The rock arch "Tsarevi porti" is formed in gneiss rocks. Geological research in the area of the village of Kovachevitsa, shows that wind and water erosion is unlikely as physical-chemical factor in the formation and development of rock arches in gneiss.

Lithological studies of the rock massif show that the physical processes of weathering and selective weathering led to the formation of the arch. Process of formation of the arch is seriously affected by disturbance of masses in the bedrock because of geomorphological change of the slope as a whole. The reason for this is cutting of the relief in the line of the river valley (Katskov and Marinova, 1992).

The rock sanctuary near the village of Kovachevitsa, Garmen Municipality is located in the composition of a large rock group with a length of 20-25 meters, oriented in a natural way, almost in the plane of the meridian. There is an excellent view of the sky and horizon. After the analysis of the structural elements of the monument the essential observational purposes who have placed ancient observers are determined. The geometric center of the observational lines connecting extreme points of sunrises with the characteristic relief of the rock arch were studied in the process of archaeoastronomical research. Survey of the rock arck has been done with an accuracy of ± 5 " and the geographic coordinates have been determined using GPS receiver (φ = 41 ° 42 '05,5", λ = 23 ° 48' 55,4", h = 1147 m.

Figure 7. The limb of the northeastern horizon seen through the opening of the arch of the sanctuary "Tsarevi porti"

3. ARCHAEOASTRONOMICAL RESEARCH

Choice of the specific rock group with a particular orientation is deliberate. All astronomical hypotheses about the described sacred territory are associated with the lintel of the arch. Projection of the light opening is connected with additional elements of the rock sanctuary, located close to the main axis of the monument (grooves, altars) (Maglova et al., 2010).

The rock arch is cut high in the mountains, in a natural rock. To the south, there is a large ground with hewn in the bedrock altar. During the summer solstice the sun rises over the opposite peak named Popova kapa and the sunlight goes through the arch and illuminates the altar (Figures 2, 3, 4, 5 and 6).

It became clear that cult activities are concentrated in two main centers respectively in position "above" and "below", similarly to the great part of the investigated megalithic sanctuaries in the region (Markov et al., 2008).

In structural and semantic terms, these are positions related to space, and the underworld. "Outpouring" of sacred fluids through the rock-cut grooves and rock arches in the rock pools gives a reason to believe that this is evidence of ritual bath as part of an ancient ritual during the summer solstice.

This ritual could be found in the ritual practice of many Indo-European peoples. Moreover, probably during the summer solstice "Feast of manhood Sun" was a popular practice in some of the largest megalithic temples in the Western Rhodopes.

Calculations show that during the summer solstice, when the sunlight passes through the opening along the axis of the facility, the outer lower edge of the lintel of the arch is projected exactly on the sacrificial altar, separate rocks and rock groups around the facility (Figures 7 and 8). This is connected with building of a system of markers and observational points along the line of sight from the place of observation to astronomically significant point on the horizon (sunrise over Mount Popova kapa at summer solstice) (Stoev and Maglova, 2002).

4. RESULTS AND DISCUSSION

Most probably, the basic geometric elements of the arch are connected with cosmogonic view of the ancient Eneolithic communities that have considered the celestial sphere in direct relation with different spatial orientations of the rock arches relative geographical directions.

This can be interpreted as reconciling the needs of economic and social life, which is found in the cult of the Great Mother Goddess. Her woship originated during the Paleolithic, where She appeared as a self-sufficient for the movement of the World cycle.

For the primitive religious consciousness Earth is taken for granted: its scope, its hardness, its diverse topography and vegetation are living and active cosmic unity (Markov, 2007).

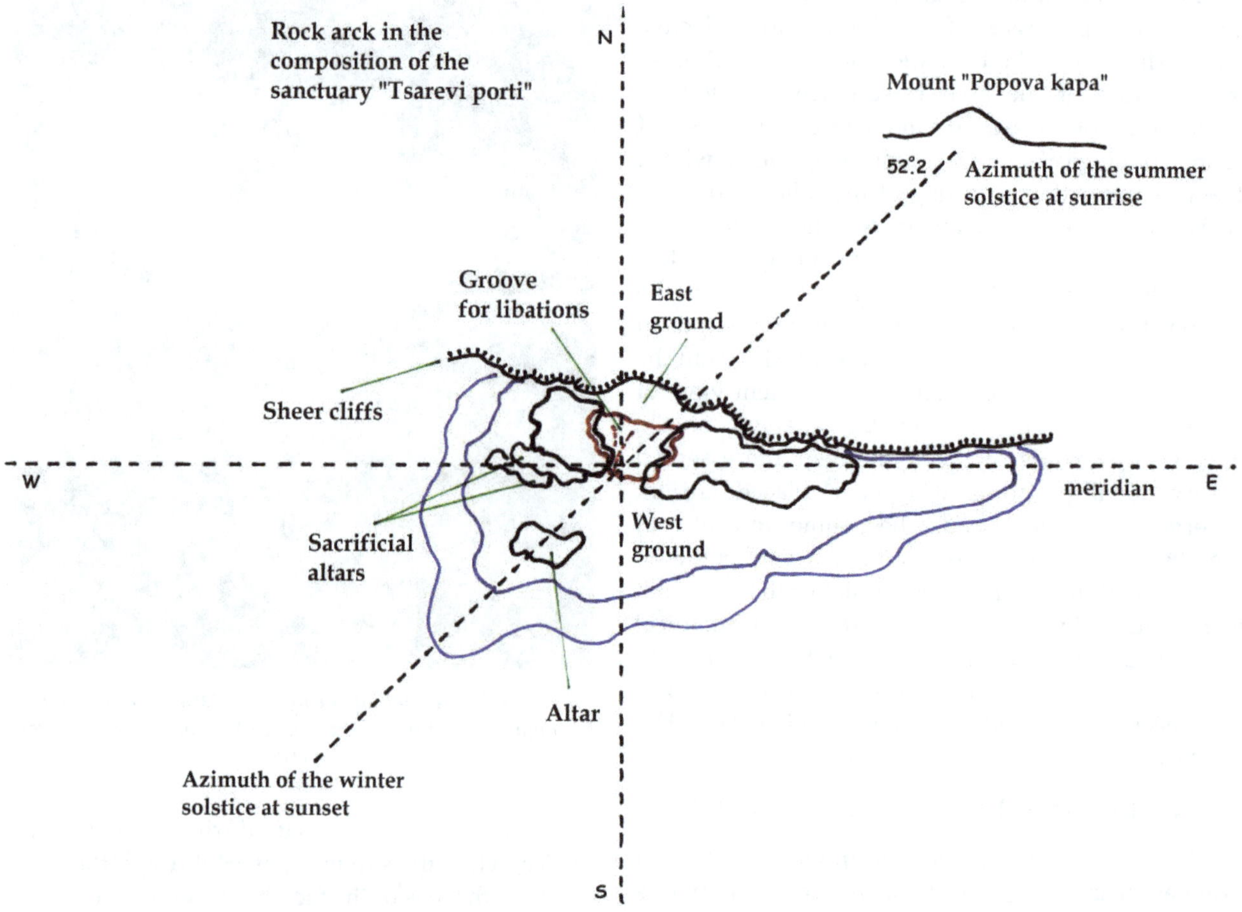

Figure 8. Orientation of the main structural elements and astronomical azimuths within the rock arch "Tsarevi porti"

5. CONCLUSIONS

The investigated megalithic monument-rock arch "Tsarevi gates" near the village of Kovachevitsa, Garmen Municipality is oriented along the axis connecting the light opening with extreme sunrise in the east horizon line that matches the characteristic peak "Popova kapa". Eastern orientation, geometry and ratios between the different functional elements of the sanctuary show that it has been deliberately constructed and continuously exploited for maintenance of solar calendar.

Archaeoastronomical hypothesis about the rock arch "Tsarevi porti" is connected with direct observations of the sunrise during the summer solstice, and solar projection, produced by the passage of sunlight through the rock hole that falls on exactly corresponding to its size entrance marker (rock altar).

The problems of more general observational practice in the space of this and other already investigated megalithic monuments, as well as probable prognostic capabilities of rock complexes associated with daily and annual solar cycle will be subject to additional archaeoastronomical research. This also applies to the modeling of spatial-temporal organization of the sanctuary from the Bronze Age to the Chalcolithic included in the value system of the society of that time and their transmission in later epochs until today. An important test for the credibility of archaeoastronomical hypothesis of this study related to the calendar projections in the axis of the facility is the attempt to implement it in the later Thracian cult architecture.

ACKNOWLEDGEMENTS

We thank the anonymous reviewers for their constructive comments. This work was partially supported by the University Research Centre for Ancient European and Easterly Mediterranean Cultures of the South-West University "Neofit Rilski", Blagoevgrad, Bulgaria. The authors would especially like to thank the students from the University for their work during the expeditions.

REFERENCES

Katskov, N. and Marinova, R. (1992) *Geological map of Bulgaria in scale 1: 100 000*, Committee for Geology and Mineral Resources Series, Enterprise geophysical surveys and geological mapping, Sofia, 9-11, 29-34.

Markov, V., Gotsev, A. and Yankov, A. (2008) *Field studies of the Western Rhodopes, Prin and Rila*, Archaeological Discoveries & Excavations Series, National Archaeological Institute with Museum, BAS, Sofia, 9-10.

Markov, V. (2007) *Cultural Heritage and continuity. Heritage from the ancient pagan sacred places of the Bulgarian folk culture*, Neofit Rilski University Press, Blagoevgrad.

Maglova, P., Stoev, A. and Stoeva, M. (2010) Spatial orientation of triliths (rock arks) from the Western Rhodopes: Archaeoastronomical interpretation. *Boundaries of culturology*, Blagoevgrad, Neofit Rilski University Press, 53-71.

Potyomkina, T. M., Maglova, P. V. and Stoev, A. D. (2006) Towards methodology of the archaeoastronomical investigations. *View of life of the ancient and traditional societies of Euroasia. Collection of papers in memory of V.N. Chernetsov*, RAN, Institute of Archeology, "TAUS", Moscow, 189-197.

Raduncheva, A. (2003) *The Late Eneolithic society in the Bulgarian lands, Excavations and surveys*, Sofia.

Stoev, A. and Muglova, P.(2002) Seasonal meridional Sun culminations in the cult practices of the archaic societies, *Proceedings of the National astronomical conference*, 31 May - 1 June 2002, Shumen, 56-64.

Stoev, A., Maglova, P. and Spasova, M. (2013) Megalithic culture of the Eneolithic societies: Archaeoastronomical aspects. *Proceedings of the First International Symposium "Ancient cultures in South-East Europe and the Eastern Mediterranean. Megalithic monuments and cult practices"*, October 11-14, 2012, Neofit Rilski University press, Blagoevgrad, 194-206.

27

THE GUNDESTRUP CAULDRON: IS THIS THE KEY TO THE ENIGMA?

Claude Maumené

Independant scholar

Corresponding author: Claude Maumené (c.maumene2@orange.fr)

ABSTRACT

The Gundestrup cauldron is one of the most emblematic and studied archaeological relics of the Celtic world but also one of the most mysterious.

Observation of the night sky, the symbolism of the cauldron and of numbers five and eight, traces of an Indo-European base-ten computation, the antiquity of the octaeteric calendar system, all suggest a new interpretation of the Gundestrup cauldron, based on the cycle of Venus.

The five inner plates would each respectively represent the structure of time, as it appears in a computation close to that of the « octaeteris », rhythmed by the synodic cycle of the planet Venus, of the Moon and of the Sun. As for the eight outer plates, as Paul Verdier had suggested, they would represent astronomical events, or in other words, the dates or the periods associated with the appearance-disappearance of celestial object(s) or constellation(s). We particularly underline the importance of the male and female arm movements seen on the different plates. The shining hands of the divinities would rise just like the diurnal body (the Sun) or the nocturnal body (Venus), rising at dawn and in return setting at twilight. The figures associated with the Deities would evoke the constellations they represent, at times under, or, at other times, just above the horizon or sometimes higher up in the night sky. So each motif would recall nocturnal celestial events, observed at dawn or at twilight, corresponding to the heliacal rising or setting of the constellations detected. The two plates portraying the female divinity, her arms folded away under her shoulder line recalling the earthly horizon, would evoke the disappearance of Venus below the horizon, each of these two plates corresponding to the two periods during which the planet is not visible.

KEYWORDS: Gundestrup cauldron, astronomy, calendar, Venus, octaeteris.

1. INTRODUCTION

The Gundestrup cauldron is one of the most emblematic archaeological testimonies of the Celtic world. It was discovered in 1891 (eighteen ninety-one) in a peat bog in Gundestrup in Denmark.

The cauldron is traditionally estimated to date from the first or second century before Christ. However, its precise dating is still to be determined (Nielsen, 2005). Its geographical origin is also subject to debate (Berquist et al., 1987 ; Kaul and al., 1991 ; Taylor, 1992 ; Nielsen, 2005). To all evidence the Gundestrup cauldron is a highly sacred relic. It is made of pure silver. The outer plates each portray a male or a female bust. The five inner plates depict scenes of figures surrounded by animals, some of which are mythical. The numbers eight and five, corresponding to the eight outer plates and the five inner plates have oriented our research in the direction of Venus.

Five synodic cycles of Venus and eight solar years do indeed correspond to an almost identical number of days, and to the number of days in ninety-nine lunar months.

Figure 1. The Gundestrup cauldron

This coincidence is the origin of the calendar system known since Antiquity as the octaeteric cycle[1]. Apparently this eight year's calendar left traces particularly in Greece, Crete, Armenia, Albania, Lithuania, and even as far away as in China, among the Yi people (Broutian, 1997 ; Gregoraides, 2009 ; Kalinowski, 1986 ; Straižys, 1997 ; Maumené, 2012 ; Panagiotes and al., 2008 ; Weibao, 1996 ; Weibao, 1997 ; Weibao and al., 2010).

The imagery depicted on the cauldron possibly enabled memorization of time structure and certain key dates, thanks to representations of the night sky (Verdier, 1998 ; Goudineau et al.,2006).

[1] Octaeteris » was described by the Greek astronomer Cleostratus (around 548 to 432 BC).

The five inner plates would materialize the elements of a calendar close to the octaeteris, in rhythm with the cycle of Venus, the Moon and the Sun. The eight outer plates would represent astronomical events, dates or periods, matching the appearance or the disappearance of remarkable celestial objects.

2. THE FIVE INNER PLATES

On Figure 2, we suggest each horse rider represents the four seasons of the Celtic year, each season presided over by one of the four opening festivals, Beltane, Lughnasadh, Samain and Imbolc.

The horsemen are linked two by two, each group composing one of the two light and dark halves of the Celtic year, just as, for example, they appear in the Coligny calendar. The rider leading the procession, wearing a bird crest, might symbolize the Beltane festival, since it takes place in the nesting season and is also when the migratory birds return[2]. The second rider's helmet, crested with a boar, could represent Lughnasadh. Many common features link the boar and Lugh, the Celtic god. We are tempted to associate the third rider's helmet bristling with horns with Cernunnos and the Samain festival. The last rider wears a mysterious helmet. Traditional Irish folklore mentions the making of a belt on the date of Imbolc, using a braided straw rope through which the cattle and the family members were made to pass (Le Roux and al., 1995). It is tempting to compare this with the thing above the helmet with its semi-circular profile and braided aspect. If the four horsemen represent the Celtic festivals and the seasons they preside over, we think the legs and the fore and rear parts of the riders could possibly represent the three lunar months of each season, and symbolize the twelve lunar months of the year.

Figure 2. The four seasons of the Celtic year

[2] This association of the cavaliers and the four seasons is reinforced by the décor engraved on a bronze sword case from Hallstatt (Naturhistorisches Museum Vienne), also portraying the four cavaliers and the procession of foot-soldiers associated with an eight-spoke wheel.

The foot-soldiers are seen to represent the eleven days required to synchronize the Moon and the Sun. A year consists in fact of 12 synodic moons + 11 days. The ground level indicated by the frieze of vegetation would represent the boundary between the terrestrial and the underworld from which the eleven foot-soldiers come, resuscitated here by the Dagda[3] (who acts here as a regulator of time). Naturally the dog, generally considered to be a chtonic animal belonging to the warrior class, finds its place in the underworld. The last foot-soldier carrying a sword on his shoulder and wearing a boar for crest could mark the end of the season of Lugh. He is followed by three foot-soldiers whose battle cry, amplified by their carnyx, seems to emerge from the underground world into the world above. They could possibly represent the three nights of Samain, marking the end of the battle season.

2.1.1. The goddess with five animals

Figure 3 represents a goddess surrounded by five animals, two elephants, a wolf and two griffins, and also two rosettes with six petals which probably stand for the twelve lunar months[4]. This divinity would represent the moon. She is crossing her arms. The gestural language of the various figures is one of the mysteries of the Gundestrup cauldron. The arms raised would, we think, be the sign of the rising and setting of the celestial body each figure represents. The gilding of their hands and of the upper part of their arms reinforces the idea of a symbolic astral associated with the arms. If, respecting the tradition, we place the deities in the North of the world, the East is on the right and the West on the left side of the observer.

Figure 3. The moon with five animals or the dark season

Supposing the shoulder line corresponds to the horizon, the moon would not be rising at this point. So we would be in the new moon period, the ideal phase to symbolize the dark season. The five animals would represent five of the months of a ten month year[5], or half a year, that is to say, the dark season placed under the sign of the moon. In the Yi calendar, the ten months of the year are each represented by an animal. Other antique calendars present a monthly code in animal forms. Françoise Bader demonstrates the use of animals in Antiquity, as a coding, designating the calendar months[6].

2.1.2. The god with five animals

Obviously this plate (figure 4) forms a pair with the preceding one. Here we would be in presence of a representation of the light season. The five animals would represent the other five months. The central figure, depicted as a bust, both arms raised[7], would

[3] The Dagda is indeed the « governor of time », « guarantor of fertility and of earthly replenishment». His height is colossal (Rabelais will name him Gargantua) and among his attributes is the cauldron of abundance, of immortality and resurrection. He is the great Lord of the elements, of the cosmos and of time, the first of the Druids, assimilated with the sun in his diurnal aspect. He perceives all things within the circle of his course. He reigns over the living and the dead, he is the master of heaven and of the underground world.

[4] Taylor T. 1992 : p.84-89 followed by De Gourcuff (to be published) have pointed out the connection of this illustration to that of the Indian Goddess Lakshmi, likewise surrounded by two elephants and two Lotus flowers, counterparts of the two rosettes of plate C6573. Indian astrology assimilates Lakshmi and the planet Venus.

[5] Some calendar vestiges suggest the existence of an ancient computation containing ten months, in some cases of 36 days, sometimes assembled in pairs, forming five seasons per year. Indeed, the Etruscan, Roman, Armenian and Lithuanian calendars take into account 10 months of 30 or 31 days, then add a period of 2 months to terminate the calculation. This period could correspond to a period outside time as it appears in the Indo-Nepalese calendar system (the latter accounting for 12 lunar months of which 2 have no spatiotemporal correspondence, being considered outside time). On their side the Greek, Minœn and Yi calendar systems account for 10 months of 36 days. An intercalation allows for adjustment of time following the solar cycle, inserting respectively 5 or 6 days per year (or 21 days every four years).
The year is cut into 2 halves in the antique Roman calendar system, as in the Yi and Indo-Nepalese calendars, as well as in the Celtic calendar of Coligny. These calendar systems sometimes join the months in groups of two. Most of the time they are the product of Indo-European cultures: Roman, Lithuanian, Armenian, Greek, and Minoan, Indo-Nepalese, and also Yi or Etruscan. In current everyday language we still recollect a ten month calendar, not just because the name December means the tenth, but also in current expressions such as «every thirty-sixth day of the month », or « to see thirty-six candles », referring perhaps to the length of a month which has disappeared.

[6] Bader, F. 1997, p. 490-515.

[7] This representation of divine busts and gestural coding seem to be echoed in a bronze pendant recently discovered in the tomb of a woman, buried in a Celtic necropolis, representing a stylized figure identified as Lugh by Daniel Gricourt and Dominique Hollard. The deity is recognizable due to his lyre-shaped legs (image linked to a solar divinity) and by his head in the form of a ring

be an avatar of the sun. The two arms lifted would signify it rises in the East in the morning and sets in the West in the evening. The portion of a wheel would evoke the half-year period and its eventual division into eight sub-period.

Figure 4. The sun with five animals or the clear season

2.1.3. Cernunnos and the eight seasons

This plate (figure 5) is probably one of the Gundestrup cauldron's most enigmatic elements. We consider it can be understood with reference to the cycle of Venus and to the octaeteric cycle. Cernunnos is surrounded by eight animals. They would represent eight seasons of 72 or 73 days forming the complete Venus cycle of 584 days. Or they might represent even more simply the eight years of the octaeteric cycle[8].

Figure 5. Cernunnos and the eight seasons

2.1.4. The sacrifice of the bull

We suggest seeing in figure 6, the constellation of the Bull, well known to the people of Antiquity, and also the Lesser and the Greater Dog, as well as Orion the warrior. All these constellations are indeed present in the same part of the sky and are close to one another. The scene would depict the sacrifice of the bull. Our hypothesis is that the three bull sacrifices were associated with the insertion of three embolismic months, based on the octaeteric cycle.

Figure 6. The sacrifice of the bull

3. THE EIGHT OUTER PLATES.

3.1. *The feminine deities.*

The goddess with her right arm raised is holding a bird which looks like a dove, known to be an attribute of Aphrodite and Astarte. So the central deity could represent the planet Venus, which is clearly associated with these deities.

On the left side, a figure corresponding to a counterpart of the central divinity (the same hairstyle, braids, and torque) could be a second representation of the goddess. From an astronomical viewpoint, seated on the shoulder of the divinity, we would symbolically attend to the rising of the planet Venus, to its first appearance in the evening sky.

The shoulders of the goddess would represent the earthly horizon. The two dogs would respectively represent the Lesser Dog (*Canis Minor*) and the Greater Dog (*Canis Major*), lying under the earthly horizon in a position evoking that of a dead dog.

Figure 7. The goddess Venus

The figure lying on the right, also dead, would represent Orion, also concealed under the horizon, so belonging to the world of the dead. On the right another feminine figure, identical to the figure on the left, is standing as if suspended in the air, braiding the hair of the goddess. The only reference known relating to hair in the sky brings us back to Berenice.

(heliacal aspect). The arms are held aloft, the same attitude as the figures of the cauldron (Gricourt D. et Hollard D., 2005, p. 37).

[8] It is also possible to see in them a sort of interlocking of the 8 days, allowing for adjustment of the 8 seasons of 72 days each to the length of the Venusian cycle (8x72+8=584). Alternatively, the 8 animals could be interpreted as the 8 divisions of the Celtic year, equinoxes and solstices, intersected by the 4 Celtic festivals, Beltane, Lughnasadh, Samain and Imbolc.

She would have sacrificed her hair to Aphrodite. Her hair was reflected in the sky in the form of a group of seven stars, an identical number to that of the braids illustrated on this plate[9]. This scene could recall, from an astronomical viewpoint, Venus and Berenice's hair at the precise moment when the planet and the constellation are visible in the East prior to the break of dawn, at the same level above the horizon.

In short, the remaining feminine plates represent the goddess with her arms crossed. The arms folded under the shoulder line would evoke the disappearance of Venus under the horizon, during the periods of conjunction (inferior and superior) of the planet with the sun.

Figure 8. Venus arms almost crossed (inferior conjunction)

Her tightly closed mouth and slightly swollen cheeks suggest the divinity could be immersed.

The surface of the water would be materialized by the dots ; the waves on top of the image and the liquid element, by the underlying stippled line.

Astronomically, Venus, during the inferior conjunction disappears from the evening sky (under the horizon, possibly in the sea) then reappears in the morning sky - after eight or nine days. So the plate would portray the passage from one spouse to another, from an extremely hirsute so necessarily aging god corresponding to the evening sun, to a youthful smooth-skinned divinity, image of the morning sun. This celestial marital alternation is echoed in mythology (Gricourt and al, 2009).

The following plate (figure 9), also represents the same feminine divinity with her arms folded, or rather crossed. So this plate could depict the second period of the conjunction (the superior conjunction) of Venus, longer than the preceding one. Her arms, this time are totally crossed. It is worthy to notice that the two nipples of the deity are not visible on those two plates, indicating possibly that the planet is not visible, neither in the east, neither in the west

On left side, on the right shoulder of the goddess, in the West, a man is wrestling with a lion. The scene specifically recalls certain Greek representations of Hercules wrestling bare-handed with the lion of Nemea[10]. The group would logically have been codified as a celestial event connected to the constellation of the Lion, probably during its last appearance in the evening sky.

Figure 9. Venus arms fully crossed (superior conjunction)

In the East, on the left shoulder of the divinity, another figure is to be seen, similar to the one depicted in the scene of the bull sacrifice, except that he is unarmed. He is Orion again. The group could mark the date of the heliacal rising of Orion, and more specifically of Betelgeuse (the left shoulder), the brightest star in the constellation.

3.2. *The masculine deities*

The masculine deities represent the sun in the different phases of its cycle[11].

3.2.1. The dragon God

The scene is mysterious, the monsters in the hands of the god can be described as winged serpents with horses heads and wings. Possibly in the Greek records the image recalls Ladon, the fantastic reptile sent by Hera to look after the apples in the garden of the Hesperides. This suggestion is

[9] Four braids on the left side and three on the right side, becoming four.

[10] As for example, a terracotta amphora dated 540 BC, housed in Rouen, inventory n° 538.1(A) ; 9820030 Rouen (n° Charbonneaux Centre).

[11] On examining the four masculine plates, both a match and a chronology can be detected. The plates representing the god with a torque would form a pair, the chronology within each pair would result from the pilosity. In figure 10, the beard is at its beginning and the moustache is absent, indicating that this is the first plate, following a chronological order. It is followed by figure 11 in which the pilosity is more developed.
In the two following plates the torque is absent (or eventually hidden), and the pilosity is well developed, organised, in the form of harmoniously tidy locks. The order could be as follows : figure 12 then 13 based on the number of strands in the beard.

backed up by the presence in the same plate of a hideous two-headed dog (in the bottom of the image), possibly representing Orthos, the bicephalous dog of Hades, conceived like Ladon by Typhon and Echidna. The figure whose legs are separated from his body remains mysterious. As far as we know, only Soslan in Ossetian mythology, is seen to be mutilated like this, not by a hideous dog, but by a magic wheel.

Figure 10. The dragon God

The star Ɣ of the Dragon, the brightest, known to the Greeks as the Dragon star, is situated on the head of the monster. At times in the East, at others in the West, the Dragon is always visible above the horizon. The constellation Draco (Serpens) circles the celestial pole, which, according to the conviction of the Indo-Europeans, is held up by the cosmic tree. In plate (figure 10), the Dragon is represented in the East and in the West, overpowered by the divinity, far above the horizon, so without any reference either to a heliacal rising or setting. Like the group of stars and constellations close to the pole, the Dragon's stars never disappear below the horizon. Perhaps one should interpret in this manner the organization of the scene, where the Dragon seems to be captured above the horizon?

3.2.2. The God with a unicorn

Figure 11, represents, the same divinity, slightly older. On his left is a figure similar to the one on the central plate: with the same clothing, the same hairstyle and the same kneeling position. On this plate the figure is unarmed. It could concern a representation of Orion. As for the little figure on the shoulder of the God, it looks like a unicorn. The Indian donkey, said by Ctesias to be a unicorn, (or monoceros), already belonged to the realm of the imagination of the people of Antiquity in Persia as well as in India. It does not belong to the repertoire of Ptolemy. Joseph Juste Scaliger, a great French erudite of the XVIth century mentions the presence of a unicorn in a celestial sphere of Persian origin. When examining a map of the night sky, this constellation is situated immediately on the East of Orion. If accepted that there are representations of the sky in the plates of the cauldron, then the constellation of the Unicorn would be known since Antiquity, and would be illustrated on the cauldron, beside other images of the sky.

On the left of the deity we again find a figure we believe to be the Herdsman. This interpretation ensues from the examination of Figure 12 (see below) where the same figure appears, brandishing a wild boar at the end of his arm. This boar is interpreted as a representation of the Great Bear. The brightest star of the constellation of the Herdsman is called Arcturus (α Boötis). It is the guardian of the bear and is situated in the prolongation of the Great Bear's « tail ».

Figure 11. The God with a unicorn

3.2.3. The God with a boar

The quadruped brandished at the end of the arm of the figure, also lifted up by the divinity, is most certainly a wild boar, recognizable due to its dorsal mane. Indeed, the European boar has short, woolly fur, mixed with bristles, forming a mane along its spine. Numerous Celtic representations of the boar depict this same characteristic feature on the back of the animal. So the resemblance between the crest of the cavalier and the foot soldier of plate figure 2 leads to recognition of the same animal.

The boar is associated among the Celts with the sacerdotal class and with the God Lugh, as well as with the warrior class. Its presence on Gallic coins, together with a circle with a globe in its middle and 7 other globular clusters linked together by an arc has made it possible to relate the boar to the constellation of the Great Bear. The figure holding the boar at arm's length could then indeed be the Herdsman, accompanied by the Great Bear. Astronomically both constellations are in the exact spatial configuration depicted by the plate of the cauldron.

Figure 12. The God with a boar

The Herdsman seems to follow the Great Bear in the sky. The two animals on either side of the central divinity might respectively represent the heliacal rising of Pegasus (α Pegasus) and the heliacal setting of the Greater Dog (Sirius), respectively corresponding to the beginning of February (the 3rd) and the end of April (April 28 to 30).

3.2.4. The God with stags

Here, the divinity seems to be the oldest one. She is holding a deer in each hand, by its back legs. In Greece, the sacrifices of deer and hunts carried out in honour of Artemis took place in the ninth Attic month Elaphebolion corresponding to the period from mid-February to mid- March, moulting period of their antlers by all cervids. In the Coligny calendar the name usually said to correspond to the tenth month Elembiu, contains the Indo-European root for the deer, elen-(bho-) in Greek elaphos (elnbhos).
The stag (figure 13) recalls a constellation of the Stag, of which very little traces are left. Indeed Ptolemy does not mention it, but it is present in Roman and Mesopotamian Antiquity. It is mentioned by Firmicus Maternus, a Latin writer of the fourth century AD, who situates it on the North constellation of Pisces. In Mesopotamia, the constellation of the Stag was assimilated to the oriental part of Andromeda, specifically to α Andromeda. In both examples this means we are in the square of Pegasus, North of Pisces. If α Andromeda is used as reference point, its heliacal rising takes place towards the beginning of February (February 3 to 5) and its setting in the beginning of March (March 5 to 7). But the stags appear on a stippled setting, identical to the one around the goddess with her arms crossed. Perhaps in this representation, the period of invisibility of the Stag, rather than the period separating its rising and heliacal setting is to be perceived?

The base plate would represent the region of the sky assembling the Bull, Orion and his Dogs. One of

Figure 13. The God with stags

4. THE CENTRAL PLATE OF THE BULL

Figure 14. The central plate of the bull and the hermaphrodite warrior

the most enigmatic elements is certainly the sex of the warrior. Indeed he seems to have a bosom, making him a woman warrior, a sort of Amazon. Astronomically, following the thread of a Venusian presence associated with the cauldron, the scene could correspond to the conjunction between Venus and the constellation of Orion, a kind of feminization of the hero, merging the bodies into a hermaphroditic form. This particular celestial event takes place every eight years. So it could have been used as a point of reference, setting off the ritual ceremonies associated with the change of the calendar cycle.

5. CONCLUSION

The precise meaning of certain images of the Gundestrup cauldron still remains unsolved. But the main signification of the vessel has been partially revealed to us.

It is a commemoration of the order of the universe. Its commissioners were most probably well-informed observers of the sky. The most ancient astronomical text we have available is a list of observations of the movements of the planet Venus, particularly important in Mesopotamia, being identified with the goddess « Inanna « (Ishtar), more than 1600 years BC. In parallel, many myths, above all Mesopotamian, have incorporated astronomical contents. Given the limited distances combined with the trade routes over the Mediterranean Sea, such observations were certainly shared by all the astronomical scholars around the Mediterranean coast and doubtless even beyond. Traces of this knowledge (lost deliberately or not) remain to be discovered. As for the cauldron, it invites us to reconsider the place and perhaps the identity of some of our feminine deities (Sirona, Rhiannon, Epona, and Rigani) or even that of certain fairies of medieval literature, and above all, Melusine. By proposing a new interpretation of our heritage, we hope to have helped towards the opening path of further new research. Remember that the most ancient astronomical text available is a list of observations of the movements of the planet Venus, more than 3600 years ago. So the incorporation of astronomical knowledge linked to that planet into an object made probably on the shores of the Black Sea during the first centuries before Christ would not be at all improbable.

ACKNOWLEDGEMENTS

We thank the anonymous reviewers for their constructive comments. Special thanks to Joël Hascoët and Joseph Monard for their support, their impartial reviews and their sound advice. The authors would also like to thank the Christian Thioc, photographe at the Service of the « Musée et sites gallo-romains de Lyon - Fourvière » for providing us the pictures of the cauldron.

REFERENCES

Bader, F. (1997) Mars, avril, mai : le pic, la louve, le sanglier et la truie, in B. Cassin et J.L. Labarrière (Eds.) *L'animal dans l'antiquité*, Paris, Antiquité VRIN, pp. 491-519

Broutian, G.H. (1997) *The Armenian Calendar*, Echmiatzin (Armenia), Echmiatzin Publishing House (in Armenian). *http://www.hyeetch.nareg.com.au/ armenians/names_p3.html*

Goudineau, C., Verdier P. (2006) : Religion et Sciences, in J.-L. Bruneaux et Al., (eds) *Religion et société en Gaule*, Ouvrage édité par le Pôle archéologique du département du Rhône à l'occasion de l'exposition *Par Toutatis ! La religion des Gaulois.*Arles Cedex, (France), Edition Errance

Gregoriades, P. D., (2009) Eneoros Minos and the Minoan Calendrical Abacus. *History of Mechanism and Machine Science*, vol. 6, 2009, pp. 319-324.

Gricourt, D., Hollard, D., (2005) Lug, dieu aux liens : à propos d'une pendeloque du Ve siècle av. J.-C. trouvée à Vasseny (Aisne), Dialogues d'histoire ancienne, 2005, vol. 31, n° 1, p. 7.

Gricourt, D., Hollard, D., (2010) *Cernunnos, le dioscure sauvage - Recherches sur le type dionysiaque chez les Celtes*, Paris, Éd. l'Harmattan.

Hatt, J.-J. (1989) *Mythes et Dieux de la Gaule*, Paris, Picard

Kalinowski, M. (1986) : L'astronomie des populations Yi du Sud-Ouest de la Chine, *Cahiers d'Extrême-Asie*, vol. 2(1), pp. 253-263.

Kaul, F., Marzov, I., Best, J., de Vries, N. (1991) *Tracian Tales on the Gunstrup Cauldron*, Amsterdam, Najade Press.

Le Roux, F., Guyonvarc'h, C.-J. (1995) *Les fêtes celtiques*, Rennes Cedex, Editions Ouest-France Université.

Maumené, C. (2012) Considérations calendaires sur la grande Troménie de Locronan et sa périodicité, *Ollodagos*, t. XXVI (2011-2012), p. 229- 264.

Nielsen, S., Andersen, J., Baker, J., Christensen, C., Glastrup, J., et al., (2005) The Gundestrup cauldron : New scientific and technical investigations, *Acta Archaeologica*, vol. 76, p. 1-58.

Panagiotes D., Gregoriades (2008) The Phaistos Disc : The Oldest Portable Calendar in Use by the Minoan Navy, International Phaistos Disk *Conference 2008, Society of Antiquaries, London 31 October - 1 November 2008, MINERVA, The International Review of Ancient Art & Archaeology.*

Sterckx C. (1998) *Sangliers père et fils : rites, myths et dieux celtes du porc et du sanglier*, coll. « Mémoire de la société Belge d'Etudes Celtiques», n° 8.

Straizys, V., Klimka, M. (1997) The cosmology of the Ancient Balts, *JHA*, vol. 28, ppS57-S81.

Taylor, T. (1992) The origin of the Gundestrup cauldron, *Scientific American*, vol. 266, p. 84-89.

Verdier, P. (1998) Celtic astronomy: the enigma of the Gundestrup cauldron *La revue Archéologue*, June-July 1998, n° 36

Weibao, L. (1996): New Evidence for the Ten-month Calendar, *Publication of the Yunnan Observatory*, 1996-01.

Weibao, L, (1997): Advance in the Research on the Ten-month Solar Calendar, *Progress in Astronomy*, 1997-01.

Weibao, L., Chen J., (2010): Examination of the "Seats of the Five Emperors" *in the Traditional Chinese Constellations, Astronomical Research & Technology*, 2010-02.

28

DARK SKY: AESTHETICS OF THE EXTRATERRESTRIAL LANDSCAPE

Gary Wells

Ithaca College, Ithaca, New York, USA
(wells@ithaca.edu)

ABSTRACT

One consequence of the exploration of the solar system is the accumulating body of images of the surfaces of planets, comets, moons, and asteroids. These images are landscapes, although the land they represent may be quite unlike Earth. Extraterrestrial landscapes may incorporate conventions of representation that provide a familiar grounding for the viewer. However, images of alien landscapes also break some of those conventions and force us to consider the nature of landscape itself. The presence of artistic conventions in the pictures of non-terrestrial landscapes taken during missions to various bodies in the Solar System suggest a counterpart in the history of earthly landscape images. Conventionalizations in these images may be the consequence of imager design, processing, and editing of images that converge around or mimic representational norms found in pictorial images of terrestrial subjects. Deviations from the norms of terrestrial representation may constitute the emergence of an aesthetics of mediation, which may be the result of the unique conditions found on the body itself or the result of human intervention in the imaging process. "Normalized" views of extraterrestrial landscapes simulate a human viewpoint in color or perspective, but certain aspects of imaging technology, and associated science goals, dissociate the earth-bound viewer from the unearthly terrain.

KEYWORDS: landscape, aesthetics, Moon, Mars, Comet 67P/Churyumov-Gerasimenko, photography

The Apollo moon landings provided the first opportunity for humans to describe aesthetic experiences on an extraterrestrial landscape based on direct observation. During his lunar excursion, Neil Armstrong described the moon's surface as having "a stark beauty all its own." Buzz Aldrin described the scene a few minutes later with "beautiful, beautiful," and as "magnificent desolation" (NASA 1969). While not necessarily the most eloquent descriptions, they were the reactions of humans in a truly virgin landscape. Like tourists in an exotic land, the astronauts had the privilege of an unmediated view of this extraterrestrial landscape. That both astronauts described what they saw in terms of beauty is interesting, if not particularly revelatory. It was not beauty in the classical sense of order and harmony, but a fascination with the new and the unfamiliar, a reaction of enthusiasm more than a statement of aesthetic theory.

Figure 1. Untitled photograph from the Apollo 15 mission, July 1971 (Museum of Modern Art, NY).

Within a decade of the first moon landing, photographs taken by astronauts were included in an art exhibition for the first time (Figure 1). In 1979 the Light Gallery in New York City featured NASA photos, curated by Peter McGill, interspersed with the works of Ansel Adams. Audiences were invited to think not only about the visual similarities between images but also to imagine that the astronauts were more than scientists doing documentary work. A fascinating review of the exhibition by Gene Thornton in the *New York Times* expressed skepticism about the context and the message: "It takes a lot of nerve to present NASA photographs as works of art" (Thornton, 1979). The reviewer was insistent that these photos were not works of art, but he did point out that they expressed their own visual language of landscape: "To come from the bare surface and black sky of the moon to the clouds, trees, snow-covered mountains and dazzling sunlight of Ansel Adams's West is not merely to come from the moon to the earth but from one esthetic to another." Today, many of the photos from that exhibition are part of the permanent collection of the Museum of Modern Art.

Figure 2. Rosetta Photograph of Comet 67P/Churyumov-Gerasimenko, September 2014 (ESA/Rosetta/Navcam - CC BY-SA IGO 3.0).

Over the last few years, a steady stream of images has come to us from bodies around our Solar System: the Curiosity rover on Mars, the Rosetta probe to Comet 67P/ Churyumov-Gerasimenko (Figure 2), the New Horizons flyby of Pluto. Along with the body of images from earlier landers, orbiters, and flybys that explored Venus, the moons of the outer planets, and our own Moon, we now have a substantial visual record of the appearance and nature of worlds beyond our own. For the scholar and student of visual culture, these are images that continue to present an opportunity and challenge. While we take the image of the earthly landscape for granted, these extraterrestrial landscapes open the possibility of reconsidering how we define and witness landscape as an experience, a visual genre, and as an idea. Because we have now a significant body of work to consult, we may be able to address the ways in which we can construct a rudimentary aesthetics of the extraterrestrial landscape. I have been thinking about the ways in which these images might serve as teachable moments in visual culture, as challenges to our descriptive language and our theories of visual meaning. The difficulty, however, is that these post-Apollo images are the products of the robotic explorer. They represent a mediated vision of places no human has actually witnessed. Any discussion of aesthetics, of beauty and its other, has to be framed by the intervening technology.

Figure 3. Curiosity rover photograph of Mount Sharp, Mars, September 2015 (NASA/JPL-Caltech/MSSS).

We naturally seek out the familiar in the terrain of an unknown place. It is an act of contextualization and orientation. We look at pictures of alien worlds and identify what we know, based on both experience and visual memory. The language of description that we use with these images, and by "we" I mean those of us who are not invested in the science being done with the images, is sometimes sprinkled with phrases that suggest resemblance or affinity. A Martian scene may "look like the American Southwest," for example (Figure 3). Our descriptions resemble the language of the tourist, because the tourist vision is comparative (Jakle, 1987). The tourist sees the landscape as an outsider. The native sees that same landscape as ordinary and prosaic, but the visitor brings a different view and supplies a different meaning. The extraterrestrial landscapes, however, have no native view. There is no intrinsic context of personal experience, and even astronauts on the Moon were temporary visitors and "outsiders," tourists, who could only provide an external context for what they saw. These are unique views of the previously unseen. The "first views" sent by spacecraft have few earthly equivalents - most every landscape of Earth has been seen long before the sight was recorded or described.

Gene Thornton's objections to seeing space photos in a gallery was that they lacked the artistic intent of the human photographer's vision. The Curiosity rover photograph of Mount Sharp is "white balanced" to aid geologists who are studying the rocks in the scene. The color shift renders the landscape as almost banal, nearly invisible because it is so similar to what we already know and see on Earth, but that was not the point of the color manipulation. The scientists processing the image have inadvertently created a simulated native view of comforting familiarity. It was incidental, or accidental, that they also created the kind of picture that nearly anyone might snap with a cell phone on a visit to Arizona.

Figure 4. Curiosity rover photograph of Mount Sharp, Mars, February 2017 (NASA/JPL/Ken Kramer/Marco Di Lorenzo).

The "true color" view of the same general landscape, on the other hand, may appear as moody and emotional, shifted enough from the common impression that we read the photograph expressively (Figure 4). Only in this second image can we speak of the differences to the familiar landscapes of Earth, and only if we acknowledge that this image is also selective and contrived. It is still the robot's vision, not our own, even if there are hints of drama that we humans bring to or impose upon the picture, manifested through the choices made in the production of the picture itself (Lynch and Edgerton, 1998). But is this Romanticism merely wishful thinking about the exotic locale? It is not the choice of the robot, nor of the controllers on Earth seeking data. There is not (yet) an "art view" built into the mechanism that allows choices, only a process of gathering images that may or may not be interesting. Chance plays an important, and poorly studied, role in the way that we remotely perceive other worlds.

Buzz Aldrin has said that his words from the surface of the Moon were both a visual observation and a conceptual label, a reaction to what he saw but also to what he knew about what he saw (Aldrin, 2014). The hostile environment, the lack of atmosphere and the fact that the landscape had not changed substantially in thousands of years were conceptual factors that shaped his sensory experience of the lunar landscape (Figure 5). But his own physical presence in that place allowed him to have a genuine emotional connection to the landscape, something that the viewer of the photographs cannot fully experience. Although the Apollo astronauts were trying to accurately describe and document the places they visited, I would argue that the impulse of emotion guided their verbal and photographic reactions. Aldrin's "magnificent desolation" was a reaction of awe (and perhaps of fear), inspired by both the sight and the idea of the place, that could only come from actually being in that place, with only his spacesuit keeping him alive in the hostile environment. And this emotional connection is what aesthetics in the traditional sense attempts to define and understand. But the photographs made by astronauts on the Moon were guided by the imperative of gathering science data. The choice of view and composition in any given photograph was situated within the tension of the immediate, human, experience and the rational act of documentation.

Figure 5. Buzz Aldrin, Photograph of the Apollo 11 landing site, 21 July 1969 (NASA).

Figure 6. Rosetta Photograph of Comet 67P/Churyumov-Gerasimenko, January 2015 (ESA/Rosetta/Navcam - CC BY-SA IGO 3.0).

Sooner or later we gravitate to the idea of the sublime when we look at the landscapes of dark skies and airless alien mountains. Our instinct toward comparison and contrast when confronted with the sight of the new and the strange is sometimes an attempt to gauge danger and elation. Edmund Burke's characterization of the sublime as the emotions of extreme experiences in the face of nature are apt for the astronauts who are standing on the surface of the Moon (Burke, 1767). As Jean-François Lyotard points out, "modern aesthetics is an aesthetics of the sublime, though a nostalgic one" (Lyotard, 1986). But just as the viewers of paintings of landscapes are never able to connect to the experience of the artist who journeyed to distant and exotic places to create them, we have little prospect of sharing the connection between participant and location that the Apollo astronauts experienced. The emotion of the experience arrives indirectly through our assessment of the context of creation and the degree to which we assign an expressive intent to the final realization of the image. For all the other bodies in the Solar System that have only been visited by machines, such as Comet Churyumov-Gerasimenko, we have no human spectator at all, and we face the problem of interpreting images of places that we cannot visit and for which we lack any direct human experience (Figure 6). The conventions utilized in the production of such images are applied after the fact, in the absence of a guiding human observer. They make visible a different kind of tension between alien-ness (Romantic "horror") and familiarity (cliché, metaphor, or touristic *indifférance*) (Dumont, 1984).

The artist Robert Smithson once described the tensions between the direct observation of a place and the photographic record made of that experience. His commentary on the beauty (or anti-beauty) of the New Jersey landscape was as a visitor to a strange but familiar land, littered with "craters," "voids," "nebulosity," and "minute particles…under a bleakly glowing sun" (Smithson, 1979). His photographs conveyed the composed, calculated views of an artist's structured vision, but the immediate reaction of the artist to the actual place was indifference, a sense of disinterest, dreariness, and dissolution. His reaction suggests a postmodern sublime, different from the excitement of astronauts on the Moon, but similar to the clichés employed in describing the artifacts of extraterrestrial exploration. He found himself experiencing the landscape through the camera as if "photographing a photograph" and "walking on an enormous photograph made of wood and steel" (Smithson, 1979).

This may be why I have been so fascinated by the Rosetta images. Their weird compositions, sharp contrasts of light and dark, and disorienting perspectives create a sense of dislocation and otherness that flirts with earthly analogues but does not precisely fit any of them. The brilliant landscape with the dark sky is strange enough, but to have that same darkness below is to pull the ground from beneath our virtual feet. The visual plunge into the void gives us pause, even as we are reassured that this is merely a picture, and a highly composed one at that. The spacecraft is an imperfect substitute spectator, guided by humans but unfettered by many of the physical limits of humans. It is we Earth-bound viewers of the image that have the emotional reaction to the scene. And for the moment I am setting aside the planning and processing done by the engineers and scientists that make this image in the first place, not to mention the constraints of the camera itself. I am not sure that what we experience in looking at an image like this is the shiver of the sublime, what 18th century landscape spectators called the awful sensation of nature, at least not as Burke knew it. But it does raise the prospect that the images of extraterrestrial landscape are unexpectedly difficult to describe without resorting to worn out analogies and comparisons. The pictures released by the ESA and NASA appear to us to be more than documentation. But we need better tools to respond to them.

Figure 7. Ansel Adams, Monolith, the Face of Half Dome, Yosemite National Park, California, April 17, 1927. (Metropolitan Museum of Art, NY, Gift of Ansel and Virginia Adams, 1979, © Photograph by Ansel Adams).

Comparisons help, but they fall short because they lack some of the essential elements that distinguish the extraterrestrial from the terrestrial. The

1979 Light Gallery exhibition built a deliberate comparison between the extraterrestrial landscape and the landscape photos of Ansel Adams. His photograph *Half Dome at Yosemite* from 1927 exemplifies the formal and aesthetic argument that the curators were making (Figure 7). The crisp detail, high contrast, and the dark sky bear a passing similarity to the Apollo lunar landscapes, but even more to some of the Rosetta pictures. The deliberate manipulations used by Adams to achieve the look of his photograph were driven by an overarching creative aesthetic rather than a documentary priority. Artfulness, for want of a better word, guided the creative process.

The dark sky of Ansel Adams' photograph is a consequence of choices of film, filter and printing by an artist. For the Rosetta images, or those of the Moon or other bodies lacking an atmosphere, the dark skies are also artifacts of imaging technology, choices of exposure time for example. But we think and react differently to these than we do to the obvious artistic or expressive aspects of earthly landscapes, perhaps because we project our expectations about science rather than art onto the images. The inherent tensions between the "art" image and the "not-art" image are contextual, but not formal. The same criteria for formal description (and aesthetic judgment) could be applied equally to both. It is our presumed knowledge of intent and the craft of image making that make them seem different.

Figure 8. James Nasmyth, Group of Lunar Mountains (Ideal Lunar Landscape), 1874.

But an argument advanced by Barbara Stafford suggests that the visual artifacts of technology intersect with the traditional views of art precisely because perception is perhaps "*the* constitutive form" of knowledge (Stafford, 1996). The nineteenth century impression of the lunar surface is based on the expectation that the Moon will look like a rougher Earth, as in James Nasmyth's 1874 image (Figure 8). It doesn't look like the moon we found when we got there. But Nasmyth's starry sky and jagged peaks are a surprising counterpart to the Rosetta images. Nasmyth assumed the starry sky would be prominent in our visualization of the extraterrestrial landscape as a direct spectator on the surface of the moon. But the images we have received are dominated by the utter blackness of space as an artifact of their technological, rather than perceptual, origins.

Aesthetics is both a theory of beauty and a broader definition and philosophy of art. But the "de-definition" of art since the early 20th century has meant that the work of art as artefact has only to be accorded an appropriate context in order to become art. Much as Gene Thornton saw the status of NASA lunar photographs changed due to their placement in an art gallery, the opportunity to see, and critique, the extraterrestrial landscapes of Rosetta on a par with landscape paintings or terrestrial art photographs could merely mean that we give them the chance to be seen and "appreciated" by the artworld. The mere exhibition of such images does not mean that we have developed a specific explanation for their beauty, any more than the exclamations of astronauts on the Moon can be seen as reasoned theory instead of immediate emotional reaction.

NASA's Art Program, in existence since 1962, was initially an attempt to suggest that the experiences of the Space Age could find expression in artefacts that were, as a class of images, unambiguously designated as art. This class of images did not contain the mission photographs themselves but works that generally presented their "art-ness" through distinctive style and media in a clear separation of art and science. But consider the more recent manifestation of aestheticizing the visual culture of space exploration, the Juno mission to Jupiter. Not only does the mission seek input from the public about what to photograph, NASA also invites members of the public to process the raw image data returned from the Juno spacecraft in ways that are both scientifically useful and aesthetically pleasing. "Some creations are works of art," according to the mission web site, and the public is encouraged to create galleries of such processed images based on the contributions of their favorite "'artist'" (in quotes on the site) (NASA, 2016). By making the mediating role of the "artist" explicit, the mission scientists imply that the Juno images have two separate identities, as data and as expressive artifact (Latour, 1998). The problem of aesthetic value is shifted away from the image makers, whose interest is in the data that the images encode. The public "artists" who manipulate the data into pretty pictures could be seen as disinterested agents who mediate between science and art, or virtual tourists who impose their aesthetic vision and upon that data as an act of interpretation. The tension between earth-bound viewer and machine explorer is left unresolved.

Art historian James Elkins has voiced skepticism about a theory of the visual that embraces both art and scientific imaging (Elkins, 2003). But we live in an era when the divisions between the two blur and merge as never before. We have internalized the digital, the virtual, and the artificial in our image-making and our image consumption. The immediacy of the tourist experience is being replaced by the acceptance of remote viewing as a viable alternative to physical presence. If, as students and scholars of the image, we are to build the language and conceptual framework to describe and interpret the images of this era, then perhaps the collaborations of humans and machines will help define an aesthetics of mediation. The experience of the extraterrestrial landscape in images, and even in person, may become the expression of a postmodern sublime.

Figure 9. View of boulder photographed during second Apollo 17 EVA, 12 December 1972 (NASA).

REFERENCES

Aldrin, B. (2014) Reddit Interview, 8 July 2014. https://www.reddit.com/r/IAmA/comments/2a5vg8/i_am_buzz_aldrin_engineer_american_astronaut_and/ (Accessed 3 January 2018).

Burke, E. (1767) *A Philosophical Enquiry in the Origin of our Ideas of the Sublime and Beautiful.* 5th edition. London: J. Dodsley, pp. 58.

Dumont, J-P. (1984) A Matter of Touristic "Indifférance." *American Ethnologist*, vol. 11, no. 1 (February 1984), pp. 139–151.

Elkins, J. (2003) *Visual Studies: A Skeptical Introduction.* New York: Routledge, pp. 85–94.

Jakle, J. A. (1987) *The Visual Elements of Landscape.* Amhert: The University of Massachusetts Press, pp. 8–12.

Latour, B. (1998) How to Become Iconophilic in Art, Science, and Religion? In *Picturing Science, Producing Art.* Ed. Caroline A. Jones and Peter Galison. New York: Routledge, pp. 418-440.

Lynch, M. and Edgerton, S. (1998) Aesthetics and Digital Image Processing: Representational Craft in Contemporary Astronomy. In *Picturing Power: Visual Depiction and Social Relations.* Sociological Review Monograph, no. 35. Ed. Gordon Fyfe and John Law. New York: Routledge, pp. 184–220.

Lyotard, J.–F. (1986) *The Postmodern Condition: A Report on Knowledge.* Manchester University Press, p. 81.

NASA (1969) *Apollo 11 Technical Air–to–Ground Voice Transcription.* https://www.jsc.nasa.gov/history/mission_trans/ AS11_TEC.PDF, pp. 379–382 (Accessed 20 February 2018).

NASA (2016) Juno Mission, Junocam Image Processing Gallery. https://www.missionjuno.swri.edu/junocam/processing/ (Accessed 14 April 2018).

Smithson, R. (1979) A Tour of the Monuments of Passaic, New Jersey. In *The Writings of Robert Smithson*. Ed. Nancy Holt. New York University Press, pp. 53–56.

Stafford, B. (1996) *Good Looking: Essays on the Virtue of Images*. Cambridge, Mass.: MIT Press, p. 39.

Thornton, G. (1979) Photography View; When Scientists Take the Shots; Pictures by NASA. *New York Times*, January 14, 1979, p. 27.

Permissions

All chapters in this book were first published in MAA, by University of the Aegean; hereby published with permission under the Creative Commons Attribution License or equivalent. Every chapter published in this book has been scrutinized by our experts. Their significance has been extensively debated. The topics covered herein carry significant findings which will fuel the growth of the discipline. They may even be implemented as practical applications or may be referred to as a beginning point for another development.

The contributors of this book come from diverse backgrounds, making this book a truly international effort. This book will bring forth new frontiers with its revolutionizing research information and detailed analysis of the nascent developments around the world.

We would like to thank all the contributing authors for lending their expertise to make the book truly unique. They have played a crucial role in the development of this book. Without their invaluable contributions this book wouldn't have been possible. They have made vital efforts to compile up to date information on the varied aspects of this subject to make this book a valuable addition to the collection of many professionals and students.

This book was conceptualized with the vision of imparting up-to-date information and advanced data in this field. To ensure the same, a matchless editorial board was set up. Every individual on the board went through rigorous rounds of assessment to prove their worth. After which they invested a large part of their time researching and compiling the most relevant data for our readers.

The editorial board has been involved in producing this book since its inception. They have spent rigorous hours researching and exploring the diverse topics which have resulted in the successful publishing of this book. They have passed on their knowledge of decades through this book. To expedite this challenging task, the publisher supported the team at every step. A small team of assistant editors was also appointed to further simplify the editing procedure and attain best results for the readers.

Apart from the editorial board, the designing team has also invested a significant amount of their time in understanding the subject and creating the most relevant covers. They scrutinized every image to scout for the most suitable representation of the subject and create an appropriate cover for the book.

The publishing team has been an ardent support to the editorial, designing and production team. Their endless efforts to recruit the best for this project, has resulted in the accomplishment of this book. They are a veteran in the field of academics and their pool of knowledge is as vast as their experience in printing. Their expertise and guidance has proved useful at every step. Their uncompromising quality standards have made this book an exceptional effort. Their encouragement from time to time has been an inspiration for everyone.

The publisher and the editorial board hope that this book will prove to be a valuable piece of knowledge for researchers, students, practitioners and scholars across the globe.

List of Contributors

Dr Nicholas Campion
Associate Professor in Cosmology and Culture, University of Wales Trinity St. David

Lionel Sims
University of East London

Xabi Otero
Jauzarrea

Juan Antonio Belmonte
Instituto de Astrofísica de Canarias and Universidad de La Laguna, Tenerife, Spain

M. Carmen Pérez Die
Museo Arqueológico Nacional, Madrid, Spain

Lucía Díaz-Iglesias Llanos
Centro de Ciencias Humanas y Sociales, CSIC, Madrid, Spain

A. Rodríguez-Antón and J. A. Belmonte
Instituto de Astrofísica de Canarias, 38200, La Laguna, Spain
Universidad de La Laguna, 38200, La Laguna, Spain

David Espinosa Espinosa and Marco V. García Quintela
University of Santiago de Compostela

A. César González-García
Institute of Heritage Sciences, Incipit-CSIC Instituto de Ciencias del Patrimonio, Incipit-CSIC, Santiago, Spain
Heritage Sciences Institute, Incipit-CSIC, Santiago de Compostela, 15704, Spain

Josef Vit
Independent Scholar, Oberbettingen, Germany

Michael A. Rappenglück
Adult Education Centre and Observatory, Gilching, Germany

César Esteban
Instituto de Astrofísica de Canarias, La Laguna, Tenerife, Spain
Departamento de Astrofísica, Universidad de La Laguna, La Laguna, Tenerife, Spain

Luis Benítez-de-Lugo-Enrich
Departamento de Prehistoria y Arqueología. Universidad Nacional de Educación a Distancia, Valdepeñas, Ciudad Real, Spain
Departamento de Prehistoria y Arqueología. Universidad Autónoma de Madrid, Madrid, Spain

Hossam M. K. Aboulfotouh
Faculty of Environmental Design, King Abdulaziz University, Jeddah, KSA
Faculty of Fine Arts, Minia University, Minia, Egypt

Juan A. Belmonte
Instituto de Astrofísica de Canarias, Tenerife, Spain

Andrea Polcaro
University of Perugia, Italy

Thomas T. Gough
Independent Scholar, Parkhead Farmhouse, Ballindalloch, Scotland AB37 9BJ

Marianna P. Ridderstad
University of Helsinki, Finland

Eshbal Ratzon
Tel Aviv University

Stanisław Iwaniszewski
Posgrado en Arqueología, Escuela Nacional de Antropología e Historia - Instituto Nacional de Antropología e Historia, Mexico

Vance R. Tiede
Astro-Archaeology Surveys

Andrea Rodríguez-Antón
Instituto de Astrofísica de Canarias and Universidad de La Laguna, Tenerife, Spain

Mosalam Shaltout
N.R.I.A.G., Helwan, Cairo, Egypt

Roslyn M. Frank
University of Iowa, Department of Spanish and Portuguese, Iowa City, Iowa, USA

Sona V. Farmanyan
M. Abeghyan Institute of Literature, National Academy of Sciences, 15 Grigor Lusavorich Str., Yerevan 0015, Armenia

Areg M. Mickaelian
Byurakan Astrophysical Observatory (BAO), National Academy of Sciences, Byurakan 0213, Aragatzotn Province, Armenia

Mosalam Shaltout
N.R.I.A.G., Helwan, Cairo, Egypt

Roslyn M. Frank
University of Iowa, Department of Spanish and Portuguese, Iowa City, Iowa, USA

Sona V. Farmanyan
M. Abeghyan Institute of Literature, National Academy of Sciences, 15 Grigor Lusavorich Str., Yerevan 0015, Armenia

Areg M. Mickaelian
Byurakan Astrophysical Observatory (BAO), National Academy of Sciences, Byurakan 0213, Aragatzotn Province, Armenia

S. Draxler and M. E. Lippitsch
Institute of Physics, Karl-Franzens-Universität Graz, Austria

M. Incerti, G. Lavoratti and S. Iurilli
Architecture Department of Ferrara (Italy)

Areg Mickaelian
Byurakan Astrophysical Observatory (BAO), National Academy of Sciences, Byurakan 0213, Aragatzotn Province, Armenia

Simone Bartolini
Italian Military Geographic Institute (Florence)

Marco Pierozzi
Italian Navy Hydrographic Institute (Genova)

Mauro Peppino Zedda
Isili, Sardinia

Giangiacomo Gandolfi
Zetema/Planetario di Roma - Sovrintendenza BBCC Comune di Roma

Alexey Stoev and Penka Maglova
Space Research and Technology Institute, Bulgarian Academy of Sciences, Stara Zagora Department

Vassil Markov
South-West University "Neofit Rilski", Blagoevgrad, Bulgaria

Mina Spasova
Institute for the Study of Societies and Knowledge, Bulgarian Academy of Sciences, Sofia, Bulgaria

Claude Maumené
Independant scholar

Index

www.ingramcontent.com/pod-product-compliance
Lightning Source LLC
LaVergne TN
LVHW060526210926
848485LV00005B/1155
* 9 7 8 1 6 3 9 8 9 0 5 7 6 *